Sustainable Practices in the Textile Industry

Scrivener Publishing
100 Cummings Center, Suite 541J
Beverly, MA 01915-6106

Publishers at Scrivener
Martin Scrivener (martin@scrivenerpublishing.com)
Phillip Carmical (pcarmical@scrivenerpublishing.com)

Sustainable Practices in the Textile Industry

Edited by
Luqman Jameel Rather
State Key Laboratory of Silkworm Genome Biology, Southwest University, Chongqing, P.R. China
Mohd Shabbir
School of Chemical Engineering and Pharmacy, Wuhan Institute of Technology, Wuhan, Hubei, P.R. China
and
Aminoddin Haji
Textile Engineering Department, Yazd University, Yazd, Iran

Scrivener Publishing

WILEY

This edition first published 2021 by John Wiley & Sons, Inc., 111 River Street, Hoboken, NJ 07030, USA and Scrivener Publishing LLC, 100 Cummings Center, Suite 541J, Beverly, MA 01915, USA

For more information about Scrivener publications please visit www.scrivenerpublishing.com.

Wiley Global Headquarters
111 River Street, Hoboken, NJ 07030, USA

For details of our global editorial offices, customer services, and more information about Wiley products visit us at www.wiley.com.

Library of Congress Cataloging-in-Publication Data

ISBN 978-1-119-81888-5

Cover image: Pixabay.Com
Cover design Russell Richardson

Set in size of 11pt and Minion Pro by Manila Typesetting Company, Makati, Philippines

MIX
Paper from responsible sources
FSC® C013604

Contents

Preface

In recent years, the textile industry has been the focus of rising interest in global markets due to varied and changing world market conditions. Increasing environmental and health concerns owing to the use of large quantities of water and hazardous chemicals in conventional textile finishing processes, has led to the design and development of new dyeing strategies and technologies. Effluents produced from the textile wet processing industry are very diverse in chemical composition, ranging from inorganic finishing agents, surfactants, chlorine compounds, salts and total phosphate to polymers and organic products. This has forced Western countries to exploit their high technical skills for the advancement of textile materials with high quality technical performances, and the development of cleaner production technologies for cost-effective and value-added textile materials. *Sustainable Practices in the Textile Industry* is a collection of the current sophisticated ways used to minimize the use of bioresource products to improve dye extraction and dyeing properties. Highlighted in this book are the innovative ways in which wet chemical processing methods are used to alleviate the environmental impacts arising from this sector. The major challenge in the textiles and fashion sector is that it requires massive sustainable innovation in terms of material and end-use products to mitigate the huge environmental impacts arising from chemical processing. Therefore, this book also contains innovations in eco-friendly methods for textile wet processes and applications of enzymes in textiles in addition to advancements in the use of nanotechnology for wastewater remediation.

The book is compiled of 13 chapters from various research areas dealing with the application of different sustainable technologies for enhancing the dyeing and comfort properties of textile materials with substantial reduction in wastewater problems. Chapter 1 deals with the sustainable extraction of natural dyes from plant sources and their subsequent applications in the textile industry. Chapter 2 deals with the advancements in non-aqueous dyeing systems. Chapter 3 gives a brief account of structural coloration of different textiles achieved as a result of scientific observations of nature. Chapter 4 deals with the use of enzymes for enhancing dyeing properties of different textiles. Chapter 5 deals with the use of sustainable processes for textile coating. Chapters 6 through 8 give a detailed account of the functional finishing properties achieved on different textiles using different dyeing methods with natural and synthetic functional finishing agents. Chapter 9 provides up-to-date information regarding sustainable development for brands and manufacturers in the textile industry. Finally, the remaining Chapters 10 through 13 deal with the advanced techniques used for wastewater remediation.

The authors who contributed to this book are specialists in fields involved in using different dyeing systems other than aqueous solvent, employing enzymes in dyeing procedures, surface modifications, sustainable developments for textile manufactures, functional finishing and different advanced techniques for wastewater remediation. Thus, the editors hope that students, researchers and academicians of various fields, such as textile dyeing, chemical engineering, environmental science, materials science among others, will find this book of great interest and useful in their curriculum. We expect it will definitely be helpful for engendering new ideas in textiles research, leading to interdisciplinary research collaborations.

Now the time has come to thank those who supported this book in any way. We acknowledge the great efforts of the eminent authors without whom this book would have been unimaginable. We also appreciate the interest shown and the support given by the publisher, which allowed us to compile this reference book.

Luqman Jameel Rather
Aminoddin Haji
Mohd Shabbir

Part 1

SUSTAINABLE DYE EXTRACTION AND DYEING TECHNIQUES

1

Extraction and Application of Natural Dyes

Sanjeeda Iqbal and Taiyaba Nimra Ansari*

Department of Botany, Govt. Holkar Science College, Indore, India

Abstract

Environmental pollution and population explosion are becoming the world's biggest issues. Eco friendly products and practices are popularizing day by day due to present national and international awareness on environmental situations. Textile industries are one of the reasons of environmental pollution and affect all forms of life adversely. Textile dyeing process generally uses chemical dyes and synthetic mordants. Chemicals in the dyeing process began with the discovery of "Mauve" by WH Perkin in 1856. These synthetic dyes are manufactured from coal tar, petrochemicals and many other chemicals, which cause allergies such as contact dermatitis, respiratory diseases, skin irritation and cancer etc. Naturally dyed textile materials are in demand globally because of harmful effects of chemical dyeing and continuous efforts of researchers in this field. Natural dyes can be obtained from natural resources like plants, minerals, insects and fungi but most of the dyes are taken from plant parts i.e. leaves, barks, flowers, fruits and roots. Natural dyes have some special properties like soothing color, biodegradable, non-hazardous, non-carcinogenic and antimicrobial resistance etc. Natural dye extraction process requires plant parts and sometimes their by-products as a raw material. Natural dyeing practices enhance the cultivation of flowering crops, which provides an extra source of income to farmers. Thus, large scale production of naturally dyed fabric in future will solve the problem of human as well as environmental health.

***Keywords*:** Natural dyes, source, fiber, mordant, dyeing, fastness, antibacterial, UV-protection

**Corresponding author*: taiyaba_ansari@rediffmail.com

Luqman Jameel Rather, Mohd Shabbir and Aminoddin Haji (eds.) Sustainable Practices in the Textile Industry, (3–42) © 2021 Scrivener Publishing LLC

1.1 Introduction

In the present scenario people are more inclined to be nature-friendly, health conscious and become aware about the environment. The meticulous environmental standards are being imposed by many countries in response to the toxic and allergic reactions associated with synthetic dyes. It has created a revolution in research and development in eco-friendly and non-toxic colorant. Environmental considerations are now becoming vital factors during the selection of consumer goods including textile as well as in cosmetic, pharmaceutical and food industry all over the world. Coloring agent or dyes play an important role in all these industries during manufacturing and other production process. Both qualitative and quantitative research investigations have been undertaken all over the world on safe coloring substance. Natural dyes are widely used in following application (Figure 1.1):

a. Textile Coloration
Coloring of textile material is called dyeing. Dyeing is a process to enhance the beauty of fiber or fabric by coloring them. The coloring compounds can be synthetic or natural and capable of being fixed to fabric defined as dye. India is a diverse country of region and culture. Dyeing practices varied immensely due to availability of local dye-yielding plants and minerals, the natural sources. Until year 1856, these natural sources were used in coloring but with the discovery of chemical dyes, the use of natural dyes decreased. The side effects of the continuous use of chemical dyes gradually began to

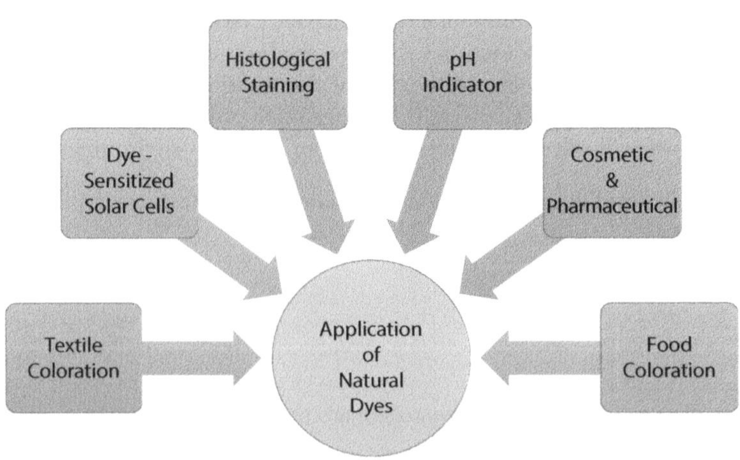

Figure 1.1 Schematic representation of applications of natural dyes.

appear as health problems and skin diseases that resulted by wearing synthetic dyed fabric for decades. Hazardous chemicals were used in the manufacture of dyes in large scale for rapid growth of textile industries. All those synthetic, hazardous chemicals disposed of in nature after dye preparation, coloring and printing, which ultimately created various pollution problems in the environment. Keeping this situation in mind, many experiments are being done by scientists and researchers for the development of new sources and techniques of natural colors, so that it can be adopted by textile industries on a large scale. There is an increasing awareness among humans about their health and nature, due to which the demand for eco-friendly clothes is also increasing. Natural clothes can be obtained at a higher price from some small scale industries and handlooms, but they are not able to meet the demand of all people. In the past few years, many research works related to natural dye have been done. In this chapter, an attempt has been made to exhibit how natural dyes are used on clothes and their significance for human health and nature.

b. Food Coloration

Human appetizer and choice of food are influenced by color. Food colors are used in processed food, drinks and condiments. They are often added to maintain and improve the appearance of the food. The addition of saffron, turmeric and other spices are reported from ancient times. Commercially available colors are made of chemicals that can be harmful to human and environment. Therefore, natural food colors are in demand again. Most commonly used natural food dyes are saffron, turmeric, annatto, beetroot and carrot etc.

c. pH Indicator

pH is measure of relative amount of free hydrogen and hydroxyl ions in water. The range of pH is in between 0 and 14, where 7 is considered neutral. pH greater than 7 indicates the base, whereas pH less than 7 indicates acidity. A pH indicator is a compound that changes color in solution over a narrow range of pH value. Small amount of indicator compound is sufficient to produce a visible color change. There are many colors of indicators in nature thus, influenced by change in pH range. Some of the natural dyes show color changes with variation of pH. Red cabbage dye is very good example of natural pH indicator [1].

d. Histology Staining

Study of the microanatomy of cells, tissue and organs is called histology. Observations of cells are performed with the help of a microscope. Staining is the technique to highlight and differentiate the structure of tissue. Stain and dyes are applied in staining on tissue and cells. Dyes also can be used to color cells, tissue, organelles as well as microorganism such as bacteria and fungi [2]. Saffron, a natural dye extracted from *Saffron crocus* was the first stain in histology used by Antonie van Leeuwenhoek, the father of microbiology [3]. Hematoxylin stain is a naturally occurring dye found in Logwood tree widely used in histological study. Researchers also found *Punica granatum*, *Curcuma longa*, *Syzygium cumini* and *Sorghum bicolor* effective in staining [4–7].

e. Cosmetics and Pharmaceutical
Cosmetic products are used to clear, improve or change the complexion, skin, hair and nails. Colors or dyes are the most important ingredient in production of cosmetics. Henna is traditionally used for coloring hands and hair. Saffron and turmeric also examples of natural dyes utilized in cosmetics. In pharmaceutical field colors of medicine are used to differentiate the dosage. Imparting color to drugs helps in their distinctive appearance. The colorant employed in pharmaceuticals is considered safe such as beet root, paprika and annatto.

f. Dye-Sensitized Solar Cells
Dye-sensitized solar cell is third generation photovoltaic (solar) cell that converts any visible light into electrical energy. It is low-cost solar cell belonging to the group of thin film solar cells. Performances of dye sensitized solar cells are mainly based on dye used as a sensitizer [8]. Godibo *et al.*, attempted the preparation of Dye Sensitized Solar Cells using flowers of *Amaranthus caudatus*, *Bougainvillea spectabilis*, *Delonix regia*, *Nerium oleander*, *Spathodea companulata* and a mixture of the extracts [9].

In addition to the above mentioned applications, there is a growing interest for using natural dyes to dye leather, stain wood, pulp, some plastics [10–14]. This chapter intends to discuss the application of natural dyes in textile.

1.2 What are Natural Dyes?

Natural dyes can be derived from natural sources such as plants, animals and minerals. A large number of herbs, shrubs, trees, insects, animals, microbes and minerals have been identified for extraction of coloring compounds [15]. Red, yellow, brown, blue, black, green and orange color can be obtained from natural dyes.

1.3 Why Natural Dyes?

Natural dyes are recommended to be applied on textile materials. Following points support the use of natural dyes on a large scale.

1. Eco-friendly: Natural dyes are extracted from natural sources therefore they are environment safe.
2. Biodegradable: These dyes are capable of being decomposed by microorganisms.
3. Renewable: Replaced by the new material obtained from nature.
4. No health hazard/Non-toxic: Natural origin of these dyes makes them harmless.
5. Variety of shades: Varieties of color, shades and hues present in nature itself.
6. Soothing, soft and lustrous color: Natural dyes are soft and relaxing.
7. Utilization of waste material: Many agriculture waste products can be used in the dyeing process.
8. Antibacterial/UV Protective: Naturally dyed fabrics get special properties like antibacterial and UV protection.

As there are many advantages in using natural dyes but they also have some drawbacks:

1. Expensive: Natural dyes are expensive due to being limited in source.
2. Faded easily: Sometimes their poor attachment on fabric makes them fade easily.
3. Difficult to produce/collect: Collection is somewhat difficult in large amounts.
4. Time consuming: The complete process like collection of dye takes long time.
5. Reproducibility of shades is difficult to control: Natural dyes produced by secondary metabolic activities of plants or by very special processes in other animals, which depend on climate conditions, age and seasonal variations. Thus, one particular shade cannot be achieved again and again by a single dye.

1.4 What are Synthetic Dyes?

Synthetic dyes are made by organic molecules. They are derived from coal tar hence also known as coal tar dyes. William H Perkin synthesized "Mauve" the first synthetic dye in 1856 in the United Kingdom. Then, a significant number of dyes were discovered and industries quickly adopted them to grow, mainly in the United Kingdom, Switzerland and Germany [16].

The Sudan I (Solvent Yellow 14) is one of the members of azo-dyes widely used in textile industry [17]. It is enzymatically transformed, through the action of the intestinal flora, into carcinogenic aromatic amines, when present in the bodies of animals or humans [18]. In the case of azo-dyes, especially, carcinogenicity can be produced by both the dye itself and its own converted compounds [19]. The study of National Toxicology Program confirmed the neoplastic liver nodules in rats by the presence of Sudan I dye [20]. The Basic Red 9 dye, used in the textile, leather, paper and ink industries [21], develops carcinogenic potential in humans [22], and high toxicity to environment [23]. Under anaerobic conditions, it breaks down into carcinogenic aromatic amines, and when disposed in water bodies can cause allergic dermatitis, skin irritation, and cancer [24]. According to the *in vivo* tests on rats, it causes local sarcomas and tumors in the liver, bladder [25], mammary glands and hematopoietic system [26].

The Crystal Violet dye, shows an intense color [27], and is a member of the cationic triphenyl methane group, and is responsible for mitotic poisoning and abnormal accumulation of metaphases [28] as well as the *in vitro* clastogenic effects observed in Chinese hamster ovules [29], which induce chromosomal damage too [30]. According to Bharagava *et al.*, this powerful carcinogenic agent promotes fish tumors [28, 31] and hepatocarcinoma, reticular cell sarcoma in various organs, such as the vagina, uterus, ovary and bladder [32] as well as hardened gland adenoma and ovarian atrophy in rats. In humans, it is capable of generating respiratory and renal failure, chemical cystitis, skin irritation and digestive tract disorder [28].

Advantages/Merits of Synthetic dyes

1. Easy preparation.
2. Available in large numbers and quantities.
3. Quality of fast colors
4. Cost effective.

Disadvantages/Demerits of Synthetic dyes

1. Production on high temperature

2. Carcinogenic
3. Hazardous to human health.
4. Problem of environmental pollution.

1.5 Sources of Natural Dyes

Natural dyes can be classified in following groups on the basis of sources (Figure 1.2) [33]:

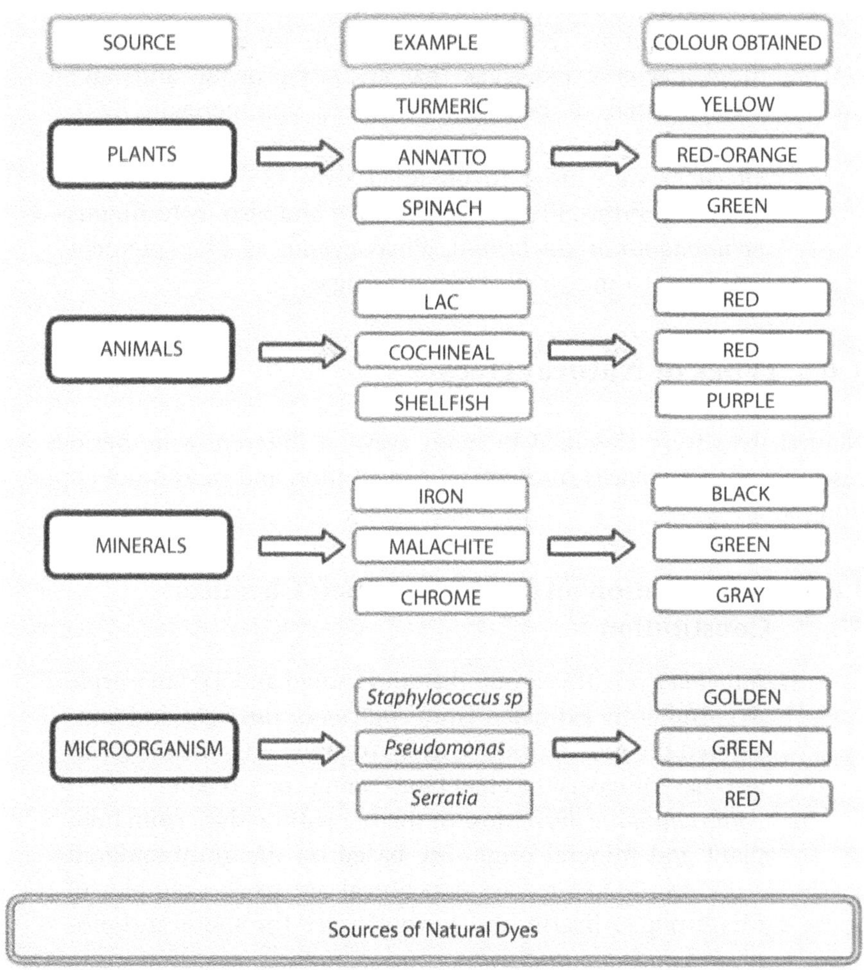

Figure 1.2 Sources of natural dyes.

1. *Plants:* Roots, leaves, fruits, flowers and barks can be used as a source of natural dyes. Different colors can be obtained from each part such as Sappan-wood tree pods give red, barks give brown and root gives yellow color. Many by-products of plants can also be used to form dyes.
2. *Animals:* Dyes can be obtained from dried body of insects for example, Lac, Cochineal and Kermes. Cochineal is a brilliant red dye produced from insects living on Cactus plants. Carmine and Tyrian purple dye derived from cochineal, shellfish (*Murex* spp.) respectively.
3. *Minerals:* Mineral dyes include iron buff, iron black, manganese bistre, chrome yellow, and Prussian blue.
4. *Microorganisms:* Natural colorant can be extracted from fungi, bacteria and algae that are fast growing and have the potential of being standardized commercially [34]. *Chitosan*, *Serratia* spp., *Trichoderma virens* and *Alternaria alternata* were used to obtained dyes [35]. Natural Red color is produced by *Monascus anka* and also from fungus *Echinodontium tinctorium.* Phycocyanin is blue pigment extracted from *Spirulina plarensis* algae.

1.6 Types of Natural Dyes

Natural dyes were classified in many ways at different time periods by researchers on the basis of chemical constitution and method of application [36].

1.6.1 Classification on the Basis of Their Chemical Constitution

1. *Indigoid dyes:* This group includes Indigo and Tyrian purple dye. Indigo is extracted from *Indigofera tinctoria* and considered the most primitive dye. Woad plant (*Isatis tinctoria*) also has indigo as the chief blue dyeing component.
2. *Anthraquinone dyes:* Most of the red natural dyes from both plant and mineral origin are based on the anthraquinoid structure. Madder, Lacs, Cochineal are some examples of this group. Alizarin and purpurin are the main chromophores in *Rubia tinctorum.*

3. *Alpha naphthoquinones:* Lawsone (henna) is a most important member of this class. Another dye is juglone, isolated from the shells of unripe walnuts.
4. *Flavonoids:* Yellow dyes obtained from this group and can be classified under flavones, isoflavones, aurones and chalcones. These yellows are found in a variety of plants, including Persian berries (*Rhamnus* spp.), young fustic (*Cotinus coggygria*), old fustic (*Chlorophora tinctoria*) and yellow wood (*Solidago virgaurea*).
5. *Di-hydropyrans:* In chemical structure, di-hydropyrans are similar to the flavones. These natural dyes give dark shades on cotton, wool and silk. Logwood and Sappan-wood are the most common examples.
6. *Anthocyanidins:* Orange dye carajurin obtained from leaves of *Bignonia chica*. Carajurin is a chemical member of this class.
7. *Carotenoids:* The class name carotene is derived from the orange pigment found in carrots. In these, the color is due to the presence of long conjugated double bonds. Usually, red, orange and yellow colors come in this category and can be obtained from different plants, e.g. yellow, orange color in sunflower [37, 38].

1.6.2 Classification Based on Method of Application/Preparation

1. *Direct Dyes:* Direct dye soluble in water can be taken up directly by the material. Direct dye also called substantive dyes because of their excellent substantivity for cellulosic material like cotton and viscose rayon. Turmeric, *Chebulic myrobalan* and Annatto used in direct dyes.
2. *Vat Dyes:* As the name suggests that the dye is prepared in a large container for storing and mixing liquids or wooden vessels commonly known as 'Vat'. This is a primitive method of dye preparation.
3. *Mordant Dyes:* Mordant dyes are attached to textile fibers by a fixing agent "mordant" which can be organic or inorganic substance. Since chromium is used extensively hence, mordant dyes are sometimes called chrome dyes.

4. *Acid Dyes:* These dyes performed in acidic medium. Sulfonic or Carboxylic groups of dye molecules can form electrovalent bonds with amino groups of wool and silk.
5. *Basic Dyes:* These dyes form an electrovalent bond with the carboxylic group of wool and silk. Berberine has been classified as basic dye.
6. *Disperse Dye:* Disperse dye have low aqueous solubility and low molecular weight. These dyes require post mordanting treatment with chromium, copper or tin salt (Figure 1.3) [36].

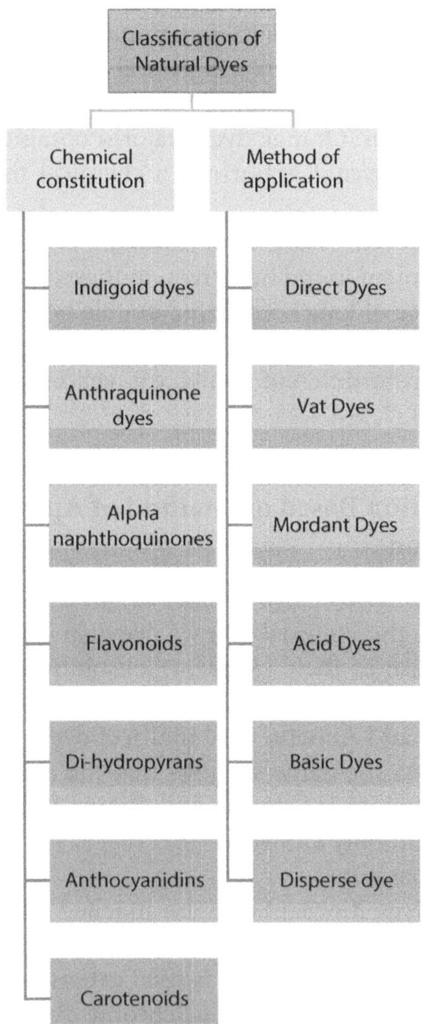

Figure 1.3 Classification of natural dyes.

1.7 Natural Dyes Need Fixing Agent (Mordants) for Bonding

Mordants (from the Latin verb "modere" meaning "to bite") are natural salts that can form a stable molecular coordination complex with both dye and fiber. Natural dyes and their use in dyeing is the most ancient art of all times. Most of the natural dyes have very low affinity towards fabric, therefore a fixing agent is required to attach dye on fabric. Mordants are substances that are able to form complexes with molecules of dyes. Mordants can be applied before dyeing, after dyeing or within dyeing mixed in a dye pot. Process of Mordanting improves the color fastness properties of dyed fabric. Mordants are classified in three categories such as Metallic mordants, Tannins and Oil mordants [39]. The fourth category is bio-mordant, which is generally obtained from natural resources.

1.7.1 Metallic Mordants

i. *Aluminum:* Potash alum is the most widely used aluminum mordant for natural dyeing. Alum does not affect the color. The shade of dye depends on the amount of mordant. If deeper shades are required on fabric a greater amount of mordant is needed. Alum forms weak sulfuric acid when dissolved in water during the mordant process. This can result in acidic fumes which are corrosive, and irritating when inhaled.

ii. *Iron:* Iron salts in the form of ferrous sulfate (also known as green vitriol) are extensively used in dyeing and printing. Mordanting with iron salts produces a black or gray color to the fabric and reduces the darkness of other colors. Repeated high exposures may lead to nausea, vomiting, stomach pain, constipation and may affect liver.

iii. *Copper:* In copper mordanting, fabric treatment is done with the help of copper sulfate (blue vitriol). It is known for improving the light fastness of various dyed materials. High temperature operations such as boiling in dyeing generate fumes that have different health effects. Long-term exposure of copper can cause irritation, burn the skin, eyes and throat. It can cause headache, nausea, vomiting, diarrhea and abdominal pain.

iv. *Tin:* Tin mordant brightens the color. Stannous and stannic chloride are used as mordants. Stannic chloride is preferred for cotton. It causes severe skin burns and eye damage. It also causes skin allergies.

v. *Chromium:* Potassium dichromate is used in mordanting procedure and referred as Chrome. It is highly toxic and quite hazardous to health. Small amounts can cause contact dermatitis. It is a known carcinogen meaning it causes cancer.

1.7.2 Tannins and Tannic Acid

Tannic acid or tannins are used as a primary mordant for cotton and cellulosic fibers which do not have much affinity for metallic mordants. A cotton fabric treated with tannic acid can absorb all types of dyes.

1.7.3 Oil Mordants

In the past, castor and til (sesame) oils were used as mordants but they were later replaced by Turkey Red Oil (TRO). Sulfated castor oil is largely used in textile industries.

Many metal salts such as chrome, copper, tin and lead are seldom used now due to research evidence of their extreme toxicity either to human health, ecological health or both. Only a small amount of these metal salts get fixed on the fabric and the rest is discharged as effluent which leads to the contamination of land and water resources. In order to make natural dye sustainable many scientific workers are developing natural mordants that can be replaced with metallic-salt-based mordants.

1.7.4 Bio-Mordants

Bio-mordants are those substances that can be obtained from natural sources. According to many researchers bio-mordants are eco-friendly and effective to use than synthetic mordants [40]. There are some examples of bio-mordants:

i. *Myrobalan:* It is one of the most important and widely used mordants in dyeing processes. It can be considered as dye and mordant both. Myrobalan mordant is obtained from fruits of *Terminalia chebula* commonly known as 'Harda'. It gives pale yellow color on fabric.

ii. *Oak gall:* Gallnuts are obtained from the oak tree. It is the earliest and richest source for natural tannin. These are collected and ground for use as a tannin mordant.

iii. *Sumac:* The leaves of sumac contain tannin which can be used in the process of mordanting cotton. *Rhus glabra* species of sumac also known as "rhubarb". Leaves of sumac are rich in tannin suitable for dyeing and their use as mordant.
iv. *Pomegranate rind:* Dried pomegranate rind (*P. granatum*) powder also used as mordant in natural dyeing. Sangeetha *et al.* applied lemon leaves extract using *P. granatum* rinds as mordant on Silk fabric [41].
v. *Catechu:* Catechu was used as a natural mordant since the ancient times as myrobalan [42, 43]. Catechu is extracted from the heartwood of *Acacia catechu.* It produces various shades of brown. Catechu mordants were applied with *Sticta coronate* a lichen that produces dye for coloring silk fabric [42].
vi. *Aloe vera:* Aloe vera leaves contain sticky substance. Researchers are working on exhibited fixing properties of aloe. Fresh leaves of aloe vera can be taken as biomordant for dyeing silk fabric [44].

Other than above mentioned names many sources have been explored for bio-mordants. Adeel *et al.* explored the fixing properties of acacia (*Acacia nilotica*), henna (*Lawsonia inermis*), turmeric (*C. longa*), pomegranate (*P. granatum*) and rose (*Rosa indica*) with natural dye extracted from cinnamon bark. New shades observed on silk fabric with improved fastness properties [45]. Bark of *Xylocarpus moluccensis* tested to be used as a biomordant, and significant improvement in the percentages of dye absorbed in the silk fabric was observed [46]. Wool yarn dyed with madder roots with gallnut (*Quercus infectoria*) extract as biomordant [47]. Rani *et al.* investigated that harda powder, pomegranate peel, orange peel and amla powder can be used as alternative copartner of metal mordants. Dyeing was done on protein fabrics with *Carica papaya* leaf natural extract [48]. Wool yarn dyeing performed with *Adhatoda vasica* extract. The effect of various metal salts and extracts of gallnut, pomegranate peel and babool bark as mordants were comparatively evaluated [49]. Aminoddin extracted Berberine from *Berberis vulgaris* wood and applied on wool fiber using the extract of roots of *Rumex hymenosepalus* as biomordant [50]. Banana pseudostem sap was applied as a biomordant with *Senegalia catechu* stem extract on wool [51]. To provide best options of synthetic mordants various scientists are exploring and developing the new and effective bio-mordants.

1.7.5 Method of Application

i. *Pre Mordanting (Chrome mordant process):* This is two bath process in which fabric is first mordanted in a dye bath then squeezed and immersed for dyeing at required temperature.
ii. *Post Mordanting (After chrome method):* The method involves first dyeing the fabric and then treating it with mordants, later on material is taken out for washing, squeezed and dried.
iii. *Simultaneous Mordanting (Meta-Chrome process):* The meta-chrome process involves only one single step because dye and mordant is mixed together to work simultaneously on fabric.

1.8 Fibers/Fabrics Used for Natural Dyeing

Dyeing is a process where dye molecules transport to a substrate surface from the dye solution. The substrates are fabric or fiber. Mainly fibers have two groups i.e. Natural and Man-made (Synthetic) fibers. These fibers/ fabrics are mainly used to perform natural dyeing [52].

1.8.1 Cellulosic Fiber

a) *Cotton:* Cotton is a natural fiber obtained from different species of *Gossypium* plant. The fiber is almost pure cellulose, soft, fluffy, staple that grows in a ball around the seeds of cotton plants.
b) *Jute:* Jute is a long, soft and shiny bast fiber. It is one of the strongest natural fibers.
c) *Linen:* Linen has a very good quality of absorption. Flax (*Linum* sp.) plants commonly known as linseed are the source of linen fiber.
d) *Hemp:* It is obtained from a variety of *Cannabis sativa* plant species as bast fiber.

1.8.2 Protein Fiber

a) *Wool:* It is an animal fiber. Wool is obtained from the sheep or other hairy mammals.

b) *Silk:* It is a natural protein fiber and made up of Fibroin. Silk fibre is obtained from the cocoons of silkworms.

1.8.3 Synthetic Fiber

a) *Nylon:* It is produced by reaction of amino acid with itself or between diamines and diacids.
b) *Acrylic:* Polymer acrylonitrile found in acrylic fiber. It is soft, light weight and warm fiber.
c) *Polyester:* They are polymer and contain ester functional groups in their main chain.

1.9 Extraction of Natural Dyes

Extraction of color from natural dye is one of the most important steps of dyeing. Raw materials of natural dyes are leaves, barks, fruits, flowers and roots of many plants as well as some animals and minerals as described earlier. Sources of natural dyes are carbohydrates, proteins, lipids, fats and many other cell inclusions, so only the desired coloring material requires an extraction process. In the extraction process of natural dyes, the cost of extraction and the yield of color affects the cost of dyeing. Following methods generally used for extraction of coloring materials:

i. *Aqueous extraction:* Aqueous extraction method is a traditional and most popular procedure in natural dyeing. Small pieces of fresh or dried dye material are ground in powdered form. That is soaked in water, boiled, filtered to obtain aqueous dye solution. Aqueous extraction depends on many factors such as time and temperature of the boiling, fresh or dried dye material and material to liquor ratio. Maryam *et al.* gave one such example i.e. extraction of color from Onion (*Allium cepa*) skin in aqueous condition as 5 g of dye dissolved in 100 ml water at the temperature of 100 °C for 60 min [53].
ii. *Alkali or acid extraction:* Glycosides can be hydrolyzed in acidic or alkaline condition. Mostly, natural dyes constitute glycosides, so extraction in acidic or alkaline medium can improve color yield. Some of the natural dyes are pH sensitive therefore; they destroy their dyeing properties in unwanted pH condition.

iii. Microwave and ultrasonic assisted extraction: The extraction of natural dye can be done by microwave and ultrasonic assistance. Microwave energy used in extraction of natural dye with a very minimum amount of solvent. Microwave increases the rate of the processes so the extraction can be completed in a shorter time with better yield. Thangabai and Kalaiarasi's studies revealed that microwave assisted extraction of Padauk (*Pterocarpus* sp.) wood are more efficient as compared to conventional extraction methods [54]. Natural dye from Sorghum husk extracted with the help of ultrasound-microwave-assistance [55].

iv. Fermentation: Indigo extraction is the best example of fermentation method of extraction. In presence of indimuslin enzyme, glucoside indican breaks into glucose and indoxyl [56]. The enzymes produced by the microorganisms present in the atmosphere or those present in the natural resources used in fermentation for assisting the extraction process.

v. Solvent extraction: Natural dyes can also be extracted depending upon their solubility by using organic solvents such as acetone, petroleum ether, chloroform, ethanol, methanol, or a mixture of solvents such as ethanol and methanol, mixture of water with alcohol, and so on.

vi. Supercritical fluid extraction: Supercritical fluid extractions have become popular in recent years to isolate the organic compound from herbs and dyes as well as dye from natural sources. In supercritical fluid extraction a dense gas as a solvent that usually has carbon dioxide above its critical temperature (31 °C) and critical pressure (74 bar) for extraction is used [14].

1.10 Dyeing Process

1.10.1 Preparation of Fabric Before Dyeing

Raw textile materials are required to prepare base before processing such as dyeing and printing etc. Natural and man-made fibers contain undesirable matter like dirt or stains which are regarded as impurities. Textile materials in this "raw" state are to be "cleaned" and "finished" to make gray yarn or gray fabric.

a) *Weighing:* The weight of textile material helps us to know about the amount of soap for washing, the quality and quantity of chemicals and dye stuff to use in the mordanting and dyeing processes. Therefore, the first step is to weigh the yarn of fabric while it is still dry [57].
b) *Soaking:* Fiber or fabric is soaked for 12 h in tap water to remove the water soluble impurities.
c) *Scouring:* Fibers contain oil and fats on their surface; they are hydrophobic in nature which affects the absorbency of the fibers. The outer hydrophobic layer has to be removed before dyeing. The process by which this water resistance layer is removed from the fabric is called "Scouring".
d) *Bleaching:* Bleaching is the process of discoloration or removal of natural and other coloring matter from fibers. This is a process of whitening fibers using oxidizing agents.

1.10.2 Mechanism of Dyeing

a) *Adsorption:* Firstly the dye molecules in the dye bath move towards the fiber and those that are nearest to the fiber get "adsorbed" on to the fiber surface. They form a very thin layer of molecules on the surface of fiber.
b) *Penetration:* Secondly the adsorbed dye molecules adhered to the outer surface of fiber gradually penetrate or infiltrate into the pores or canals of the structure.
c) *Fixation:* The final step is one where the dye molecules find suitable locations according to dye size where they get "fixed" or "anchored".

1.10.3 Process of Dyeing

a) *Pre-Soaking the Material:* Textile stuff, whether it is fabric, yarn or loose fiber is thoroughly wet in water before dyeing begins. Such wetting is achieved by soaking for hours. A thoroughly wet textile dyes well.
b) *Enzyme Assisted Dyeing:* Enzyme assisted dyeing is also performed for textile coloration [58]. Ultrasonic dyeing on cotton and silk fabric is performed with *Terminalia arjuna*, *Punica granatum* and *Rheum embodi* dye. In pretreatment enzyme protease, amylase, diastase and lipase

are complexed with tannic acid. Both fabrics showed rapid dye adsorption kinetics and total higher adsorption [59]. Raja and Thilagavathi demonstrated that alkaline protease enzyme process improve the quantity of natural dye exhausted [60].

c) *Sonicator Assisted Dyeing and Plasma Treatment:* Ultrasonic dyeing technique is also called Sonicator dyeing that improves the penetration of dye in fiber or fabric and increases color strength. It is a rapid dyeing process and can be run under mild conditions and low temperatures. Dyeing of wool fabrics carried out with natural dye "lac" through conventional and ultrasonic techniques [61]. In another study *Eclipta* leaves were taken as natural dye for cotton fabric using both conventional and sonicator methods. Results revealed that Ultra-sonication method showed higher color strength values [62]. Vankar *et al.* demonstrated the sonicator dyeing method to improve dye uptake on cotton, silk and wool [63]. Plasma treatment in dyeing is conducted for improving the dye uptake of fabric. It is a surface modification technique that performs before dyeing on textile materials [64]. Low-temperature plasma is widely used in non-destructive surface modification of textiles where a wide range of properties can be obtained. Plasma treatment performed on silk fabric and dyeing done with natural dye extracted from *Phytolacca decandra* [65]. Plasma treatment was conducted to improve the adhesion of chitosan on cotton fibers. After that the cotton is dyed with natural dye extracted from pomegranate rinds. The results exhibited that plasma treatment can enhance the color strength of the dyed sample [66].

d) *Printing:* Printing on textile in India has been a part of India's cultural identity for thousands of years [67]. Printing produces more colorful effect on the fabric. Printing is a process where colorful designs are created which can be done by hand or machine. Hand printing is done by two methods viz., block printing and screen printing. Boruah and Kalita revealed that turmeric dye produced various soft and stable natural print on eri silk. Three different mordants alum, stannous chloride and ferrous sulfate were selected for printing [68]. Kavyashree investigated the efficiency of natural dye in screen printing on cotton and silk fabrics. Three natural dyes indigo, madder, and sappanwood were selected for screen printing. The results revealed that these dyes can be considered as the recommendable alternative to harmful synthetic dyes [69]. Jimmy *et al.*, investigated the color resistant material from flour of *Colocasia esculenta* using *Acacia catechu* as natural dye for batik technique [70].

e) Dyeing Condition

i. Dyeing Condition for Cellulosic Material

Cotton is the most popular textile material. Many researchers have attempted to dye on cotton with the natural dyes. Each fabric performs differently in dye bath on the basis of their chemical structure. Dyeing parameters such as dyeing time, temperature, pH, material liquor ratio, dye and mordant concentration play an important role in dyeing. Several studies standardized the dyeing condition for cotton and reported the results as dyeing temperature, 70–100 °C, dyeing time, 60–120 min, material to liquor ratio, 1:20–1:100, and pH, 10–12 may be required for natural cellulosic material.

Vankar *et al.* used *Eclipta* as natural dye for dyeing cotton fabric by conventional and sonicator methods [62]. Teli and Paul attempted to dye cotton fabric with extraction of coffee seed coat. Dyeing was done by pre, meta and post-mordanting methods using various mordants. The results showed that coffee seed extract can develop a range of shades with good fastness properties [71]. In another study only coffee seeds were used for dyeing purposes. Some mordants such as $FeSO_4$, $CuSO_4$ and $SnSO_4$ were applied for improvement of color strength of cotton fabric [72]. Shanker and Vankar applied dye extracted from *Hibiscus mutabilis* using 1:40 for M:L ratio on cotton fabric. Dyed cotton fabric exhibited good fastness properties [73]. Dayal *et al.* isolated dye from *Parthenium hysterophorus* and employed on cotton fabric. The dyeing done with M:L ratio 1:100 at 95–98 °C for 60 min on dyebath [74]. Indi and Chinta the fruits of *Phyllanthus reticulatus* utilized for dye extraction and application. Premordanting was done with alum (8%) and tannic acid (4%) at the temperature 80 °C for 60 min. Same treatments were performed for Post mordanting. Dyeing was carried out for 10% shade at 80 °C for 45 min at pH from 3–7 [75]. Vankar and Shanker dyed cotton with aqueous extraction of *N. oleander* flowers. Mordanting was done with metal salt i.e. $FeSO_4$, $SnCl_2$, $CuSO_4$, $SnCl_4$, $K_2Cr_2O_7$ and alum at 60 °C for 30 min. Then, dye is applied on cotton while keeping the M:L ratio as 1:30 and pH was set at 4 [76]. A study has been conducted for improvement of washing and light fastness by Mukherjee *et al.*, where pre mordanting was carried out with aluminum sulfate, zinc sulfate, copper sulfate, magnesium sulfate and sodium dichromate. Dyeing was done with M:L ratio 1:20 at the temperature of 90 °C for 45 min. Natural dyes obtained from *Curcuma longa*, *Butea monosperma*, *Tagetes erecta* and *Nyctanethes arbor-tristis* were taken for experiment by different researchers [77].

Kulkarni *et al.* attempted dyeing cotton with natural dyes isolated from Pomegranate peel. Copper sulfate and ferrous sulfate were applied in ratios for mordanting. About 4% dye extraction was applied at 80 °C for 60 min with M:L ratio 1:40 [78]. Srivastava *et al.* studied the dyeing capability of Lichi peels on cellulosic fabrics. Many experiments were performed to determine the dyeing parameters, such as extraction medium, optimum concentration of dye material, extraction time and concentration of mordants and mordanting methods. One such example revealed that 5 g of dye material with mordant like $FeSO_4$, alum and tannic acid at 60 °C for 1 h produced good results in dyeing after experimentation [79]. A report by Jain presented that three natural mordants Anar, Arjun and Babul bark were applied on cotton fabric for better results. However, on the other hand colorant extracted from Jamun leaves, bark, bark peel and fruit in pre mordanting method dyed for 60 min at 60 °C temperature gave good results too [80]. Single jersey cotton knitted fabric that has been mordanted with some natural extract like pomegranate peel seeds, pomegranate peel bark and some of Gymnosperm leaves *Thuja orientalis* and *Araucaria excelsa* gives significant results at 95 °C temperature for 60 min in exhaust method. Then, dyeing of samples was done with natural dyes extracted from mango seed kernel (*Mangifera indica* L.) after above mordanting. The dyeing was carried out at 100 °C temperature for 60 min [81]. In a series of studies eco-friendly garments, inner wears, child clothing and home furnishing materials were prepared by dyeing cotton material with Myrobalan (*T. chebula*) and Turmeric (*C. longa*). Compared to the synthetic dyed cotton fabric, the above dyed fabrics showed excellent results in terms of fastness properties. Herbal Textile is finished entirely with herbal extractions, without using any chemicals [82]. Chandel *et al.* attempted to extract organic dye from *Brassica oleracea* Var. *botrytis* (Cauliflower) and applied it on 100% pure cotton. It revealed that different shades from cauliflower can be prepared using different mordants [83].

Singam *et al.* studied natural dyes based on *Lawsonia inermis*, *Azadirachta indica* and *Curcuma longa* were used to produce eco-friendly and non-toxic fabric for the people. The extraction process of natural dyes is an aqueous technique and then proceeded to hot bath dyeing later. The aim was to find the optimum concentration of natural dyes and super hydrophobic coating removal from cotton fabric for the green technology dyeing process [84]. Pan and his colleague explored that extract of Deodara, Jackfruit and Eucalyptus leaves yield light brown and light mustard shades on jute fabric. Fastness properties toward washing showed good in all manners [85].

ii. Dyeing Condition for Protein Material

Wool and silk fibers both have complex chemical structure and are susceptible to alkali treatment. They respond very well in acidic conditions. Mehtab *et al.* have utilized neem bark (*A. indica*) for dyeing of wool yarn. They optimized dyeing conditions such as pH 4.5, dye concentration 0.05 g per gram of wool, dyeing time 60 min and temperature 97.5 °C indicated good light and wash fastness properties [86]. Bechtold *et al.* isolated colorant from ash-tree bark (*Fraxinus excelsior* L.) for dyeing on wool. Meta mordanting process with $FeSO_4.7H_2O$ was applied, which revealed that 1-2 gm extraction of bark is sufficient to dye 1 g wool yarn [87]. A study has been conducted by Jayalakshmi and Amsamani for application of Annatto and Catechu using bio-mordants to dye wool. Mordanting and dyeing were conducted by then at room temperature for 30 min. Myrobalan and Karavelum (Babul) bark were used in 1% concentration as biomordant while liquor of Tamarind and Green tea were used for fixing treatment. The experiment concluded that use of natural mordant and fixing agents improve color fastness of wool [88]. Mohammad *et al.* extracted colorant from Henna leaves for dyeing woolen yarn. Dyeing was conducted by using 1, 5, 10 and 20% of dye concentrations with 1:40 Material to liquor (M:L) ratio at 30 °C. Thirty six shades were obtained by varying concentrations of dye and mordants [89].

Uddin evaluated the performance of dyes taken from Mango leaves for silk dyeing. Dye extraction conditions were optimized such as the temperature, time, and material-to-liquor ratio found to be 98 °C, 60 min, and 1:10, respectively. Dyeing was done on silk with the ferrous sulfate, alum (potassium aluminum sulfate), and tin (stannous chloride) as fixing agents individually and using four different combinations of these mordants at 60 °C for 60 min keeping M:L ratio of 1:30 [90]. Swamy observed the coloring potential of one Gymnosperm *Casuarina equisetifolia* leaves and applied on silk fabric where color strength of dye was improved by using different mordants [91]. Banerjee *et al.* studied natural dyes isolated from *Camellia sinensis*, *A. cepa*, *Laccifer lacca* and Iron ore. Dyes employed on Eri silk yarn with different mordants [92]. Shukla *et al.* collected eleven species of lichens from different regions of Himalaya to extract dye and applied it on Silk, Tussar silk and Absorbent cotton. The lichens can produce orange, yellow, blue grey, purple and brown color dyes. However, the author recommended that due to small size of lichens and slow growth they could serve for local handlooms but cannot fulfill the requirements of textile industries [93]. Khan *et al.* demonstrated that Myrobalan, Gallnut and Pomegranate extract can be used as a dyeing agent on wool yarn. Three different mordants like Alum, Copper sulfate and Ferrous sulfate combined with above agents improved fastness properties [94].

iii. Dyeing condition for synthetic fiber
Nylon is a synthetic polymer containing amide link known as synthetic fiber. Lokhande and Dorugade attempted dyeing of nylon fabric with two different techniques viz. open bath and HTHP (high temperature and high pressure) dyeing. Natural dyes extracted from Onion (*A. cepa*), Lac (*L. lacca*) and Turmeric (*C. longa*) were applied with various mordants on Nylon Fabric. HTHP Dyeing has been found to give better results as compared to the open bath dyeing [95]. A study has been conducted by Miah *et al.* on nylon fabric dyed with onion extract using various mordants such as Alum, Copper sulphate and Potassium dichromate by HTHP dyeing methods [96].

Exhaust dyeing method is commonly used for the application of natural dyes on polyester fabrics. The dyeing of polyester is conducted using material: liquor ratio in 1:15–1:50, temperature above 90 °C and pH ranges from 4–8 for 60–90 min [97]. Elnagar *et al.* reported UV/ozone pretreatment was employed to activate fiber and improve dye ability of polyester and nylon. Fabrics were pre mordanted by ferrous sulfate 6% (owf) keeping material to liquor ratio 1:15 at 60 °C for 60 min. Dye isolated from Curcumin and Saffron and applied on nylon and polyester [98]. Shahin *et al.* studied the process of dyeing polyester fabric with Chinese Rhubarb "Dolu" (*Rheum officinale*) after optimization. The dyeing process was performed with 50% dye extract at temperature 100 °C for 60 min and M:L ratio 1:100 [99]. Guizhen reported that *Rhizoma coptidis* colorant can be used on acrylic fiber successfully. Dyeing carried out different concentrations of dye at 60 and 95 °C for 5 h keeping material liquor ratio 1:200 and pH adjusted at 6.5 [100].

1.11 Evaluation of the Dyed Fabric

1.11.1 Color Strength or K/S Value

Color strength or K/S value is the most important parameter to test the quality check of a sample in terms of depth of the color dyed on fabric.

Strength of any colorant is related to absorption property. Reflectance (%) of the dyed fabric samples can be measured by using a spectrophotometer. The Kubelka–Munk theory gives the following relation between reflectance and absorbance:

$$K/S = [(1 - R)^2/2R]$$

where R is the reflectance, K is absorbance and S is the scattering.

The above equation is used for detection of color strength. Color strength of the dyed fabric influenced by the increase of reflectance value. When shade percent increases, reflectance percent & color strength (K/S) decreases. If shade percent is more but color depth is less than quality of textile does matter a lot for customers and selling manufacturers as well. Color strength of a textile is a quality control assurance system. It is also important to know about the color combination by which color is produced. So, it makes the necessity of color strength testing in textile industries. Color strength or K/S value were carried out by different scientists for natural dyed fabrics [50, 51, 101].

1.11.2 Color Fastness Properties

Colorfastness can be defined as a property of any dyed textile material to retain its original color without fading, changing or running during washing, wetting, cleaning, exposure to light, heat or any influences. The color fastness standards or protocol generally used by the American Association of Textile Chemists and Colorists (AATCC) and the International Organization for Standardization (ISO). After testing, the color of the sample is compared by "Gray Scale for Color Change" and a "Gray Scale for Staining". Color Fastness uses a rating system from1 to 5, where 5 indicate excellent and 1 shows very poor results. The main color fastness tests are as follows:

1. Colorfastness to Washing: the ability of dye to resist during washing of fabric called washing fastness.
2. Colorfastness to Light: light fastness is resistance of dye during the exposure to light or heat.
3. Colorfastness to Rubbing: the ability to sustain original color of dyed fabrics by abrasion or rubbing.
4. Colorfastness to Perspiration: the ability to not fade when dyed fabric is perspired.

Natural sources provide a large range of color with various shades on textile materials. To use dyed material every day, the color should remain on the fabric in any situation. Therefore, researchers and scientists attempted several trials and found very good results for fastness [102, 103].

1.12 Some Special Characteristics of Naturally Dyed Fabric

1.12.1 Antimicrobial Properties

Textile materials are the carriers of microorganisms such as pathogenic, odor generating bacteria, mold and fungi. Fabric surface is a good growth medium for microorganisms. Microbial growth causes multiple problems to fabric as discoloration, fiber damage, staining and unpleasant odor as well as many skin problems to the weaver. All these problems decrease the quality of fabric and demand of customers. Thus, research led to develop the antimicrobial textile material which carried out extensive work for the sake of seller and buyer. Antimicrobial agent inhibits the growth of microorganism and can be applied to the textile substrates by coating, exhaust, pad–dry–cure, spray and foam techniques. The substances can also be applied by directly adding into the fiber spinning. All the above mentioned procedures are applicable generally on synthetic dyed materials. Although natural dyes are known to have medicinal and protective character, they show remarkable antimicrobial properties. They prevent the growth of bacteria and have been proven environment friendly and health protecting even preventing diseases. Antibacterial efficiency of dyed textiles could be tested qualitatively and quantitatively. Antimicrobial properties of dyed fabric can be performed on solid medium (diffusion tests) or liquid culture (suspension tests) by standard methods [35, 50, 104].

1.12.2 UV Protection

The sunlight that reaches the earth has ultraviolet rays that cause damage to human skin. It is common to use clothes to avoid the scorching heat of the sun, but nowadays the demand for solar ultraviolet radiation (UVR) protected clothes has increased to circumvent skin problems. Textiles such as hats, accessories and umbrellas propose the safest protection from ultraviolet (UV) radiation exposure [105]. Their UV rays blocking properties depend on fabric types, fabric construction and dyeing technique etc. For this reason, the interest of scientists and researchers is moving towards developing such a dyeing system. Standard test method is applied to determine the ultraviolet radiation transmitted or blocked by textile fabrics. Several researchers reported that dyeing with natural dyes increases UV protection properties of fabric [106, 107].

1.12.3 Deodorizing Finishing

Sweating and grown bacterial colonies are responsible for generating odors in garments. In present scenario due to priority of health and hygiene consumer demands have been increased for deodorizing fabrics. Researchers are also working on improving deodorizing properties with antibacterial and UV protection properties.

1.12.4 Moth Resistant and Insect Repellent

In advances textile products insect repellent finishes are being serve as protective clothing. Many insects and moths like mosquitoes, bugs, bees, and ants are cause of many diseases. Therefore, these textile materials act as barrier and repellent for insects. Natural dyes extraction contains many active compounds which work against insects. Several methods are available to evaluate the insect repellency however the most used techniques are the cage test, cone test, and the excito chamber test.

1.13 Conclusion

The contribution of textile industry is spread globally with 7% of the total world exports and employs 35 million workers, generating around 1 trillion dollars economy around the world [108]. The textile industrial sector is one of the biggest global polluters that consume high amounts of fuels and chemicals [109] despite its undeniable importance. The special emphasis is placed on the enormous use of drinking water in various operations of its production chain such as washing, bleaching, dyeing among others [110]. The textile industry is responsible for an extensive list of environmental impacts [111]. Some of the research studies are given below to access the impact of synthetic dyes on overall nature. Air water and soil pollutions are major environmental hazards of synthetic dyeing industries. Where air pollution caused by particulate matter, dust, oxides of nitrogen, sulfur and volatile organic compounds.

The water is an essential resource for life on the planet and for human development. The textile industries discharge untreated effluents into water bodies [109]. Most of the residual waters of textile industry create high levels of biochemical oxygen demand (BOD) and chemical oxygen demand (COD) [112]. The effluent sample collected from Karur, a textile city in Tamil Nadu (India) showed dark violet color, objectionable odor, high TDS, alkaline PH, high amount of chlorides, nitrate, sulfates, heavy metals, and low amount of dissolved oxygen (DO), all parameters exceeded the BIS (Bureau of Indian Standard) limits [113]. The active, direct, basic and acid dyes are soluble organic compounds [114]. Their high solubility in water makes it difficult to remove them by conventional methods [115]. They impart color to a given substrate due to the presence of chromophoric groups in its molecular structures [116].

The reduction in the rate of photosynthesis and dissolved oxygen levels affecting the aquatic biota are due to the color present in textile dyes that not only causes aesthetic damage to the water bodies, but also prevents the penetration of light through water [112, 115, 117]. The textile dyes are toxic, mutagenic and carcinogenic agents that act as environmental pollutants and cause biomagnifications by entering in food chain. Organisms at higher tropic levels show higher levels of contamination compared to their first level [30, 118–120]. Here it is necessary to mention that around 15–50%, of azo-type textile dyes do not bind to the fabric during the dyeing process, and are released into wastewater which is commonly used in developing countries for the purpose of irrigation in agriculture [121]. The use of these azo compounds affects negatively to germination and growth of plants and to soil microbial communities [117, 121]. The textile dyes, along with a large number of industrial pollutants, are highly toxic and potentially carcinogenic which makes them environmentally degrading and are the cause of various diseases in animals and humans [122, 123]. Whereas, the textile sludge reveals problems related to surplus volumes and unwanted composition, high loads of organic matter, micronutrients, heavy metal cations and pathogenic microorganisms [109]. Often sludge is sent to the landfills, as a result the toxic chemicals present in synthetic dyes percolate polluting ground water resources.

According to studies conducted by Clark, the acute toxicity of textile dyes is caused by oral ingestion and inhalation, especially by exposure to dust causing triggering irritations to the skin and eyes [19, 124]. The workers are prone to contact dermatitis, allergic conjunctivitis, rhinitis, occupational asthma or other allergic reactions due to producing or handling reactive dyes [125]. A conjugate forms between human serum albumin and the reactive dye, which acts as an antigen producing immunoglobulin E (IgE) antibodies, which combine with histamine and causes all these diseases [19, 125]. The greatest potential long-term hazard to human health is the genotoxicity of textile dyes [19, 126]. For example, the strong genotoxic effects of textile dyes, pointed out by Tiwari *et al.* in their study on *A. cepa* root cells exhibiting chromosomal aberrations. The carcinogenesis is comprised of multiple stages favored initially by mutagenic factors [125, 127]. The textile dyes, especially those of the azo and nitro type, may offer carcinogenicity, and its effects show themselves over time [128].

1.13.1 Overview

Globalization of the world market encouraged the textile industries for large scale production and application of synthetic dyes. These dyes are non-degradable, complex in nature, and a major reason of environmental contamination and serious public health concern. All above research investigations highlighted the impact of environmental, contamination and health hazards of flora, fauna and human beings by enormous use of synthetic dyes. Generally, the long list of synthetic dyes are toxic to humans and environment. In this discussion only three examples of dyes are given here with research evidences as Sudan I, Basic Red 9 and Crystal Violet dye to show their impact on organisms and ecological balance. Many scientists investigated the harmful effect in micro and macro-organisms as well as human beings to be aware of the production, exposure and disposal of these dyes. The synthetic dyes and their effluent discharge are responsible for little danger of skin irritation to cancer of different body parts. During textile processing, inefficiencies in dyeing result in large amount of the dye stuff being directly lost to the wastewater which ultimately finds its way into the environment. This is the picture of most of the developing countries. Approximately 10–15% dyes are released into environment during dyeing process making the effluent highly colored, unpleasant and toxic. As far as human health and environmental hazards are concerned, important prerequisites for risk should be assessed. Nowadays authorities, scientists and general public is aware to potential risk of production and exposure of synthetic dyes. Textile industries are liable to establish the treatment plants for wastewater before being disposed into the environment. But efficiency of treatment plants, honesty of workers, reliability of industry and implementation by government affect the extent of residual amounts reach to the environment. Export demands associated with low cost labor prevalent in India; determine the existence of small-scale textile factories that clandestinely release toxic dyes into water bodies [129].

The effluent disposal of synthetic dyes in water resources causes bio magnification, effects aquatic life where toxic chemicals enter in their bodies later goes to humans via food chain. Another bitter example of wastewater release from dyeing industry is their uses in irrigation by poor, illiterate, innocent farmers in agriculture enable the harmful chemicals of synthetic dyes reach to both animals and human beings through food chains.

1.13.2 Legislative Regulations for Synthetic Dyes

Though, stringent environmental legislations do exist in most of the countries, the will to implement these laws faithfully is needed to overcome the human health and environmental hazards of synthetic dyes [130]. The ministry of environment and forests, government of India has prohibited the handling of 42 Benzidine based dyes from 1993 onward. In January 1997 the use of Azo dyes was banned in India under the clause of Environmental (Protection) Act 1986. Indian legislation prohibited import of hazardous Azo-dyes on 31 March 2002. In Europe Sudan I an Azo-dye was banned in 2009. However Sudan I, III and IV have been classified as category 3 carcinogens. Many Benzidine based dyes are prohibited but Basic Red 9 dye is still in industrial use, which is carcinogenic and genotoxic to humans other mammals and aquatic fauna. As recent research studies showed the harmful effects of Crystal Violet dye on humans, rats and fishes with dangerous carcinogenic disorders. This alarming situation suggests us to ban the use of Crystal Violet dye with immediate effect. Despite of prohibition of many dyes, they are still in use.

1.13.3 Sustainability Aspects of Natural Dyes

Sustainability can be defined as the processes and actions through which mankind avoids the depletion of natural resources to keep an ecological balance so that society's quality of life doesn't decrease. Sustainability is the foundation for today's leading global framework for international cooperation- the 2030 agenda for Sustainable Development Goals (SDGs). The 17 SDGs were adopted by all United Nation Member States in 2015, with 169 targets to reach 2030. International Institute for Sustainable Development (IISD) focus areas are mainly climate, economies, resources and some of the global goals are clean water, economic growth, industry, responsible consumption and production, climate action, life on land and below water etc. Use of natural dyes on large scale leads to Sustainable Development which meets the needs of the present without compromising the ability of future generation to facilitate their own requirements. Three pillars of sustainability are economy, society and environment. Therefore; if we discuss the connection between use of natural dyes and sustainability then we come to the highlight of profit, people and planet the three principles of sustainable development. There are certain precious points of natural dyes for environment, economy and society to combat threats of climate and restoration of sustainability.

i. Environment

The textile industry is the second most polluting industries in the world. Synthetic dyes contribute to a major part of this pollution, with nearly 20% of global water pollution being linked to the textile dyeing processes.

Benefits of natural dyes are as follows

- Renewable resources.
- Non-hazardous
- Biodegradable waste, easily decompose
- No negative impact on food supply or water.

ii. Economy

The worst scenario of clothing industry is over consumption and unsafe manufacturing. Hence, the economy can be boosted by use of natural dyes because

- Some of the natural dyes are obtained from waste and by-products of plants and vegetables.
- They will provide extra source of income for rural farmers.
- Natural dyes collection can provide income source for locals.
- The process of collection of dyes also requires skilled workers.

iii. Society

Sustainability is also about being socially responsible. And over all, the textile industry isn't a very responsible one with large scale utilization of synthetic hazardous chemical dyeing and printing. Thus, natural dyeing only can be a safest, everlasting, soothing way of life due to their nature of

- Cheap or cost effective
- Variety of shades can be obtained from one dye with changing concentrations and combinations.
- Non-toxic during the process of production.
- No negative health effect on wearing.
- Other quality such as antibacterial, UV-Protection, Deodorizing.

1.13.4 Practicality of Natural Dyes

The earlier discussion shows that, the use of synthetic dyes in textile and other industries is hazardous for the life of all organisms and environmental health. Therefore, utilization of natural dyes in large extent is necessary to overcome the toxic effects of synthetic dyes. The advantages of natural dyes are as they come from natural sources therefore, not harmful to the environment, so appealing to the consumers. Natural dyes are obtained from renewable sources, therefore, impose no harm to the environment. When dyes are taken from plant parts, the large scale production of flower crops and other species definitely increase the income of farmers and can generate employability. They give soothing shades and soft hue color pay off to textile so safe and secure for all of us. When fabric is made by natural dyes, it is also protective to the users from harmful rays of sunlight. Antibacterial properties found in the naturally dyed fabric that protect the wearer from diseases.

However, there are some disadvantages of natural dyes as they are expensive, tend to fade quickly, less availability of sources, low sustainability of color and shades etc. Because of these reasons the natural dyes are less popular in use among people. On the other hand natural dyes are vibrant in colors, nontoxic, non-allergic, non-carcinogenic, safer for kids, eco-friendly, lowering human dependence on harmful products, produce no waste due to origin from nature. Thus, the importance of natural dyes is because of their utilitarian uses, aesthetic satisfaction and harmless qualities, demand us to make them popular in society.

The use of natural dyes will make us feel proud to be closely connected with nature and to recognize the importance it plays in our lives. Gradually with the help of organized plantation and farming the cost effective production and dyeing will definitely be possible in future to serve the purpose of large scale trade and commerce. Hence, the application, usage and popularization of natural dyes may be a way to conserve our environment and pay off the obligation of nature.

Acknowledgement

The authors are grateful to University Grants Commission (UGC), Delhi for the Maulana Azad National Fellowship (Taiyaba Nimra Ansari) and Dr. S.T. Silawat, Principal, Govt. Holkar Science College, Indore for providing necessary facilities.

References

1. Kapilraj, N., Keerthanan, S., Sithambaresan, S., Natural plant extracts as acid-base indicator and determination of their pka value, *J. Chem.*, 2019, 6, 2019. https://doi.org/10.1155/2019/2031342.
2. Chukwu, O., Odu, C., Chukwu, D., Hafiz, N., Chidozie, V., Onyimba, L., Application of extracts of henna (*Lawsonia inarmis*) leaves as a counter stain. *Afr. J. Microbiol. Res.*, 5, 3351, 2011.
3. Tiford, M., The long history of hematoxylin. *J. Biotechnic Histochem.*, 80, 73, 2005.
4. Gharravi, A.M., Golarlipour, M.J., Ghorbani, R., Khazaei, M., Natural dye for staining astrocytes and neurons. *J. Neuro. Sci. (Turkisk)*, 23, 215, 2006.
5. Avwioro, O.G., Onwuka, S.K., Moody, J.O., Agbedahunsi, J.M., Oduola, T., Ekpo, O.E., Oladele, A.A., *Curcuma longa* extract as a histological dye for collagen fibres and red blood cells. *J. Anat.*, 210, 600, 2007.
6. Suabjakyong, P., Romratanapun, S., Thitipramote, N., Extraction of natural histological dye from black plum fruit (*Syzygium cumini*). *J. Microsc. Soc. Thailand*, 4, 13, 2011.
7. Muhammed, A.O., Olutunde, O.A., Benard, S.A., Muhammad Ahmad, A.T., Omoowo, B.T., Hibiscus-Shorgum: A New Morphological Stain in Neuro-Histology. *Int. J. Health Res. Innovation*, 4, 31, 2016.
8. Richhariya, G., Kumar, A., Tekasakul, P., Gupta, B., Natural dyes for dye sensitized solar cell: A review, *Renew. Sustain. Energy Rev.*, 69,705, 2017.
9. Desalegn, J.G., Sisay, T.A., Teketel, Y.A., Dye sensitized solar cells using natural pigments from five plants and quasi-solid state electrolyte. *J. Braz. Chem. Soc.*, 26, 1, 2015. http://dx.doi.org/10.5935/0103-5053.20140218.
10. Sivakumar, V., Vijaeeswarri, J., Anna, J.L., Effective natural dye extraction from different plant materials using ultrasound. *Ind. Crops Prod.*, 33, 116, 2011.
11. Goktas, O., Duru, M.E., Yeniocak, M., Ozen, E., Determination of the color stability of an environmentally friendly wood stain derived from laurel (*Laurus nobilis* L.) leaf extracts under UV exposure. *Forest Prod. J.*, 58, 77, 2008.
12. Savvidou, M.I. and Economides, D.G., Colour gamut produced by applying mixtures of natural dyes on de-inked mechanical pulp. *Color Technol.*, 123, 2, 119–23, 2007.
13. Van Dam, J.E., Den, van, Oever, M.J., Keijsers, E.R., van der Putten, J.C., Anayron, C., Josol, F., Peralta, A., Process for production of high density/high performance binder less boards from whole coconut husk: Part 2: Coconut husk morphology, composition and properties. *Ind. Crops Prod.*, 24, 96, 2006.

14. Mansour, R., Natural dyes and pigments: Extraction and applications. In: *Handbook of Renewable Material for Coloration and Finishing*, M. Yusuf (Ed.), pp. 75-102, Scrivener Publishing, Massachusetts, 2018.
15. Samantha, K. and Agarwal, P., Review article on application of natural dyes on textiles. *Indian J. Fibre Text. Res.*, 34, 384, 2009.
16. Kumar, R. and Tripathi, Y. C., Natural dye from forest biomass. Training manual on extraction technology of natural dyes & aroma therapy and cultivation value addition of medicinal plants, 1, 51, 2011.
17. Petrakis, E.A., Cagliani, L.R., Tarantilis, P.A., Polissiou, M.G., Consonni, R., Sudan dyes in adulterated saffron (*Crocus sativus* L.): Identification and quantification by 1H NMR. *Food Chem.*, 217, 418, 2017.
18. Piatkowska, M., Jedziniak, P., Olejnik, M., Żmudzki, J., Posyniak, A., Absence of evidence or evidence of absence? A transfer and depletion study of Sudan I in eggs. *Food Chem.*, 239, 598, 2018.
19. Christie, R.M., *Environmental aspects of textile dyeing*. Woodhead Publishing, Cambridge, 2007.
20. National Toxicology Program. Carcinogenesis bioassay of CI solvent yellow 14 (CAS No. 842-07-9) in F344/N rats and B6C3F1 mice (feed study). *National Toxicology Program Technical Report Series*, 226, 1, 1982.
21. Duman, O., Tunc, S., Polat, T.G., Adsorptive removal of triarylmethane-dye (Basic Red 9) from aqueous solution by sepiolite as effective and low-cost adsorbent. *Microporous Mesoporous Mater.*, 210, 176, 2015.
22. Lacasse, K. and Baumann, W., *Textile chemicals: Environmental data and facts*, Springer, Dortmund, 2012.
23. Foguel, M.V., Ton, X.A., Zanoni, M.V., Maria Del Pilar, T.S., Haupt, K., Bui, B.T.S., A molecularly imprinted polymer-based evanescent wave fiber optic sensor for the detection of basic red 9 dye. *Sens. Actuators B*, 218, 222, 2015.
24. Sivarajasekar, N. and Baskar, R., Adsorption of basic red 9 on activated waste *Gossypium hirsutum* seeds: Process modeling, analysis and optimization using statistical design. *J. Ind. Eng. Chem.*, 20, 2699, 2014.
25. Pohanish, R.P., *Sittig's Handbook of Toxic Hazardous chemicals and carcinogens*, William Andrew, Amsterdam: Elsevier, Cambridge, 2017.
26. National Toxicology Program. NTP Toxicology and Carcinogenesis Studies of CI Basic Red 9 Monohydrochloride (Pararosaniline)(CAS No. 569-61-9) In F344/N Rats and B6C3F1 Mice (Feed Studies). *National Toxicology Program Technical Report Series*, 285, 1, 1986.
27. Ali, H.M., Shehata, S.F., Ramadan, K.M.A., Microbial decolorization and degradation of crystal violet dye by *Aspergillus niger. Int. J. Environ. Sci. Technol.*, 13, 2917, 2016.
28. Mani, S., Bharagava, R.N., Exposure to crystal violet, its toxic, genotoxic and carcinogenic effects on environment and its degradation and detoxification for environmental safety. Reviews of environmental contamination and toxicology, P. de Voogt (Ed.), Vol. 237, pp. 71–104, Springer, Cham, 2016. https://doi.org/10.1007/978-3-319-23573-8_4.

29. Azmi, W., Sani, R.K., Banerjee, U.C., Biodegradation of triphenylmethane dyes. *Enzyme Microb. Technol.*, 22, 185, 1998.
30. Newman, M.C., *Fundamentals of ecotoxicology: The science of pollution*, CRC Press, Boca Raton, 2015.
31. Bharagava, R.N., Mani, S., Mulla, S.I., Saratale, G.D., Degradation and decolourization potential of an ligninolytic enzyme producing *Aeromonas hydrophila* for crystal violet dye and its phytotoxicity evaluation. *Ecotoxicol. Environ. Saf.*, 156, 166, 2018.
32. Littlefield, N.A., Blackwell, B.N., Hewitt, C.C., Gaylor, D.W., Chronic toxicity and carcinogenicity studies of gentian violet in mice. *Toxicol. Sci.*, 5, 902, 1985.
33. Chakraborty, J. N., Fundamentals and practices in colouration of textiles, pp. 241–244, Woodhead Publishing India Pvt. Ltd., New Delhi, 2014.
34. Sharma, D., Gupta, C., Aggarwal, S., Nagpal, N., Pigment extraction from fungus for textile dyeing. *Indian. J. Fibre Text.*, 37, 73, 2012.
35. Kasiri, M.B. and Safapour, S., Natural dyes and Antimicrobials for textiles, in: *Green Materials for Energy, Products and Depollution*, E. Lichtfouse, J. Schwarzbauer, D. Robert (Eds.), p. 229–286, Springer, New York, 2013.
36. Gulrajani, M.L. and Gupta, D., *Natural dyes and application to textiles*, Department of textile technology. Indian Institute of Technology, New Delhi, India, 1992.
37. Dedhia, E., Natural Dyes. *Colourage*, 45, 3, 45, 1998.
38. Gupta, K.C., Gupta, P., Singh, P., Singh, S.V., Agarwal, S., Chemistry of natural dyes, in: *Natural Dyes: Scope and Challenges*, 3rd Ed, M. Daniel, S.D. Bhattacharya, A. Arun, M.R. Vinay (Eds.), p. 7–34, Scientific Publishers, Jodhpur, India, 2006.
39. Vankar, P.S., Handbook on natural dyes for industrial applications, pp. 36–42, NIIR Project Consultancy Services, New Delhi, 2016.
40. Samanta, A.K., Bio-dyes, bio-mordants and bio-finishes: scientific analysis for their application on textiles [Online First]. *IntechOpen.*
41. Sangeetha, K., Gomathi, R., Bhuvaneshwari, M., Dyeing of silk fabric using lemon leaves extract with the effect of different mordants. *Int. J. Innovative Res. Sci. Eng. Technol.*, 4, 6, 2015.
42. Mansour, H.F. and Heffernan, S., Environmental aspects on dyeing silk fabric with *Sticta coronata* lichen using ultrasonic energy and mild mordants. *Clean Techn. Environ. Policy*, 13, 207, 2011.
43. USA Antique Rugs of the Future (2004–2009) Azerbaijan Rugs, AATCC (2002) Technical manual. *AATCC*, North Carolina, 2002.
44. Teklemedhin, T.B. and Gopalakrishnan, L.H., Environmental friendly dyeing of silk fabric with natural dye extracted from *Cassia singueana* plant. *J. Textile Sci. Eng.*, 3, 1, 2018.
45. Adeel, S., Habib, N., Arif, S., Rehman, F., Azeem, M., Batool, F., Amin, N., Microwave-assisted eco-dyeing of bio mordanted silk fabric using cinnamon

bark (Cinnamomum verum) based yellow natural dye, *Sustain. Chem. Pharm.*, 17,1, 2020.
46. Abdul Rahman, N.A., Tajuddin, R., Tumin, S.M., Optimization of natural dyeing using ultrasonic method and biomordant. *Int. J. Chem. Eng. Appl.*, 4, 3, 2013.
47. Yusuf, M., Khan, S., A., Shabbir, M., Mohammad, F., Developing a shade range on wool by madder (*Rubia cordifolia*) root extract with gallnut (*Quercus infectoria*) as biomordant. *J. Nat. Fibre*, 14, 4, 2017.
48. Rani, N., Jajpura, L., Butola, B.S., Ecological dyeing of protein fabrics with *Carica papaya* l. leaf natural extract in the presence of bio-mordants as an alternative copartner to metal mordants. *J. Inst. Eng. India Ser.*, 101, 19, 2020. https://doi.org/10.1007/s40034-020-00158-1.
49. Rather, L.J., Islam, S., Shabbir, M., Bukhari, M.N., Shahid, M., Khan, M.A., Mohammad, F., Ecological dyeing of Woolen yarn with *Adhatoda vasica* natural dye in the presence of biomordants as an alternative copartner to metal mordants. *J. Environ. Chem. Eng.*, 4, 3041, 2016.
50. Aminoddin, H., Functional dyeing of wool with natural dye extracted from *Berberis vulgaris* wood and *Rumex hymenosepolus* root as biomordant. *Iran. J. Chem. Chem. Eng.*, 29, 55, 2010.
51. Ansari, T.N., Iqbal, S., Barhanpurkar, S., Ecofriendly dyeing with *Senegalia catechu* using biomordant. *Int. J. Creative Res. Thoughts*, 6, 1351, 2018.
52. Yusuf, M., Green dyes and pigments: Classes and applications, pp. 30–41, Laxmi Book Publication, Solapur, 2016.
53. Maryam, A., Khan, S.I., Riaz, A., Ali, S., Noreen, S., Effect of mordants with the application of natural dye extracted from *Allium cepa* on natural fabric. *Int. J. Home Sci.*, 5, 283, 2019.
54. Thangabai, T. and Kalaiarasi, K., Microwave-assisted natural dye extraction from *Pterocarpus soyauxii*.. *Int. J. Eng. Technol. Sci. Res.*, 5, 4, 2018.
55. Wizi, J., Wang, L., Hou, X., Tao, Y., Ma, B., Yang, Y., Ultrasound-microwave assisted extraction of natural colorants from sorghum husk with different solvents. *Ind. Crops Prod.*, 120, 203, 2018.
56. Merdan, N., Eyupoglu, S., & Duman, M. N., Ecological and sustainable natural dyes. In: *Textile and Clothing Sustainability*, S. S. Muthu (Ed.), pp. 1–42, Springer, Singapore, 2017.
57. Furry, M. S. and Morrison, B. V., Home dyeing with natural dyes, pp. 5–6, U.S. Department of Agriculture Miscellaneous publication no. 230, Washington, 1935.
58. Yusuf, M., Shabbir, M., Mohammad, F., Natural Colorants: Historical, Processing and Sustainable Prospects. *Nat. Prod. Bioprospect.*, 7, 123, 2017.
59. Vankar, P.S. and Shankar, R., Ecofriendly ultrasonic natural dyeing of cotton fabric with enzyme pretreatments. *Desalination*, 230, 62, 2008.
60. Raja, A.S.M. and Thilagavathi, G., Influence of enzyme and mordant treatments on the antimicrobial efficiency of natural dyes on wool materials. *Asian J. Text.*, 1, 138, 2011.

61. Kamel, M.M., El-Shishtawy, R.M., Yussef, B.M., Mashaly, H., Ultrasonic assisted dyeing: III. Dyeing of wool with lac as a natural dye. *Dyes Pigm.*, 65, 103, 2005.
62. Vankar, P.S., Shanker, R., Srivastava, J., Ultrasonic dyeing of cotton fabric with aqueous extract of *Eclipta alba*. *Dyes Pigm.*, 72, 33, 2007.
63. Vankar, P.S., Shanker, R., Dixit, S., Mahanta, D., Sonicator dyeing of cotton, wool and silk with the leaves extract. *JTATM*, 6, 1, 2009.
64. Haji, A., Naebeb, M., Cleaner dyeing of textiles using plasma treatment and natural dyes: A review, *J. Clean. Prod.*, 265, 1, 2020.
65. Habip, D., Nigar, M., Mehmet, K., Dilek, K., The effect of plasma treatment on the dyeability of silk fabric by using *Phytolacca decandra* L. natural dye extract. *Tekst. ve Konfeksiyon*, 26, 262, 2016.
66. Haji, A., Improved natural dyeing of cotton by plasma treatment and chitosan coating. Optimization by response surface methodology. *Cellulose Chem. Technol.*, 51, 975, 2017.
67. Saxena, S. and Raja, A., Natural Dyes: Sources, Chemistry, Application and Sustainability Issues, in: *Roadmap to Sustainable Textiles and Clothing. Textile Science and Clothing Technology*, S. Muthu (Ed.), p. 37–80, Springer, Singapore, 2014, https://doi.org/10.1007/978-981-287-065-0_2.
68. Boruah, S. and Kalita, B., Eco-printing of eri silk with turmeric natural dye. *International Journal of Textile and Fashion Technology (IJTFT)*, 5, 27, 2015.
69. Kavyashree, M., Printing of textiles using natural dyes: A global sustainable approach [Online First]. *IntechOpen*.
70. Jimmy, K., Sung, W.P., Chen, R., Silk fabric painted with natural dye from *Acacia catechu* Willd-By using flour of wild taro (*Colocasia esculenta* (L.) Schott) as resist printing paste. *Adv. Mater. Res.*, 1030, 434, 2014.
71. Teli, M.D. and Paul, R., Novel natural dye from coffee seed coat. *Int. Dyer*, 191, 29, 2006.
72. Lee, Y.H., Dyeing, fastness, and deodorizing properties of cotton, silk, and wool fabrics dyed with coffee sludge (*Coffea arabica* L.) extract. *J. Appl. Polym. Sci.*, 103, 251, 2007.
73. Shanker, R. and Vankar, P.S., Dyeing cotton, wool and silk with *Hibiscus mutabilis* (Gulzuba). *Dyes Pigm.*, 74, 464, 2007.
74. Dayal, R., Dobhal, P.C., Kumar, R., Onil, P., Rawat, R.D., Natural dye from *Parthenium hysterophorus*. *Colourage*, 55, 75, 2008.
75. Indi, Y.M. and Chinta, S.K., Application and properties of natural dye on cotton *Phyllanthus reticulatus*. *Colourage*, 55, 52, 2008.
76. Vankar, P.S. and Shanker, R., Ultrasonic Dyeing of cotton and silk with *Nerium oleander* flower. *Colourage*, 55, 90, 2008.
77. Mukherjee, A., Mitra, S., Konar, P., Mukherjee, A., Modification of natural yellow dyes for improvement of washing and light fastness. *Colourage*, 56, 74, 2009.

78. Kulkarni, S.S., Gokhale, A.V., Bodake, U.M., Pathade, G.R., Cotton dyeing with natural dye extracted from pomegranate (*Punica granatum*) Peel. *Universal J. Environ. Res. Technol.*, 1, 135, 2011.
79. Srivastava, A., Priyanka, Parmar, M.S., Eco-friendly dyeing of natural fabrics using discarded litchi fruit peel., 2012. https://www.fibre2fashion.com/industry-article/6402/eco-friendly.
80. Jain, H., Dyeing of cotton fabrics with extracts of jamun tree and its products using natural mordants. *Colourage*, 60, 40, 2013.
81. Mohammad, S.I., Sarker, P., Mamun, A.A., Iqbal, S.M.F., Influence of natural mordants in coloration of cotton knitted fabric with mango seed kernel extract dyes. *Int. J. Curr. Eng. Technol.*, 10, 358, 2020.
82. Rachel, D.A. and Hussain, B.M.Z., Healthcare textile dyed natural socks. *Int. J. Res. Trends Innovation*, 4, 79, 2019.
83. Chandel, S., Patel, J., Chauhan, A., Organic dye extraction from *Brassica oleracea var. botrytis* (cauliflower). *Int. Res. J. Eng. Technol.*, 6, 498, 2019.
84. Singam, R.T., Marsi, N., Mamat, H., Rus, A.Z.M., Nasir, S.H., Shaari, M.F., Physical properties of natural dyes based on *Lawsonia inermis*, *Azadiractha indica* and *Curcuma longa* coated with superhydrophobic coating for textile applications. *Int. J. Adv. Res. Eng. Technol.*, 11, 301, 2020.
85. Pan, N.C., Chattopadhay, S.N., Day, A., Dyeing of jute with natural dyes. *Indian J. Fibre Text. Res.*, 28, 339, 2003.
86. Mathur, J.P., Mehta, A., Karnawat, R., & Bhandari, C.S.., Use of neem bark as wool colourant Optimum conditions of wool dyeing. *Indian J. Fibre Text. Res.*, 28, 94-99, 2003.
87. Bechtold, T. and Mahmud-Ali, A., Mussak, Reuse of ash-tree (*Fraxinus excelsior* L.) bark as natural dyes for textile dyeing: Process conditions and process stability. *Color. Technol.*, 123, 271, 2007.
88. Jayalakshmi, I. and Amsamani, S., Dyeing wool using biomordants. *Colourage*, 55, 102, 2008.
89. Mohammad, F., Yusuf, M., Shahid, M., Khan, S.A., Khan, M.I., Shahidul, I., Khan, M.A., Dyeing of wool with the extract of henna leaves using mixed metal mordants. *Colourage*, 59, 51, 2012.
90. Uddin, M.G., Extraction of eco-friendly natural dyes from mango leaves and their application on silk fabric. *Text Cloth Sustain*, 1, 7, 2015.
91. Swamy, V.N., Assessment of calorimetric, antibacterial and fastness properties of silk fabric dyed with *Casuarina equisetifolia* L. leaf extract. *Indian J. Tradit. Knowl.*, 16, 714, 2017.
92. Banerjee, A.N., Kotnala, O.P., Maulik, S.R., Dyeing of eri silk with natural dyes in presence of natural mordants. *Indian J. Tradit. Knowl.*, 17, 396, 2018.
93. Shukla, P., Upreti, D.K., Nayaka, S., Tiwari, P., Natural Dyes from Himalayas Lichens. *Indian J. Tradit. Knowl.*, 13, 195, 2014.
94. Khan, M.A., Khan, M., Srivastava, P.K., Mohammad, F., Extraction of natural dyes from myrobalan, gallnut and pomegranate, and their application on wool. *Colourage*, 52, 53, 2005.

95. Lokhande, H.T. and Dorugade, V.A., Dyeing nylon with natural dyes. *Am. Dyest. Rep.*, 88, 29, 1999.
96. Miah, M.R., Zakaria, Hossain, M.A., Dipto, A.I., Telegin, F.Y., Quan, H., Eco-dyeing of nylon fabric using natural dyes extracted from onion outer shells: assessment of the effect of different mordant on color and fastness properties. *Int. J. Sci. Eng. Res.*, 7, 1043, 2016.
97. Purwar, S., Application of natural dye on synthetic fabrics: A review. *Int. J. Home Sci.*, 2, 283, 2016.
98. Elnagar, K., Elmaaty, T.A., Raouf, S., Dyeing of polyester and polyamide synthetic fabrics with natural dyes using ecofriendly technique. *Journal of Textiles*, 2014, 8, 2014.
99. Shahin, M.F., Ahmed, R.M., Marie, M.M., Optimizing the dyeing process of alkali-treated polyester fabric with dolu natural dye. *Int. J. Eng. Res. Appl.*, 4, 35, 2014.
100. Guizhen, K., Dyeing properties on natural dye extracted from *Rhizoma coptidis* on acrylic fiber. *Indian J. Fiber Text. Res.*, 39, 102, 2014.
101. Pruthi, N., Chawla, G.S., Yadav, S., Dyeing of silk with Barberry bark dye using mordant combination. *Nat. Prod. Radiance*, 7, 40–44, 2008.
102. Iqbal, S. and Ansari, T.N., Natural dyes: Their sources and ecofriendly use as textile materials. *J. Environ. Res. Dev.*, 8, 683, 2014.
103. Shabbir, M., Islam, S.U., Bukhari, M.N., Application of *Terminalia chebula* natural dye on wool fiber–Evaluation of color and fastness properties. *Text. Cloth. Sustain.*, 2, 1, 2017.
104. Susmitha, S., Vidyamol, K.K., Ranganayaki, P., Vijayaragavan, R., Phytochemical extraction and antimicrobial properties of *Azadirachta indica* (Neem). *Global J. Pharmacol.*, 7, 316, 2013.
105. Feng, X.X., Zhang, L.L., Chen, J.Y., Zhang, J.C., New insights into solar UV-protective properties of natural dye. *J. Cleaner Prod.*, 15, 366, 2007.
106. Grifoni, D., Bacci, L., Zipoli, G., The role of natural dyes in the UV protection of fabrics made of vegetable fibres. *Dyes Pigm.*, 91, 279, 2011.
107. Mongkholrattansit, R., Krystufek, J., Wiener, J., Vikova, M., Dyeing, fastness and UV protection properties of silk and wool fabrics dyed with eucalyptus leaf extract by the exhaustion process. *Fibres Text.*, 86, 94, 2011.
108. Desore, A. and Narula, S.A., An overview on corporate response towards sustainability issues in textile industry. *Environ. Dev. Sustainability*, 20, 1439, 2018.
109. Bhatia, S. C., Pollution control in textile industry. WPI Publishing. New Delhi, 2017.
110. Hossain, M.S., Das, S.C., Islam, J.M., Al Mamun, M.A., Khan, M.A., Reuse of textile mill ETP sludge in environmental friendly bricks–effect of gamma radiation. *Radiat. Phys. Chem.*, 151, 77, 2018.
111. Muthu S.S., Introduction. In: *Sustainability in the Textile Industry*, Muthu S. (Ed.). pp. 1–7, Textile Science and Clothing Technology. Springer, Singapore, 2017.

112. Setiadi, T., Andriani, Y., Erlania, M., Treatment of textile wastewater by a combination of anaerobic and aerobic processes: A denim processing plant case, in: *Southeast Asian Water environment (Biodiversity and Water environment)*, S. Ohgaki, K. Fukushi, H. Katayama, S. Takizawa, C. Polprasert (Eds.), p. 159–166, IWA Publishing, Oxford, 2006.
113. Vigneshpriya, D. and Shanthi, E., Physicochemical Characterization of Textile Wastewater. *Int. J. Innovative Res. Dev.*, 4, 48, 2015.
114. Mahapatra, N.N., *Textile dyes*, p. 1–7, CRC Press, Woodhead Publishing India Pvt, Boca Raton, 2016.
115. Hassan, M.M. and Carr, C.M., A critical review on recent advancements of the removal of reactive dyes from dye house effluent by ion-exchange adsorbents. *Chemosphere*, 209, 201, 2018.
116. Shamey, R. and Zhao, X., *Modelling, simulation and control of the dyeing process*, p. 31–53, Elsevier, Amsterdam, 2014.
117. Imran, M., Crowley, D.E., Khalid, A., Hussain, S., Mumtaz, M.W., Arshad, M., Microbial biotechnology for decolorization of textile wastewaters. *Rev. Environ. Sci. Biotechnol.*, 14, 73, 2015.
118. Aquino, J.M., Rocha-Filho, R.C., Ruotolo, L.A., Bocchi, N., Biaggio, S.R., Electrochemical degradation of a real textile wastewater using β-PbO2 and DSA® anodes. *Chem. Eng. J.*, 25, 138, 2014.
119. Khatri, J., Nidheesh, P.V., Singh, T.A., Kumar, M.S., Advanced oxidation processes based on zero-valent aluminium for treating textile wastewater. *Chem. Eng. J.*, 348, 67, 2018.
120. Sandhya, S., Biodegradation of azo-dyes under anaerobic condition: Role of azoreductase, in: *Biodegradation of Azo-dyes*, H.A. Erkurt (Ed.), p. 39–57, Springer, Berlin, 2010.
121. Rehman, K., Shahzad, T., Sahar, A., Hussain, S., Mahmood, F., Siddique, M.H., Siddique, M.A., Rashid, M.I., Effect of reactive black 5 azo-dye on soil processes related to C and N cycling. *Peer J.*, 6, e4802, 2018.
122. Sharma, B., Dangi, A.K., Shukla, P., Contemporary enzyme based technologies for bioremediation: A review. *J. Environ. Manage.*, 210, 10, 2018.
123. Khan, S. and Malik, A., Toxicity evaluation of textile effluents and role of native soil bacterium in biodegradation of a textile dye. *Environ. Sci. Pollut. Res. Int.*, 25, 4446, 2018.
124. Clark, M., (Ed.), Handbook of textile and industrial dyeing: Principles, processes and types of dyes. Woodhead Publishing, Philadelphia, 2011.
125. Hunger, K., *Industrial Dyes: Chemistry, Properties and Applications*, Willey-VCH, Weinheim, 2003.
126. Thakur, I.S., *Environmental Biotechnology: Basic Concepts and Applications*, I.K. International Pvt, New Delhi, 2006.
127. Tiwari, S., Tripathi, A., Gaur, R., Bioremediation of plant refuges and xenobiotics, in: *Principles and Applications of Environmental Biotechnology for a Sustainable Future*, R.L. Singh (Ed.), p. 85–142, Springer Science, Singapore, 2016.

128. Mondal, S., Purkait, M.K., De, S., *Advances in dye removal technologies*, Springer, Singapore, 2018.
129. Asthana, V. and Shukla, A.C., *Water security in India: Hope, despair, and the challenges of human development*, Bloomsbury Publishing USA, 2014.
130. Chavan, R.B., *Health and Environmental Hazards of Synthetic Dyes*, https://www.fibre2fashion.com/industry-article/6909/health-and-environmental-hazards-of-synthetic-dyes, 2003.

2

Recent Advances in Non-Aqueous Dyeing Systems

Omer Kamal Alebeid[1,2]*, Elwathig A.M. Hassan[3] and LiujunPei[1,4]

[1]*Engineering Research Center for Eco-Dyeing and Finishing of Textiles, Zhejiang Sci-Tech University, Hangzhou, China*
[2]*Cleaner Production Institute, Industrial Research and Consultancy Center, Khartoum, Sudan*
[3]*Faculty of Industries Engineering and Technology, University of Gezira, Gezira, Sudan*
[4]*School of Fashion Engineering, Shanghai University of Engineering Science, Shanghai, China*

Abstract
Dyeing processes consume a huge volume of water and hazardous chemicals, creating serious environmental problems. The chemicals uses in this process are extremely toxic and cannot biodegradable easily. Consequently, effluents need appropriate disposal treatment in accordance with the wastewater nature and discharge limits; hence, different physical, chemical, and biological processes have been applied for this treatment. Recently, some processes employing non-aqueous dyeing systems have been suggested with the aim of reducing water and chemical consumption and waste discharge. This chapter will discuss the latest techniques of non-aqueous dyeing process applied in textile dyeing i.e.: supercritical fluid dyeing, reverse micelle system, solvent dyeing, and silicone non-aqueous dyeing.

***Keywords*:** Supercritical carbon dioxide, reverse micelle dyeing, solvent dyeing, silicone non-aqueous dyeing

2.1 Introduction

Textile dyeing is a wet process that consumes a large volume of clean water, which puts high pressure on the global water resource [1]. As an example,

**Corresponding author*: omer_z22@hotmail.com

Luqman Jameel Rather, Mohd Shabbir and Aminoddin Haji (eds.) Sustainable Practices in the Textile Industry, (43–74) © 2021 Scrivener Publishing LLC

the conventional aqueous dyeing demands enormous amount of water (0.1–0.18 L of water for dyeing 1 g of fiber), which is dependent on the machinery, chemical structure of dyestuff, and types of fiber [2]. The aqueous dyeing process has inherent environmental problems because at the end of the dyeing process the water involves a considerable quantity of dyestuff and a large amount of salts and alkali [3, 4]. Most of these chemical mixtures represent a direct or indirect risk to human health, marine life, and cause soil and water pollution. In order to overcome these obstacles from the point of view of environmental aspects and save time and water during dyeing processing, the textile industry is continuously searching for alternative ecological dyeing techniques [5]. Therefore, the most remarkable approaches that could fulfill these requirements are supercritical fluid dying, reverse micelle dying, solvent dyeing, and silicone non-aqueous dyeing.

2.2 Supercritical Fluid Dyeing System

Recently, the supercritical carbon dioxide ($scCO_2$) is considering as one of a substantial and significant developments in green chemistry. The $scCO_2$ approach was first offered to the textile industry by E. Schollmeyer in 1988 [6, 7]. This process is a largely waterless dyeing technology utilizes only a tiny amount of recycled carbon dioxide (CO_2) at a certain pressure and temperature. CO_2 is a gas at ambient atmospheric pressure and normal temperature, when the pressure increases the CO_2 converts into a liquid form and due to the increasing intermolecular forces converts into a solid form. At a certain pressure and temperature (32 °C and 73 MPa), CO_2 is in the supercritical state, which known as supercritical carbon dioxide [8]. Mostly, disperse dyes are the paramount applicable dyes in the $scCO_2$ dyeing process because their molecular structures display low polarity, thus, can ensure an excellent solubility in $scCO_2$. Subsequently, dye molecules are fixed to the surface of the fiber by absorption, and diffusion into the polymer matrix. The combination of chemical, and physical interactions between the polymer macromolecule and the dye molecules significantly improved the excellent fixation properties [9]. A schematic diagram of the $scCO_2$ approach was presented in Figure 2.1. Before treatment, the dye powders were filled into a dye tank cylinder, and fixed into a dyestuff cylindrical vessel. The CO_2 hoarded in a gas cylinder was firstly injected into purifier for filtering and cleaning, and then melted by using a refrigerator system. The liquid of CO_2 forms was heated to an above critical temperature (31.30 °C) with an internal heat exchanger at pressure

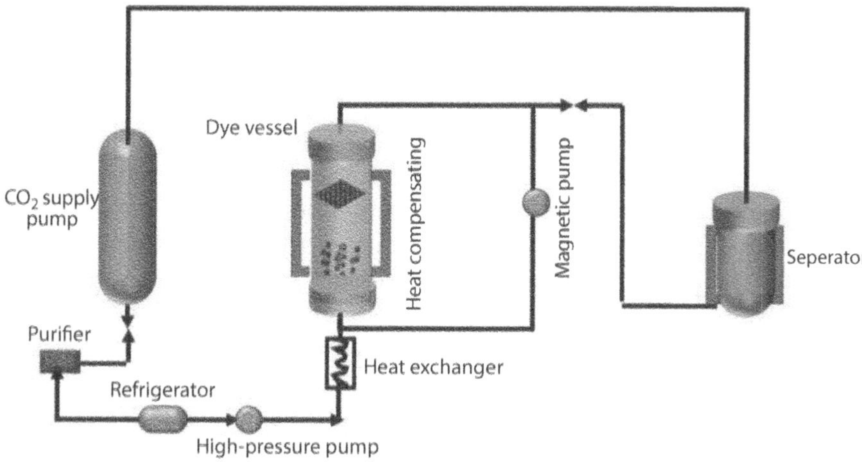

Figure 2.1 Schematic diagram of the supercritical CO_2 apparatus equipped.

above the critical point (7.30 MPa) using a high-pressure syringe pump. Subsequently, in a supercritical state form, CO_2 fluid was pumped into the dyestuff vessel in which dyes molecules could be handled. The magnetic drive pump was started to form an exemplary scCO_2 cycle, when the demand treatment temperature and pressure were reached. After the treatment, the CO_2 liquid was separated sufficiently in a pressure vessel and separator under the low temperatures (25 to 42 °C). The precipitated dye molecules were kept onto the bottom of the cylinder separator, whereas the depressurized CO_2 gas was cooled down and recovered back into the gas tank.

Supercritical CO_2 is the most extensively used supercritical fluid due to many advantages such as inexpensive, essentially nontoxic, recyclable, abundant, non-flammable, a waterless process, environmentally friendly, solvent zero discharge, easy dye recycling, wide availability, and chemically inert under many conditions [10–13]. Also, the low mass transfer coefficient and higher rate of diffusion of scCO_2 facilitate the permeation of dye molecules into polymer fibers resulting in reducing dyeing time. Moreover, since the scCO_2 dyeing is considered as waterless process, the drying stage saves time and power. Further, the dye molecules cannot be hydrolyzed; as result, most dye molecules are ready to anchor with the fiber [14]. Other significant benefits of the supercritical CO_2 dyeing system, the concentration of dye required for a certain shade is anticipated to be smaller with minimizes the use of auxiliary chemicals in dyeing and remarkably decreases air emissions.

2.2.1 Application of Supercritical CO_2 on Synthetic Fabric

Dyeing of synthetic fibers is different from natural fibers because they possess a high degree of crystallinity, thereby reducing the accessibility of the dyestuff molecules to the amorphous regions within the fiber. The success of dyeing synthetic fibers is mainly depending on the plasticization action of $scCO_2$ on polymers [5]. This phenomenon boosts the polymer chain's mobility and improves the migration of semi-crystalline polymers, thus facilitates the permeation and diffusion of dye molecules into the fiber. Take polyester fabrics as an example, it is often dyed with disperse dyes; these types of dye molecules have an extremely low solubility in water. Therefore, chemical auxiliaries such as dispersing agents, surfactants are required to stable dispersion, maintain a fine during the whole dye process. After the dyeing process is finished, however, unused dyestuffs and auxiliaries are not adsorbed onto polyester, and remain in the liquor cause environmental pollution. The idea of dyeing in $scCO_2$ has proven its efficiency to overcome these environmental problems.

Nowadays, researches on the dyeing of synthetic fabrics with disperse dyes in the $scCO_2$ have been done extensively [15–17]. The effect of the dyeing process conditions (*i.e.*, concentration, dye temperature, pressure, and time) and dyeing properties were examined and compared with conventional dyeing processing. The results indicated that the chemicals and auxiliaries such as dispersing agents, salt have no significant effect on dyeing properties. It also was found the color strength (K/S) values of the dyed fabric improved by increasing the dye content from 2 to 6% owf. (on the weight of fabric) [18]. The leveling properties of polyester fabric in a supercritical CO_2 system were investigated by Long [17]. Hou *et al.* investigated the effect of dyeing pressure and dyeing temperature on the K/S value and color yield. The results exhibited that when the pressure up to 20 MPa and the temperature increase to above 110 °C, the color yield was clearly increased, and the color strength of dyed fabric remarkably improved with increasing dye concentration [19]. Elmaaty *et al.* found that the optimized dyeing temperature and dyeing pressures are 120 °C and 15 MPa, respectively. The colorfastness properties and color strength (K/S) of the fabric dyed in $scCO_2$ were greater than that of traditional aqueous dyeing [15].

The effect of $scCO_2$ dyeing process conditions on the morphology of fibers, crystallinity, and chemical structure was investigated [20]. The result demonstrated that exposing to $scCO_2$ could not cause any chemical structure changes in the polyester fabric. The crystallite size of the polyester polymer after dyeing in supercritical CO_2 was not significantly changed and the crystallinity degree reduced. However, the dyeing in $scCO_2$ at higher temperatures could cause surface morphology changes due to increased migration of oligomer. Also, the effect of the $scCO_2$ on the morphologic structure, supramolecular, and chemical of fibers were examined [21]. The results suggested that CO_2 gas in the supercritical state absorbed into the polymer fiber and boost the oligomer migration from the polyester polymer to the surface, leading to change in the crystal index [22–24]. The influence of CO_2 temperature on the crystal structures, surface morphology, optical absorption properties, and thermal property chemical of disperse dyes were assessed with CO_2 temperature increasing from 120 to 150 °C. FT-IR spectrum of disperse dye was indicated a small changes in the intensities of the infrared bands after $scCO_2$ treatment. Moreover, the β-crystal form of the dye converted into α-crystal form due to the melting crystallization, and dissolution when the CO_2 gas temperature was greater than 115 °C. Thermal stability of the dye were moved to a lower temperature with rising CO_2 temperature [25]. It is proved that the presence of small amounts of polar or non-polar solvents can dramatically improve the solubility of dye molecules in $scCO_2$. The influence of various alcohols (ethanol, methanol, etc.) at various contents and under the working conditions of 25 MPa, 120 °C was investigated. It has appeared that the presence of a modifier content increased the polyester swelling comparing to that of pure $scCO_2$, this endows excellent dyeing properties results to be obtained [26].

Banchero *et al.* verified that the dye-uptake gradually increased by adding methanol as co-solvent, resulting in increased the dye concentration in the fibers and the solubility of the dye molecules in the supercritical CO_2 phase [27]. In respect to understand the exact adsorption behavior of the supercritical CO_2 dyeing of polyester fibers, the rate of dye-uptake and the adsorption behavior of disperse dye were studied. As a consequence, the dye diffusion within the polyester fiber could take place at temperatures below that needed for dyeing in water baths. New hydrazono propanenitrile dye with the possibility of improving antibacterial activity was successfully applied to dye polyester fibers with supercritical CO_2 as a dyeing medium. The result proved that the K/S value was higher than those of the aqueous-based dyeing approach [15]. One of the most valuable advantages of natural dyes is environment friendly, thus, they could not cause any environmental pollution during the processing stage. Moreover, like other natural dyes, applying of natural dyes to polyester fabric using the traditional dyeing method has some obstructions such as poor levelness, poor colorfastness, and difficulty of shade reproducibility. A simple and efficient dyeing process was performed using a curcumin dye molecule which was applied directly to polyester fabric in $scCO_2$ dyeing without including fabric pre-treatment procedures and mordant chemicals. Results demonstrated that the K/S increased when the concentration of dye increased and higher K/S value was obtained using low curcumin content (ca. 0.75% owf). Also, the dyed polyester fabrics showed desirable antioxidants, UV-protection, and antimicrobial activities with reasonable fastness to rubbing and washing [28].

2.2.2 Application of Supercritical CO_2 on Natural Fabric

Unfortunately, conventional synthetic dyes for natural textile fabric possess sulfonic acid functional groups thus, difficult to be dissolved in $scCO_2$ media because of its low permittivity [9, 29]. Additionally, hydroxyl and amino groups of solute molecules are demonstrated to reduce solubility in $scCO_2$. Thus, hydroxyl groups in cotton fiber and amino groups in wool fiber lead to effect negatively on the swelling of fabrics [30]. Due to these reasons, $scCO_2$ dyeing is restricted for dyeing of synthetic textile fibers which use $scCO_2$-soluble disperse dyes. Therefore, the dyeability of natural fibers such as silk, wool, and cotton fabric in the $scCO_2$ dyeing medium is known to be big challenging. To perform the dyeing of natural fibers with $scCO_2$ technology, water-soluble dyes like acid and reactive dyes in $scCO_2$ are synthesized [9, 29, 31–34]. The designed dye known as reactive disperse dyes, they have the ability to form strong covalent chemical bonds with the amine or hydroxyl groups that exist on the natural fibers, resulting in excellent fixation of the dye on the fiber. Three types of mordant dyes that possess chelating ligand properties for wool fabric dyeing were performed in supercritical CO_2, and it demonstrated promising results with good levels of wash fastness [35]. Researchers investigated the possibility of dyeing cotton and wool fabrics in $scCO_2$ using conventional disperse dyes without any modification or pre-treatment, but the K/S value of the dyed fiber was dissatisfying for the cotton and wool fabrics as well as cotton fabrics were harmed by the excess dyeing conditions [7, 36, 37]. Also, the dyeing probability of wool fabric was examined with various disperse dyes in $scCO_2$. The results indicated that the greater K/S values and dye fixation rates could be achieved with disperse blue 148 and disperse Red 153 with respect to other disperse dyes [37].

Many various reactive groups have been widely investigated for dyeing natural fibers. The chemical structures of the most reactive disperse dyes are commonly based on the anthraquinone chromophores or azo-benzene due to their versatility and excellent color strength when modifying reactive of disperse dyes. The superior reactivity of fluorotriazines as reactive groups compared to chlorotriazines was also explained. Fernandez *et al.*

synthesized reactive dyes with fluorotriazine and then they obtained dye was successfully applied on dyeing cotton fabric in supercritical $scCO_2$. It was proved that the fluorotriazines could react with cotton fabric by adding small quantities of acids, as result, K/S value up to 30 was obtained [38]. A series of fluorotriazine reactive dyes were used for dyeing treated cotton fabric by presoaked in an aprotic solvent. Methanol co-solvent were applied during the dyeing with $scCO_2$, the treatments are claimed to give excellent color strength and good fixation properties. Dyes with monofluorotriazine as a reactive functional group was found to be the best preferable modified dyes for dyeing the cotton fabric than monochlorotriazine [33].

Long investigated the dyeing behavior of cotton fabric using reactive disperses dye contain a vinyl sulfone group in $scCO_2$ by employing the catalyst of TEDA. The results suggested that the adsorption of the reactive disperse dyes containing vinyl sulfone group was remarkably dependent on the temperatures. The improvements of color strength and dye fixation efficiency on the both dry and wet cotton fabric were obtained by applying the catalyst under different dying conditions [39]. Gao *et al.* incorporated halogenated acetamide group into reactive disperse dyes and the obtained dye was applied to dye cotton fabric in $scCO_2$ without co-solvent and pre-treatment. The results demonstrated that the color strength, dye-uptake, and fastness property of dyed cotton fiber improved remarkably when increasing time and temperature. The modified disperse dyes contain bromoacetyl functional group was better than chloroacetyl functional group to fix to cotton fiber in $scCO_2$. The existence of $-NO_2$ functional group in the chromophores of dyes might be responsible to enhance the affinity of the dye to the cotton fabric due to better reactivity. Also, the surface morphology of cotton fiber after dyeing was examined, the cotton fiber was slightly swelled, and the surface became somewhat rough, seemed dehiscence (Figure 2.2), probably due to the bond split caused by dyes [34].

Luo *et al.* carried out the dyeing process of cotton and wool fabric in $scCO_2$ with water pre-treatment using azo-based reactive disperse dye containing a pendant vinyl sulfonyl reactive group. The result indicated that water was effect positively on the solubility of the dye in $scCO_2/H_2O$. Moreover, water could act as a swelling agent for the hydrophilic fabric, allowing the dye to be diffused inside the fiber [5]. Yang *et al.* developed reactive disperse dyes based on bi-acyl and mono fluoride reactive functional groups for dyeing cotton fabric in the $scCO_2$ system. On account of the hydroxyl and fluoride could form a strong hydrogen bond as shown Figure 2.3, which indicates that the bi-function group containing fluoride

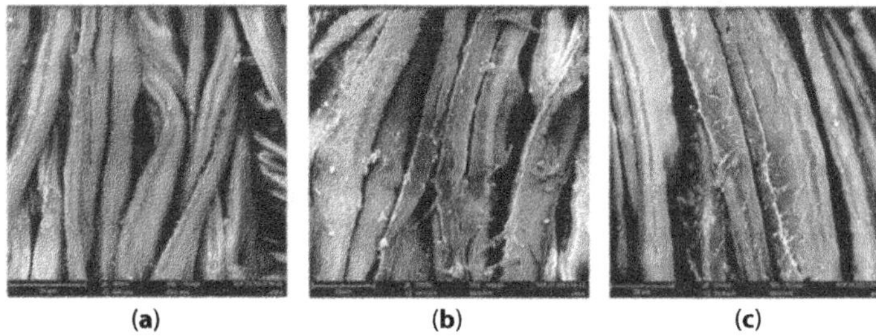

Figure 2.2 SEM photographs, (a): undyed fabric, (b & c): cotton fabric dyed with disperse dye [34].

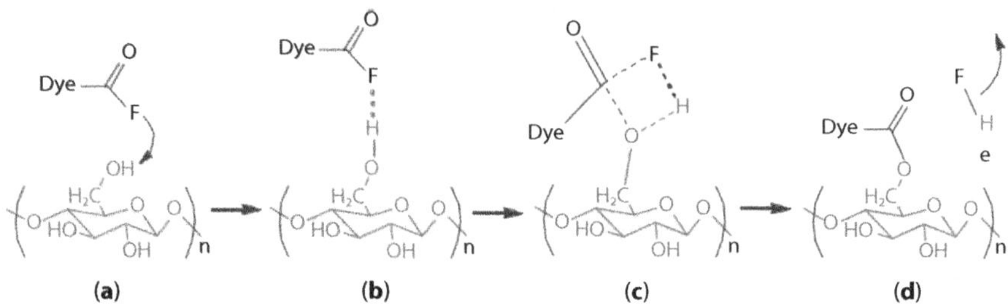

Figure 2.3 The mechanism of the reaction between reactive disperse dyes and cellulose [40].

showed a faster dyeing rate than that of mono function group but with lower color strength [40].

Diazo disperse reactive dyes containing dichlorotriazine reactive groups were profitably synthesized and then applying in the coloration of wool fabric with $scCO_2$. The obtained dyes could be efficiently used for the waterless dyeing of wool fabric with $scCO_2$ media alternative to traditional dying processes for clean production in the wool industries [9].

Wool and silk fabrics were colored with disperse reactive dyes attached with either dichlorotriazine or vinyl sulfone reactive groups in $scCO_2$. The dyeing was done with varying quantities of water in the dyeing vessel, to optimum water contents. The result showed that the reaction with and disperse reactive dyes and textile amino functional groups of silk and wool fabrics in $scCO_2$ resulting in improving the properties of dyeability and water, distributed over the $scCO_2$ water displayed a positive impact on the coloration and dye fixation properties for silk, wool (Figure 2.4) [41]. A new disperse reactive SCFX-AnB_3L dye containing three moieties of an anthraquinonoid matrix, a dichlorotriazine, and a bridge group reactive functional group were effectively developed. The result showed a very strong interaction would occur between the polar groups in the dye molecules,

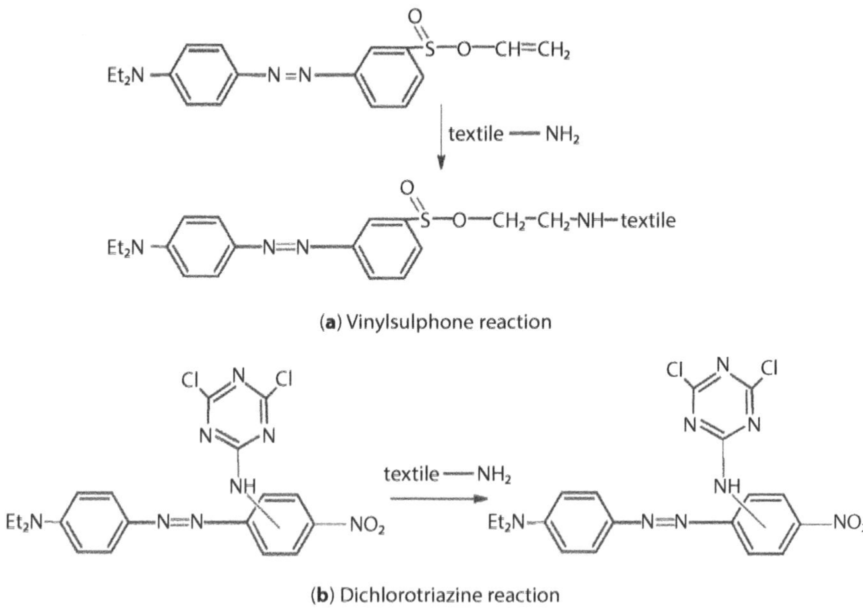

Figure 2.4 Reaction of disperse reactive dyes with textile amino groups, for vinylsulfone (a) and dichlorotriazine (b) dyes.

and the silk chains, which lead to improve the adsorption, and uptake of the dye molecules onto the silk fabrics, accompanied with a subsequent rapid diffusion of dye molecules into the amorphous regions of the fiber.

In contrast, the cotton fiber involves plenty hydroxyl groups (-OH), but their hydrogen reactivity with the dye molecules is significantly lower than that of amino ($-NH_2$) functional groups in silk fabric in the same acid-base condition [42]. An obvious limitation of natural dyes from plants such as curcumin is its incompetence to obtain a deep color depth due to curcumin's low $scCO_2$ solubility. The alkyl groups (ethyl, butyl, hexyl or octyl) were attached to curcumin to obtain better solubility and color strength. The results demonstrated that both the dyeing ability and solubility of curcumin dye improved after alkyl attaching. The best dyeability was attained by using butyl curcumin to dye silk fabric in the existence of dimethylsulfoxide (DMSO) solvent. The improvement of dyeability is related to the excellent butyl curcumin solubility and the high specific surface area of the silk fabric resulting from the synergistic effects of $scCO_2$ and DMSO. Moreover, DMSO also played a co-solvent important role in the $scCO_2$ dyeing process to increase the dyeability of curcumin dye for wool and cotton fabric. Also, methanol was found to be a promising solvent for the pre-treatment of natural fabric [32]. Table 2.1 showed the

Table 2.1 Dyeing conditions and fastness properties of wool, cotton and silk fabric dyed with different types of reactive disperse dyes.

Reactive disperse dyes	Chemical structure	Pretreatment	Fabric type	Dyeing conditions				K/S	Fastness properties			Fixation properties	Ref.
				Temp. (°C)	Pressure (MPa)	Time (h)	Conc. (% o.w.f.)		Wash	Rub	Light		
Fluorotriazine	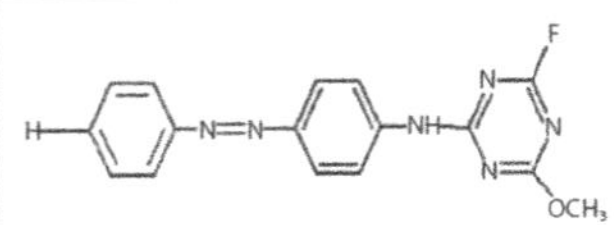	H_3PO_4, Acetic acid, Methanol	Cotton	120	35	1-3	10	26.5	-	-	-	77	[38]
	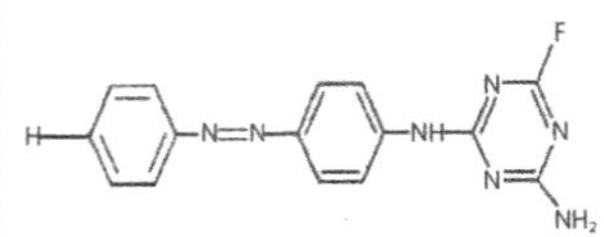		Cotton	120	35	1-3	10	25.6	-	-	-	35	[38]
Chloroacetyl	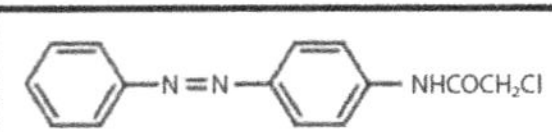	-	Cotton	80 -120	20	1-3	0.	9.8	4-5	4-5	-	-	[34]
Bromoacetyl	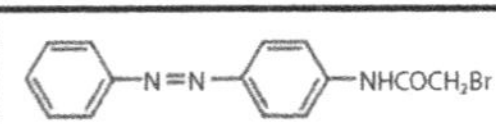	-	Cotton	80-120	20	1-3	0.5	11	4-5	4-5	-	-	[34]
Chloroacetyl	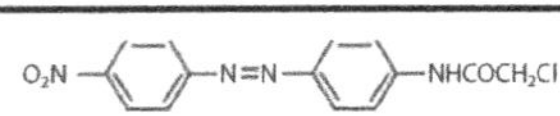	-	Cotton	80-120	20	1-3	0.5	12.6	4-5	4-5	-	-	[34]
Bromoacetyl	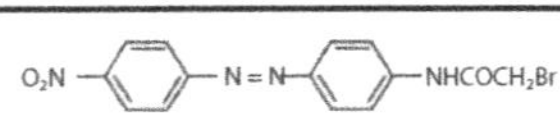	-	Cotton	80-120	20	1-3	0.5	13.2	4-5	4-5	-	-	[34]

(*Continued*)

Table 2.1 Dyeing conditions and fastness properties of wool, cotton and silk fabric dyed with different types of reactive disperse dyes. (*Continued*)

Reactive disperse dyes	Chemical structure	Pretreatment	Fabric type	Dyeing conditions				K/S	Fastness properties			Fixation properties	Ref.
				Temp. (°C)	Pressure (MPa)	Time (h)	Conc. (% o.w.f.)		Wash	Rub	Light		
Chloroacetyl	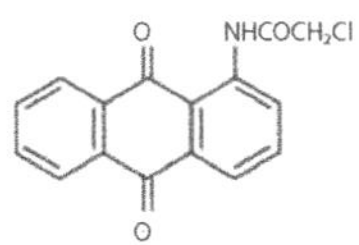	-	Cotton	80–120	20	1–3	0.5	14	4–5	4–5	-	-	[34]
Bromoacetyl	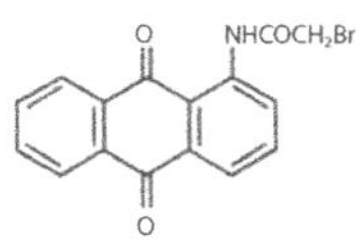	-	Cotton	80–20	20	1–3	0.5	14.3	4–5	4–5	-	-	[34]
Vinylsulfone	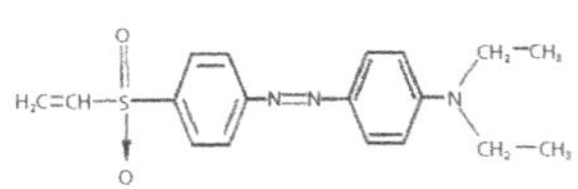	Triethylenediamine (TEDA)	Cotton	50.0–140.0	25.0	2	0.30	8.5	4-5	4-5	-	-	[39]
Vinyl sulphonyl reactive group	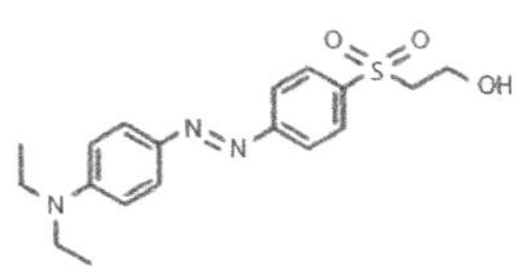	40% omf of pre-treatment water	Cotton	90	14	1	5	-	5	5	5–6	65.8	[5]

(*Continued*)

Table 2.1 Dyeing conditions and fastness properties of wool, cotton and silk fabric dyed with different types of reactive disperse dyes. (*Continued*)

Reactive disperse dyes	Chemical structure	Pretreatment	Fabric type	Dyeing conditions				K/S	Fastness properties			Fixation properties	Ref.
				Temp. (°C)	Pressure (MPa)	Time (h)	Conc. (% o.w.f.)		Wash	Rub	Light		
Vinyl sulphonyl	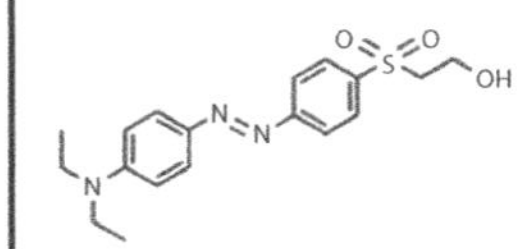	40% omf of pre-treatment water	Wool	90	14	1	5	–	5	5	5–6	99.4	[5]
Mono- and bi- acyl fluoride reactive groups	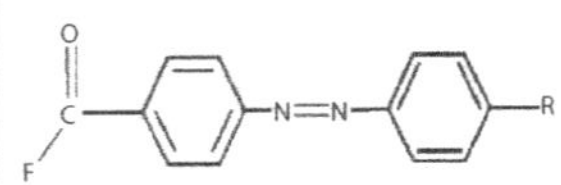 $(CH_3CH_2)_2N$- dye 1; $(CH_3COOCH_2CH_2)_2N$- dye 2; -COF dye 3	-	Cotton	80	20	2h	0.5	2.25	4-5	4-5	-	-	[40]
								2.00					
								0.75					
Vinylsulphone	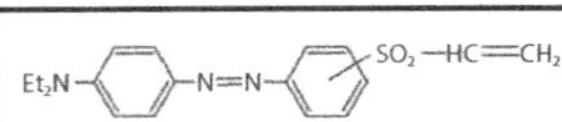	water	Wool	112	23	4h	-	91	-	-	-	-	[41]
			Silk					30					
Dichlorotriazine	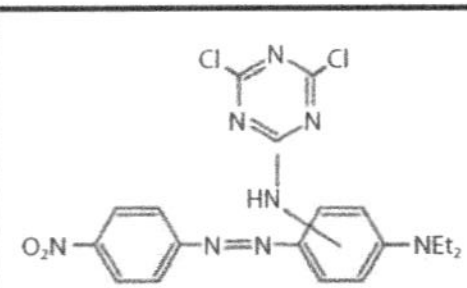	water	Wool	100	25	2h	-	4.40	-	-	-	73	[41]
			Silk					4.73					

(*Continued*)

Table 2.1 Dyeing conditions and fastness properties of wool, cotton and silk fabric dyed with different types of reactive disperse dyes. (*Continued*)

Reactive disperse dyes	Chemical structure	Pretreatment	Fabric type	Dyeing conditions				K/S	Fastness properties			Fixation properties	Ref.
				Temp. (°C)	Pressure (MPa)	Time (h)	Conc. (% o.w.f.)		Wash	Rub	Light		
Dichlorotriazine	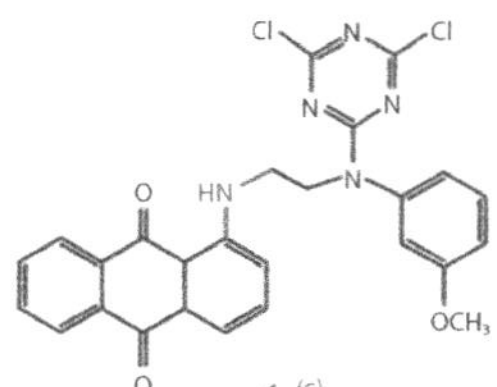 (c)	-	Cotton	–	–	–	–	–	4–5	5	–	58	[42]
			Wool	–	–	–	–	–	5	5	–	91	[42]
			Silk	–	–	–	–	–	5	5	–	95	[42]

dyeing conditions; fastness properties and color strength, of wool, cotton, and silk fabric dyed with various types of reactive disperse dyes.

2.2.3 Dyes Solubility in Supercritical Fluids

The solubility of different dyestuffs in $scCO_2$ is one a critical challenge in supercritical processing because most dyes are prepared especially for water-based dyeing; they exhibit extremely low solubility in $scCO_2$. Moreover, the solubility plays a vital role in the selection of dye, process design, color matching, and the improvement of process efficiency. Recently, many kinds of researcher have been carried out to determine the solubility of different dye solution $scCO_2$ medium to facilitate the application of this technology. Alwi and coworkers examined the solubility of 1,4-diaminoanthraquinone and 1,4-bis(ethylamino) anthraquinonein $scCO_2$ over the temperature range of 50–110 °C and the pressures from 10 to 25 MPa. It was found that the solubility improves with the rise of pressure and temperature of $scCO_2$ and the solubility of 1,4-bis(ethylamino) anthraquinone was greater than 1,4-diaminoanthraquinone. Moreover, there are many categories developed by various different researchers to enhance the solubility of dye molecules in the hydrophobic solvent.

The most powerful and important strategy is mix a small quantity of co-solvent into the $scCO_2$ medium. This approach was proved to be a readily and efficient strategy to remarkable improve the solubility of some polar and/or complexed molecules structural of the dye solutes. The co-solvent such as dimethylsulfoxide, ethanol, acetone, and methanol, etc. could act as both hydrogen bond acceptor and donor for dye solutes to enhance the solubility influence by forming some union complexes. Lee *et al.* investigated the solubility of disperse dyestuff and modified reactive disperse dye in $scCO_2$ with ethanol as a co-solvent. The result showed that the solubility significantly improves up to 20-fold, by adding only 5 mol% of ethanol solvent to the $scCO_2$ fluids. Tsai *et al.* reported that the solubility of disperse yellow 54 in $scCO_2$ with ethanol or dimethyl sulfoxide as co-solvent effectively improved in the presence of co-solvents. Bae *et al.* studied the effect of co-solvent on dyestuff molecules solubility in $scCO_2$, and their findings demonstrated that the solubility of disperse dye in the supercritical fluid was dramatically improved by mixing a small quantity of co-solvent into the fluid.

Banchero *et al.* also proved that addition of ethanol caused a huge improvement in the solubility of different types of disperse dyes in the CO_2. The improvement was strongly relevant to the dyestuff molecular weight. The mixed co-solvents of water and acetone in different contents were utilized to improve the solubility behavior of the developed reactive disperse (SCF-AOL$_2$) dyes solute in the hydrophobically scCO_2. The solubility was remarkably enhanced with increase the pressure, and temperature and reached a maximum at a temperature of 120 °C and pressure of 21 MPa with an equilibrium time of 1 h [48]. The disperse dyes solubility in scCO_2 and the influence of scCO_2 on morphological changes and crystal growth of dye powders have also been investigated [43–47]. It was found that the solubility improves remarkably with rising pressure and is not affected with temperature change. The crystalline of the disperse dye, crystal grains grown in the scCO_2, and Tm also increased with increasing the pressure and temperature. Moreover, the re-crystallized from scCO_2 exhibited stick-like morphology and the crystalline structure of the dye was not appreciable changed after treated with scCO_2. Draper *et al.* found that the solubility behavior is affected by the polarity of the dye molecule; greater solubility was showed by low polar molecules. It was also reported that the dyes with halo or amino functional groups substituent are showed more solubility than those with nitro functional groups in the same aromatic ring positions [49].

2.3 Reverse Micelle Systems

Reactive dyes react with the water molecule through the formation of the hydrogen bond, resulting in weak fastness properties of the cotton fabric [50]. Moreover, the hydrophilicity of both and water the cotton fabric leads to unavoidable impacts on the efficiency of dyeing because water molecules possess the same reactive functionality with cellulose fabric [51]. Moreover, the effective dyeing of natural fabric has not been wieldy investigated with reactive dyes in the scCO_2 because these dyes cannot be well dissolved in the CO_2 medium. Furthermore, high-pressure equipment with strict safety requirement tools need to be used in scCO_2 dyeing, leading to increase the production cost [52]. Reverse micelle methods are non-aqueous solvent media, which considered as one of the more favorable solutions to minimize the generation of impacts of the solvent applied in the dyeing processes [53].

2.3.1 Mechanism and Formation of Reverse Micelle

Reverse micelles are self-assembled structures that employ surfactants as building block agents that simplify the formation of nanoscale pool of water structures in the core region [54]. A small quantity of surfactants could be dissolved as individual molecules according to their behavior and nature like typical electrolytes. On the other hand, in a large scale showed very different behavior after a certain concentration, known as the critical micelle concentration (CMC), via forming an aggregation of a wide range of molecules called a micelle [55] (Figure 2.5a). Reverse micelles are consisting of the smaller nanoparticle sizes (<8 nm) dye-surfactant aggregates at micelles in water where hydrophobic (water-hating) tails of surfactants are directed outward of the micellar core keep in close contact with the non-polar solvent by dispersing forces [56]. The reverse micelles have the property to solubilize small amounts of polar molecules in their interior regions, which water molecules interact with the hydrophilic head of the surfactant by hydrogen bond, dipole–dipole, and ion–dipole interactions, providing a highly stable aqueous micro-environment, known as "water-pool", in a non-aqueous dyeing medium such as an organic solvent.

Commonly, an organic solvent is employed to separate these micelles from an aqueous bulk-like water core (aqueous phase); thus, the method is also called the solvent extraction method. According to the principle, the dye molecule is encapsulated and aggregated by the surfactant's polar head

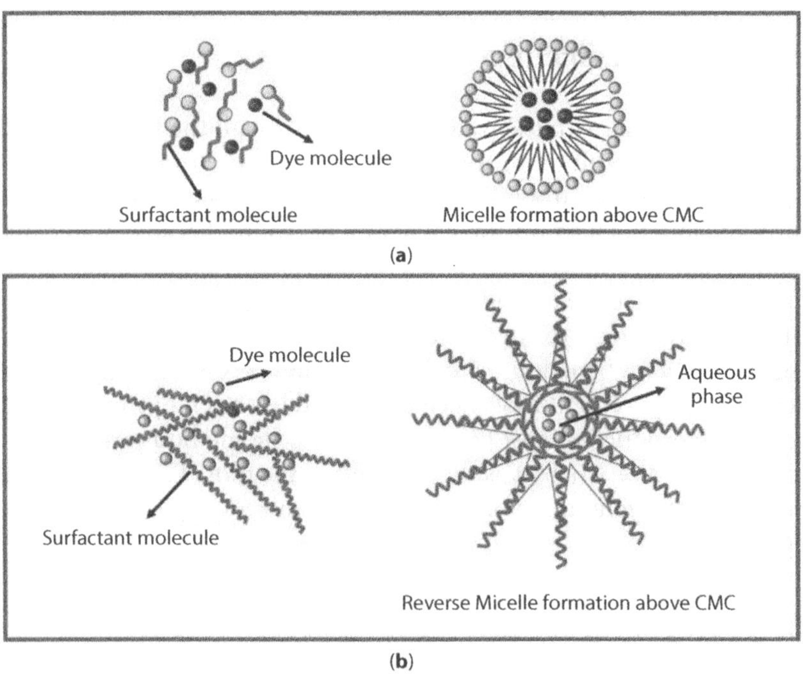

Figure 2.5 Diagram of (a): the micelle, (b): reverse micelle formation.

groups in the reverse micelles via the electrostatic interaction between surfactants and the anionic dyes (Figure 2.5b). The dye molecules separation from the aqueous phase to the micelles depends highly on the physical interaction between the surfactant's polar head groups and dye molecules.

2.3.2 Application of Reverse Micelle Dyeing System

The large amount of water used in the wet processing of cotton fabric may cause negative effects on the environment. The alternative ways to switch the dyeing medium from aqueous to non-aqueous solvent dyeing need to be investigated, so that the dyeing process could be more environmentally friendly and reusable. For similar reasons, the cotton fabric was dyed with non-aqueous reactive dyes, using a chlorinated hydrocarbon as solvent and $scCO_2$. However, poor uniformity was the main constraint associated with these dyeing systems because of inadequate swelling of cotton fabric and largely limited solubility of the reactive dyestuff. The water-soluble dyes are exhibited excellent dispersed in a non-aqueous polar solvent with the assist of reverse micelles. The reverse micelle system was first examined with the application of sodium bis-2-ethyl hexyl sulphosuccinate (Aerosol-OT) as an anionic surfactant [57–59]. Nevertheless, the polar head functional group of the surfactant in the reverse micellar system probable cause a profoundly negative impact on the polarity of the water pool, resulting in an unstable micro-environment during the process of dyeing [60, 61]. It has previously been proved that the ionic header functional group of the surfactant molecules in the water-pool could play a vital role as a dyeing auxiliaries, and the dye adsorption capacities on textile fabric is greatly affected by the surfactant molecules charge [57]. Xie *et al.* proved that the dyeing properties of the micelle dyed fabric with the reactive dye at a low liquor ratio (LR) were similar to the traditionally dyed samples. The fixation properties, reactivity, dye exhaustion, and fastness properties such as washing, perspiration fastness, and wet crocking were comparable to the traditionally dyed fabric. Moreover, the reverse micelle system could save 60% of water, 60% electrolyte saving, and 50% saving in steam during process of dyeing [62]. Non-ionic surfactant (Triton X-100) with rigid aromatic hydrocarbon structure was applied to assist the dyeing of cotton fabric in the reverse micelle system [63–65]. The result demonstrated the Triton X-100 surfactant was not completely dispersed in an aliphatic hydrocarbon solvent and according to K/S values and adsorption kinetic curves did not exhibit an obvious positive effect on dyeability. A fluorinated surfactant are developed to improve the thermodynamically stability reverse micelle, however, the fluorinated surfactants are not considered to be commercial use and caused environmental hazards [66].

Sawada *et al.* evaluated the depth and color fastness properties of dyed wool and silk fabric with acid dye, reactive dye, and direct dye in a reverse micelle system [59]. It was clearly concluded that the adsorption properties of anionic dye on silk fabric in the reverse micelle system is comparable to that in an aqueous dyeing system. Moreover, the dyeing properties of wool fabric with anionic dyes were proved to be less than that of silk fabric because of the low fiber swelling. The reactive dyes demonstrated a large ability to adsorb onto fiber surface compared with other dyes. Moreover, the fixation properties of reactive dyes on fabric could be achieved in the reverse micellar system. The dyeing mechanism of wool fabrics with acid dyes using a reverse micellar dyeing process in a $scCO_2$ as a dyeing medium was examined. A reverse micelle consisted of perfluoro 2, 5, 8, 11-tetramethy l-3,6,9,12-tetraoxapentadecanoic acid ammonium salt/water/CO_2 is showed a high possibility to solubilize acid dyes and dyeing wool fabrics in this system. The dyeability of the wool fabric with acid dyes in this system was not influenced by the CO_2 densities [67]. A non-aqueous alkane medium of octane and heptane was characterized using a reverse micellar system consisting of polyethylene glycol (PEG)-based surfactant. The result exhibited that the reactive dye was competently encapsulated within the inner gap of the reverse micelle system under the optimized parameters in dyeing process. The dyeing properties of cotton fabric with reactive dyes from the solution of reverse micellar were enhanced even without adding any salt electrolytes [52]. Tang *et al.* investigated the dyeing properties of cotton fabric with reactive dyes using PEG-based reverse

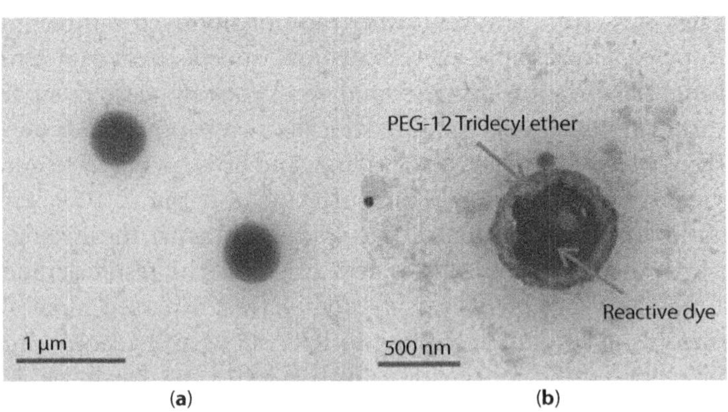

Figure 2.6 TEM images of reverse micelle-encapsulated Levafix-CA dye [63].

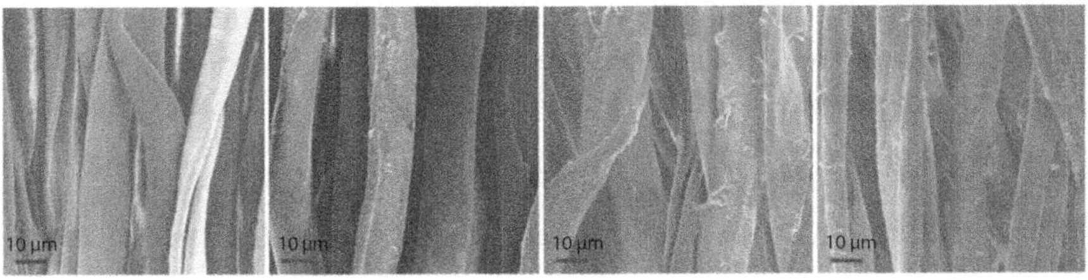

Figure 2.7 SEM photographs of the surface of cotton fiber, (a): before dyeing; (b) dyeing in aqueous solution, (c): dyeing in reverse micelle system in heptane, (d): dyeing in reverse micelle system in octane [52].

micelle in non-aqueous alkane medium of nonane (C9H20) solvent. It was indicated that employing nonane solvent media to enhance the dyeing of cotton fabric in reverse micellar can attain good color matching and excellent fixation properties similar with that of the traditional aqueous dyeing system [54].

Figure 2.6 presents the transmission electron microscope (TEM) images of reverse micelles in which the dyes are wrapped by the non-ionic surfactant (TX-100). It was observed that that the cotton fiber diameter after dipping in the surfactant reverse micelle was much higher than that after dipping in water, suggesting that the TX-100 reverse micelle can remarkably improve the swelling of the cotton fiber than water under the same processing conditions [63]. The scanning electron microscopic (SEM) images of cotton fabric proved that the surface of the fiber was not damaged from octane or heptane, and some microfibrils were observed (Figure 2.7).

2.4 Solvent Dyeing

The literature on solvent dyeing from organic solvents such as tetrachloroethylene (perchloroethylene) is extensive; in 1969 Milićević stated that around 642 patents had been granted over a 20-year period. In comparison with water, perchloroethylene can be heated up more rapidly, but the presence of 15.8% water in the system can form an azeotrope, limiting the boiling point to 87.8 °C, although pure perchloroethylene has a higher boiling point (121.1 °C) than water (100 °C) [68]. Most studies on dyeing from perchloroethylene focused on disperse dyeing of polyester, because dyeing of cellulosic fibers etc., with ionic dyestuffs required the presence of water in the system. However, the major problem with solvent dyeing of polyester (PET) dyed with disperse dyes in perchloroethylene was that the solubility parameters of disperse dyes were closer to that of perchloroethylene than of polyester, resulting in the disperse dyes retaining preferential solubility in the solvent [68–70].

One interesting solution to this problem of disperse dyes being too soluble in perchloroethylene was developed, using a silicone-fluid perchloroethylenemixture in the DACSOL procedure, which is improved by J&P Coats Ltd, and was designed for batch dyeing of polyester yarns with disperse dyes and fluorescent brightening agents. The process used a low-viscosity silicone fluid (75–80% by volume) and perchloroethylene. This increased dye bath exhaustion by reducing the dye solubility and decreased the effective liquor ratio, thereby changing the partition coefficient in favor of dye in the fiber. Lover *et al.* [68, 69] developed 2-ethoxyethanol/water system on the basis of dacsol which water substituted silicide role in perchloroethylene. Commodity dyestuff available in this dyeing process and dyeability of blended fabric also satisfied, while request higher quality water.

The removal of unfixed dye and residual solvents was then accomplished by rinsing in acetone, followed by steam drying. However, the dramatic rise in oil prices in the early 1970s rendered solvent dyeing from perchloroethylene had to be recovered for reuse, and any solvent left within the fibers after dyeing constituted a solvent loss that also adversely affected the process economics. Chavan [70, 71] studied the dyeing of alkali swollen cotton in polar/nonpolar solvent mixtures, using CI Reactive Red-31, acetone was nonpolar solvent and the polar solvent was water. At appropriate ratio of solvent mixture and alkali concentration used for cotton preswelling, it was possible to get 100% exhaustion rate and dye fixation of 79–82%. The proportion of polar-nonpolar solvent required attaining the highest exhaustion rate and dye fixation depends on the dielectric constant of the polar solvent [71].

2.5 Silicone Non-Aqueous Dyeing

In general, dyeing different fibers in solvent could obtain good dyeing performances, but there still exist more problems. Although many scholars have studied the swelling process of cotton fabrics in nonpolar solvent polarity, the dyeing mechanism is not clear yet. Special dyes and machinery and equipment are required, as well as costly of solvent recovery. Therefore, find a compatible eco-friendly medium and set a feasible dyeing technique is becoming the key of successful development of non-aqueous dyeing technology. Wang [72, 73] and his team employed an ecological environmental protection solvent D5 instead of water as the dyeing medium in non-aqueous dyeing, and the research has made noteworthy achievements. By testing the washing ability of D5, the surface morphology, structure and performance of fabric, Gaoxianyun *et al.* [73] found that the D5 washing is a multi-purpose washing way that has smaller effect on the environment and textile performance. Wei [74] studied the solubilization properties of D5 reverse micelle system, the results showed that it was feasible to use nonpolar D5 alternatives traditional paraffin in reverse micelle system, which was also friendly to the ecological environment system. Shenzheng *et al.* [75] dyed polyester fiber in D5 medium, the studies showed that adding a small amount of water in the D5 system can remarkably improve the dyeing properties of polyester. Compared with the conventional dyeing method, the same dyeing performances were obtained without adding any dispersant [76].

Miao [77, 78] and Fu [79] dyed silk fabrics with C.I. Reactive Red 195 in reactive dye/D5 suspension system. The result indicated that the dyeing equilibrium adsorption in dye/D5 suspension system was much higher than traditional water bath, and the half dyeing time is significantly reduced.

Recently, D5 non-aqueous system was used in dyeing of cotton fiber with reactive dye. As demonstrated in Figure 2.8, reactive dye solution is dispersed in the D5 medium. Since the reactive dye solution is immiscible in D5, however, it can rapidly diffuse to the cotton fiber surface, resulting that almost all dye can be adsorbed on cotton fiber surface under mechanical force. Consequently, the dye uptake is nigh to 100%, which is relatively higher than the conventional dyeing system (60–70%). After the completion of dyeing, the D5 can be recycled and reused [80–87].

Pei *et al.* [81] studied the dyeing behavior and hydrolysis reaction of dichlorotriazine dye in D5 non-aqueous dyeing. The obtained results

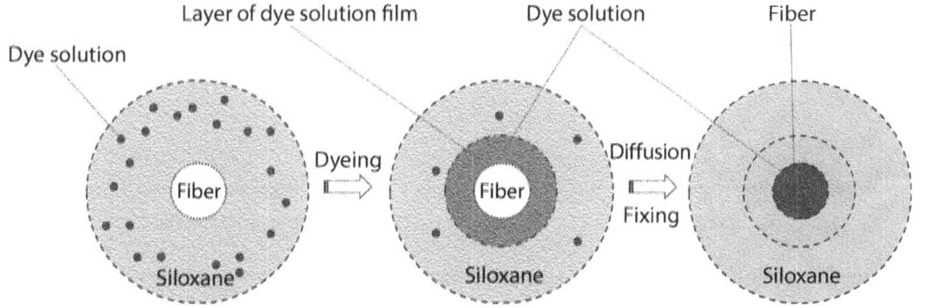

Figure 2.8 Diagram of reactive dye in D5 medium.

disclosed that color and fastness properties were much higher in D5 reverse micro-emulsion as compared with the conventional dyeing method (Figure 2.9). In D5 reverse dyeing system, the lower hydrolysis of dye is probably due to the diffusion rate of dye and the grasping water ability of surfactant (Figure 2.10). The hydrolysis model illustrates that the hydrolysis ratio constant in traditional water-based dyeing system was about 3 to 4 times faster than the one in D5 reverse emulsion system. The chemical composing of the solution might have an impact on the fiber swelling, media polarity and the molecular size. In addition, the molecular structure of D5, paraffin, and isooctane do not contain any polar groups.

Though, the molecules of ethanol and water contain some polar groups (-OH) thus, hydrogen bonds can be formed with the cellulose fibers. In addition, it was observed that (Figures 2.11a, b, e, f) Rhodamine B diffused into the viscose and cotton, demonstrating the diffusing of water molecule into the fiber. In case of the dyeing in D5 medium, Coumarin 6 only diffused on the surface or interval fiber (Figures 2.11c, d, g, h), signifying that

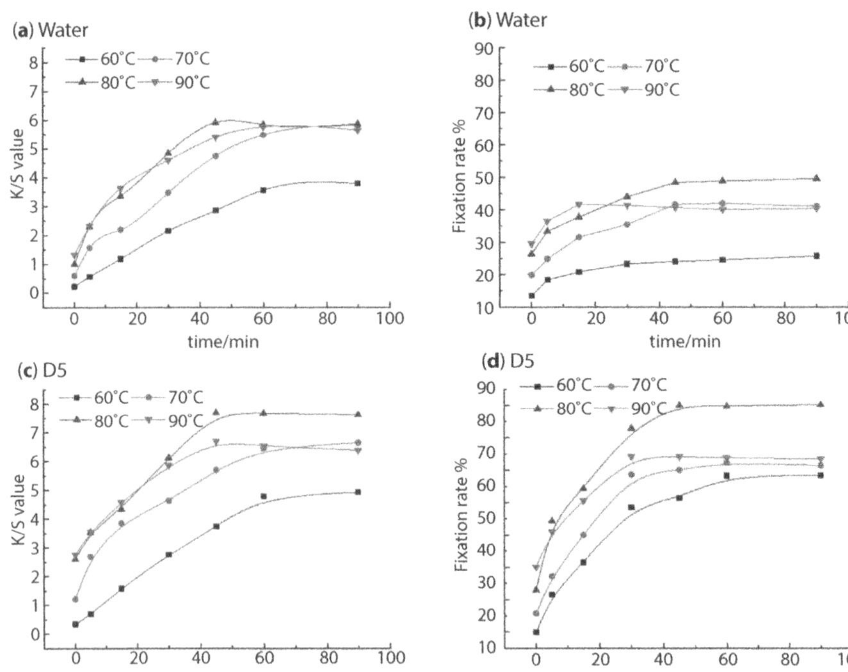

Figure 2.9 Color strength (K/S) values and fixation (%) of dichlorotriazine dye in D5 reverse micro-emulsion and aqueous medium [81].

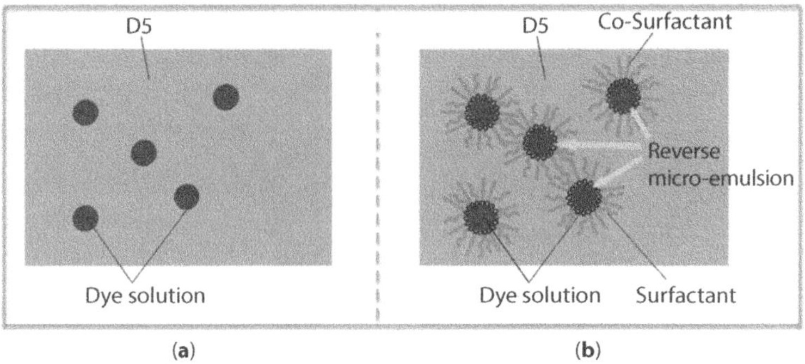

Figure 2.10 Interaction between dye and surfactant in D5 medium and D5 reverse micro-emulsion, (a): dyeing without surfactant; (b): dyeing using surfactants [81].

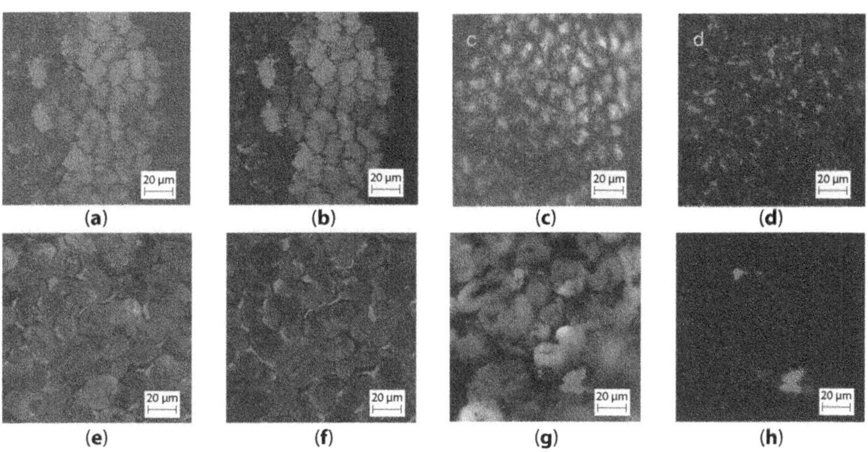

Figure 2.11 Confocal fluorescence images of distribution of water (a, b, e, f) and siloxane (c, d, g, h) in viscose (a, b, c, d) and cotton (e, f, g, h); the merged channel (a, c, e, g) and the laser channel (b, d, f, h) [82].

the D5 molecule could not permeate into the fiber. This might be explained the swelling of the fibers in the aqueous dyeing [82].

Pei [81, 82] studied the adsorption kinetic of reactive dye on cotton yarns using pseudo-first-order kinetic model and a pseudo-second-order kinetic model, the dyeing process was conducted in aqueous and D5 medium (Figure 2.12). The results showed, in D5 non-dyeing system, the dye diffusion increased with the increasing of the wettability of yarn,

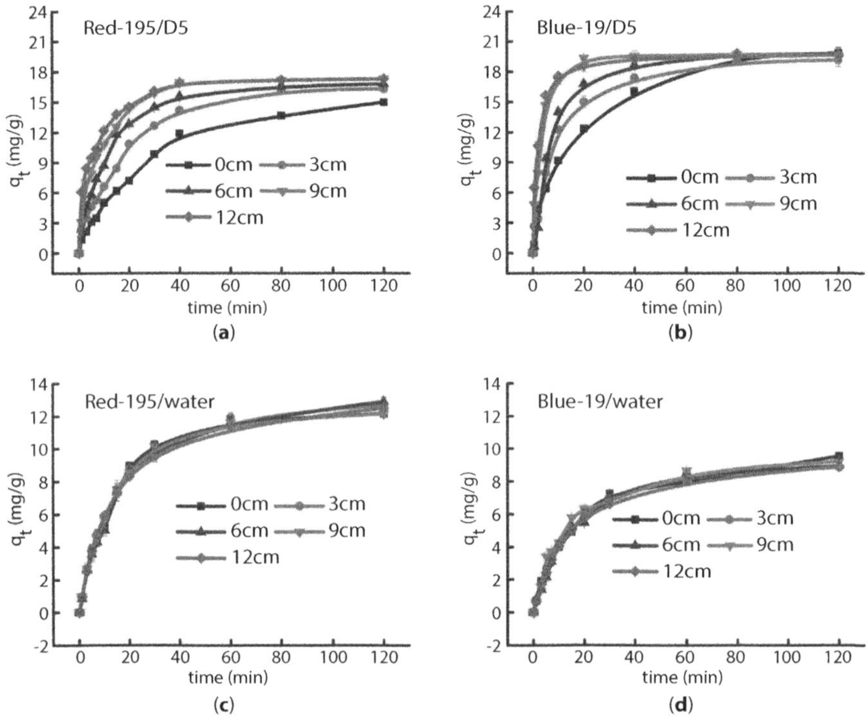

Figure 2.12 Adsorption of reactive dyes at isothermal conditions (20 °C), (a & b): Reactive Red 195 and Reactive Blue 19 in D5 medium; (c & d) Reactive Red 195 and Reactive Blue 19 in water medium, respectively [83].

however, the dye uptake did not change. While the dye uptake was more remarkably increased in the D5 medium.

Luo [84, 85] studied cotton fibers dyed with indigo dye in D5 medium, the industrial application of 100 kg/d (Figure 2.13) provides compelling proof of indigo dyeing in D5 non-aqueous dyeing system, because this technique can replace the complex process of traditional indigo dyeing, drab style, and the serious wastewater pollution.

Recently, Omer *et al.* [86, 87] applied natural dyes in D5 non-aqueous system, where the dyeing behavior of wool fibers in siloxane medium was studied using Henna dye. The obtained results demonstrated that; samples dyed in D5 medium attained 77–83% exhaustion rates compared with

64–71% in the samples dyed in aqueous system. Moreover, the SEM and EDX tests (Figure 2.14) showed that dyeing in the siloxane medium did not illustrate any adverse impact on the fiber surface in terms of morphology and elemental composition (Figure 2.14).

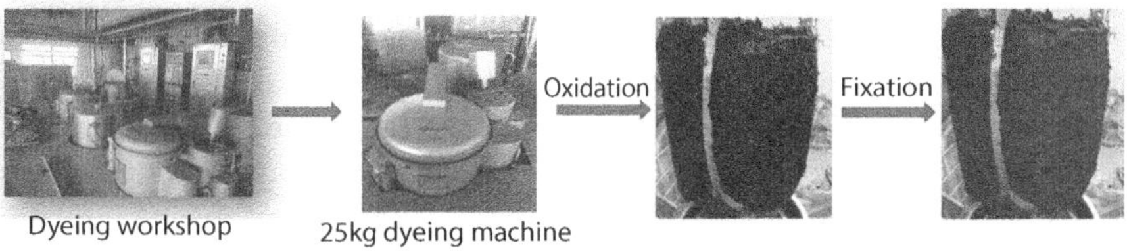

Figure 2.13 The dyeing pilot plant of 25 kg [84].

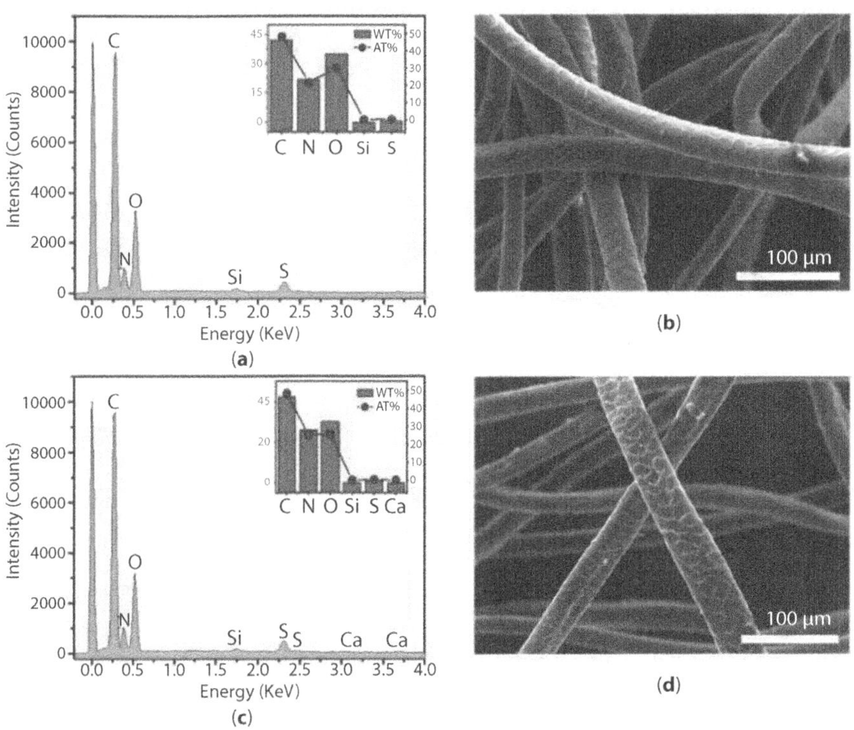

Figure 2.14 EDX Spectrum and SEM images of samples dyed in water medium (a), (b) and samples dyed in D5 medium (c) and (d) respectively [86].

2.6 Conclusion

Recently, there was a notable improvement in diverse treatment techniques for textile effluents. However, the wastewater reducing methods are greatly effective than treating the producing wastes. The use of advanced types of machinery, improved processes, and eco-friendly chemicals would assist textile factories to minimize the produced wastes and toxic emissions. Over the past few years, some processes have adopted non-aqueous dyeing techniques to minimize water and chemicals consumption.

This chapter discussed the latest advances in non-aqueous dyeing systems, such as supercritical fluid, reverse micelle, solvent dyeing, and silicone non-aqueous dyeing systems. The development of these techniques has been the focus of many researchers. However, the main challenge is to bridge the gap between the researchers' efforts and the trend of the industry. It is important to improve these techniques from lab to large scale, as any commercial progresses into these technologies will lead to an environmentally sustainable future.

References

1. Saxena, S., Raja, A., Arputharaj, A. Challenges in sustainable wet processing of textiles, in: *Textiles and clothing sustainability*, p. 43–79, Springer, Singapore, 2017.
2. Zheng, H. *et al.*, An industrial scale multiple supercritical carbon dioxide apparatus and its eco-friendly dyeing production. *J. CO_2 Util*, 16, 272–281, 2016.
3. Xie, K., Liu, H., Wang, X., Surface modification of cellulose with triazine derivative to improve printability with reactive dyes. *Carbohydr. Polym*, 78, 3 538–542, 2009.
4. Bechtold, T. and Turcanu, A., Electrochemical reduction in vat dyeing: Greener chemistry replaces traditional processes. *J. Clean. Prod.*, 17, 1669–1679, 2009.
5. Luo, X. *et al.*, Novel sustainable synthesis of dyes for clean dyeing of wool and cotton fibres in supercritical carbon dioxide. *J. Clean. Prod.*, 199, 1–10, 2018.
6. Zheng, H. *et al.*, CO_2 Utilization for the dyeing of yak hair: Fracture behaviour in supercritical state. *J. CO_2 Util.*, 18, 117–124, 2017.
7. Schmidt, A., Bach, E., Schollmeyer, E., The dyeing of natural fibres with reactive disperse dyes in supercritical carbon dioxide. *Dyes Pigm.*, 56, 27–35, 2003.
8. Hori, T. and Kongdee, A., Dyeing of PET/co-PP composite fibers using supercritical carbon dioxide. *Dyes Pigm.*, 105, 163–166, 2014.

9. Zhang, Y.-Q., Wei, X.-C., Long, J.-J., Ecofriendly synthesis and application of special disperse reactive dyes in waterless coloration of wool with supercritical carbon dioxide. *J. Clean. Pr.*, 133, 746–756, 2016.
10. Ahmed, N.S. and El-Shishtawy, R.M., The use of new technologies in coloration of textile fibers. *J. Mater. Sci.*, 45, 1143–1153, 2010.
11. McClain, J. *et al.*, Design of nonionic surfactants for supercritical carbon dioxide. *Science*, 274, 2049–2052, 1996.
12. Banchero, M., Supercritical fluid dyeing of synthetic and natural textiles–A review. *Color Technol.*, 129, 2–17, 2013.
13. Liao, S.-K. and Chang, P.-S., Special Issue–Supercritical Fluids. Literatures on dyeing technique of supercritical fluid carbon dioxid. *Am. J. Analyt. Chem.*, 3, 923–930, 2013.
14. Abou Elmaaty, T. and Abd El-Aziz, E., Supercritical carbon dioxide as a green media in textile dyeing: A review. *Text. Res. J.*, 88, 1184–1212, 2018.
15. Abou Elmaaty, T. *et al.*, Facile bifunctional dyeing of polyester under supercritical carbon dioxide medium with new antibacterial hydrazono propanenitrile dyes. *Ind. Eng. Chem. Res.*, 53, 15566–15570, 2014.
16. Özcan, A.S. and Özcan, A., Adsorption behavior of a disperse dye on polyester in supercritical carbon dioxide. *J. Supercrit. Fluids*, 35, 133–139, 2005.
17. Zheng, L.-J. *et al.*, Supercritical CO_2 for color graphic dyeing: Theoretical insight and experimental verification. *Therm. Sci.*, 19, 1287–1291, 2015.
18. Elmaaty, T.A. *et al.*, Eco-friendly disperse dyeing and functional finishing of nylon 6 using supercritical carbon dioxide. *Fibers*, 3, 309–322, 2015.
19. Hou, A. and Dai, J., Kinetics of dyeing of polyester with CI Disperse Blue 79 in supercritical carbon dioxide. *Color Technol.*, 121, 18–20, 2005.
20. Hou, A., Xie, K., Dai, J., Effect of supercritical carbon dioxide dyeing conditions on the chemical and morphological changes of poly (ethylene terephthalate) fibers. *J. Appl. Polym. Sci.*, 92, 2008–2012, 2004.
21. Schmidt, A., Bach, E., Schollmeyer, E., Damage to natural and synthetic fibers treated in supercritical carbon dioxide at 300 bar and temperatures up to 160 °C. *Text. Res. J.*, 72, 1023–1032, 2002.
22. Aigin, H., Kongliang, X., Jinjin, D., Effect of supercritical carbon dioxide dyeing conditions on the chemical and morphological changes of poly(ethylene terephthalate) fibers. *J. Appl. Polymer Sci.*, 29, 2008–2012.
23. Terada, D. *et al.*, Structural modification of polyester fibers in supercritical carbon dioxide fluid. *Text. Res. J.*, 74, 1126, 2004.
24. Zheng, H. *et al.*, An investigation for the performance of meta-aramid fiber blends treated in supercritical carbon dioxide fluid. *Fibers Polym.*, 16, 1134-1141, 2015.
25. Zheng, H. *et al.*, CO_2 utilization for the waterless dyeing: Characterization and properties of Disperse Red 167 in supercritical fluid. *J. CO_2 Util.*, 24, 266–273, 2018.
26. Kazarian, S.G. *et al.*, Partitioning of solutes and co solvents between supercritical CO_2 and polymer phases. *J. Supercrit Fluids*, 13, 107–112, 1998.

27. Banchero, M., Ferri, A., Manna, L., The phase partition of disperse dyes in the dyeing of polyethylene terephthalate with a supercritical CO_2/methanol mixture. *J. Supercrit. Fluids*, 48, 72–78, 2009.
28. Abate, M.T. *et al.*, Colouration and bio-activation of polyester fabric with curcumin in supercritical CO_2: Part II–Effect of dye concentration on the colour and functional properties. *J. Supercrit. Fluids*, 157, 104703, 2020.
29. Long, J.-J. *et al.*, Clean fixation of dye on cotton in supercritical carbon dioxide with a heterogeneous and phase transfer catalytic reaction. *Dyes Pigm.*, 115, 88–95, 2015.
30. Draper, S. *et al.*, Solubility relationships for disperse dyes in supercritical carbon dioxide. *Dyes Pigm.*, 45, 177–183, 2000.
31. Cid, M.F. *et al.*, Excellent dye fixation on cotton dyed in supercritical carbon dioxide using fluorotriazine reactive dyes. *Green Chem.*, 7, 609–616, 2005.
32. Fernandez Cid, M. *et al.*, Novel process to enhance the dyeability of cotton in supercritical carbon dioxide. *Text. Res. J.*, 77, 38–46, 2007.
33. Cid, M.F. *et al.*, A significant approach to dye cotton in supercritical carbon dioxide with fluorotriazine reactive dyes. *J. Supercrit. Fluids*, 40, 477–484, 2007.
34. Gao, D. *et al.*, Synthesis of reactive disperse dyes containing halogenated acetamide group for dyeing cotton fabric in supercritical carbon dioxide. *J. Supercrit. Fluids*, 86, 108–114, 2014.
35. Guzel, B. and Akgerman, A., Mordant dyeing of wool by supercritical processing. *J. Supercrit. Fluids*, 18, 247–252, 2000.
36. Gao, D. *et al.*, Supercritical carbon dioxide dyeing for PET and cotton fabric with synthesized dyes by a modified apparatus. *ACS Sustain. Chem. Eng.*, 3, 668–674, 2015.
37. Zheng, H. *et al.*, An ecofriendly dyeing of wool with supercritical carbon dioxide fluid. *J. Clean. Prod.*, 143, 269–277, 2017.
38. Fernandez Cid, M.V. *et al.*, Excellent dye fixation on cotton dyed in supercritical carbon dioxide using fluorotriazine reactive dyes. *Green Chem.*, 7, 609–616, 2005.
39. Long, J.-J. *et al.*, Dyeing of cotton fabric with a reactive disperse dye in supercritical carbon dioxide. *J. Supercrit. Fluids*, 69, 13–20, 2012.
40. Yang, D.-f. *et al.*, Dyeing of cotton fabric with reactive disperse dye contain acyl fluoride group in supercritical carbon dioxide. *Dyes Pigm.*, 139, 566–574, 2017.
41. van der Kraan, M. *et al.*, Dyeing of natural and synthetic textiles in supercritical carbon dioxide with disperse reactive dyes. *J. Supercrit. Fluids*, 40, 470–476, 2007.
42. Yan, K. *et al.*, Development of a special SCFX-AnB3L dye and its application in ecological dyeing of silk with supercritical carbon dioxide. *J. CO_2 Util.*, 35, 67–78, 2020.

43. Shinoda, T. and Tamura, K., Solubilities of CI disperse orange 25 and CI disperse blue 354 in supercritical carbon dioxide. *J. Chem. Eng. Data*, 48, 869–873, 2003.
44. Bao, P. and Dai, J., Relationships between the Solubility of CI Disperse Red 60 and Uptake on PET in Supercritical CO_2. *J. Chem. Eng. Data*, 50, 838–842, 2005.
45. Lin, H.-m. *et al.*, Solubilities of disperse dyes of blue 79, red 153, and yellow 119 in supercritical carbon dioxide. *J. Supercrit. Fluids*, 21, 1–9, 2001.
46. Hou, A. and Dai, J., The crystal morphology of CI Disperse Blue 79 in supercritical carbon dioxide. *Dyes Pigm.*, 82, 71–75, 2009.
47. Bach, E. *et al.*, Correlation of solubility data of azo disperse dyes with the dye uptake of poly (ethylene terephthalate) fibres in supercritical carbon dioxide. *Color Technol.*, 117, 13–18, 2001.
48. Cui, C.-L., Shi, W., Long, J.-J., Solubility and data correlation of a reactive disperse dye in a quaternary system of supercritical carbon dioxide with mixed cosolvents. *J. Taiwan Inst. Chem. Eng.*, 91, 213–223, 2018.
49. Draper, S.L. *et al.*, Solubility relationships for disperse dyes in supercritical carbon dioxide. *Dyes Pigm.*, 45, 177-183, 2000.
50. Lee, C.H. *et al.*, Effect of reverse micelle-encapsulated reactive dyes agglomeration in dyeing properties of cotton. *Dyes Pigm.*, 161, 51–57, 2019.
51. Wang, Y. *et al.*, Dyeing cotton in alkane solvent using polyethylene glycol-based reverse micelle as reactive dye carrier. *Cellulose*, 23, 965–980, 2016.
52. Perrut, M., Supercritical Fluid Applications: Industrial Developments and Economic Issues. *Ind. Eng. Chem. Res.*, 39, 4531–4535, 2000.
53. Tang, A.Y.L. *et al.*, Dyeing properties of cotton with reactive dye in nonane nonaqueous reverse micelle system. *ACS Omega*, 3, 2812–2819, 2018.
54. Göktürk, S. and Tunçay, M., Spectral studies of safranin-O in different surfactant solutions. *Spectrochim. Acta A Mol. Biomol. Spectrosc.*, 59, 1857–1866, 2003.
55. Melo, E., Aires-Barros, M., Cabral, J., Reverse micelles and protein biotechnology. *Biotechnol. Annu. Rev.*, 7, 87–12, 2001.
56. Sawada, K. and Ueda, M., Adsorption and fixation of a reactive dye on cotton in non-aqueous systems. *Color Technol.*, 119, 182–186, 2003.
57. Sawada, K. and Ueda, M., Adsorption behavior of direct dye on cotton in non-aqueous media. *Dyes Pigm.*, 58, 37–40, 2003.
58. Sawada, K. and Ueda, M., Dyeing of protein fiber in a reverse micellar system. *Dyes Pigm.*, 58, 99–103, 2003.
59. Sawada, K. and Ueda, M., Characteristics of aqueous microenvironments in non-ionic surfactant reverse micelles and their use for enzyme reactions in non-aqueous media. *J. Chem. Technol. Biotechno.*, 79, 369–375, 2004.
60. Sawada, K. and Ueda, M., Enzyme processing of wool fabrics in a non-ionic surfactant reverse micellar system. *J. Chem. Technol. Biotechno.*, 79, 376–380, 2004.

61. Xie, K. *et al.*, Micelle dyeing with low liquor ratio for reactive dyes using dialkyl maleic acid ester surfactants. *J. Clean. Prod.*, 19, 332–336, 2011.
62. Yi, S. *et al.*, Adsorption and fixation behaviour of CI Reactive Red 195 on cotton woven fabric in a nonionic surfactant Triton X-100 reverse micelle. *Color Technol.*, 128, 306–314, 2012.
63. Yi, S., Deng, Y., Sun, S., Adsorption and dyeing characteristics of reactive dyes onto cotton fiber in nonionic Triton X-100 reverse micelles. *Fibers Polym.*, 15, 2131–2138, 2014.
64. Yi, S. *et al.*, Dyeing properties of CI reactive violet 2 on cotton fabric in non-ionic TX-100/Span40 mixed reverse micelles. *Fibers Polym.*, 16, 1663–1670, 2015.
65. Clarke, M.J. *et al.*, Water in supercritical carbon dioxide microemulsions: Spectroscopic investigation of a new environment for aqueous inorganic chemistry. *J. Am. Chem. Soc.*, 119, 6399–6406, 1997.
66. Jun, J., Sawada, K., Ueda, M., Application of perfluoropolyether reverse micelles in supercritical CO_2 to dyeing process. *Dyes Pigm.*, 61, 17–22, 2004.
67. Love, R.B. and Robson, A., The use of non-aqueous solvents in dyeing. *J. Soc. Dye. Colour.*, 94, 486–492, 1978.
68. Bohrer, E., Automation in bleaching and dyeing with reference to the thies-duomat. *Color Technol.*, 81, 258–261, 2010.
69. Chavan, R.B. and Subramanian, A., Dyeing of Alkali Swollen and Alkali Swollen Solvent Exchanged Cotton with a Reactive Dye. *Text. Res. J.*, 52, 733–737, 1982.
70. Chattopadhyay, D.P., Chavan, R.B., Sharma, J.K., Salt-free reactive dyeing of cotton. *Int. J. Cloth Sci. Tech.*, 19, 99–108, 2007.
71. Pei, L., Liu, J., Cai, G., Wang, J., Study of hydrolytic kinetics of vinyl sulfone reactive dye in siloxane reverse micro-emulsion. *Text. Res. J.*, 87, 2368–2378, 2017.
72. Pei, L., Wu, P., Liu, J., Wang, J., Effect of nonionic surfactant on the micro-emulsifying water in silicone media. *J. Surfactants. Deterg.*, 20, 247–254, 2017.
73. Xian-Yun, G., Jin-Qiang, L., Yong-Qiang, L.I., The impact of d5-washing on the fabric performance. *J. Zhejiang Sci-Tech University*, 24, 7–10, 2007.
74. Weia, W., Shaa, L.I., Feng-Minga, L.U., Jin-Qianga, L., Study on the reactive dyeing in d5 reverse micelle system. *J. Zhejiang Sci-Tech University*, 27, 263–269, 2010.
75. Li, S., Liu, J., Li, Y., Li, M., Study on the dyeing of PET fiber with disperse dyes in D5 media. *Adv. Mater. Res.*, 331, 334–337, 2011.
76. Liu, J., Miao, H., Li, S., Non-aqueous dyeing of reactive dyes in D5. *Adv. Mater. Res.*, 441, 138–144, 2012.
77. Miao, H., Liu, J., Li, Y., Fu, C., Zhang, Y., Study on hydrolysis kinetics of reactive dyes in dye/D5 suspending system. *J. Tex. Res.*, 34, 77–82, 2013.
78. Fu, C., Wang, J., Shao, J., Pu, D., Chen, J., Liu, J., A non-aqueous dyeing process of reactive dye on cotton. *J. Tex. Inst.*, 106, 152–161, 2015.

79. Pei, L., Liu, J., Wang, J., Study of Dichlorotriazine Reactive Dye Hydrolysis in Siloxane Reverse Micro-emulsion. *J. Clean Prod.*, 165, 994–1004, 2017.
80. Pei, L., Luo, Y., Gu, X., Dou, H., Wang, J., Diffusion Mechanism of Aqueous Solutions and Swelling of Cellulosic Fibers in Silicone Non-Aqueous Dyeing System. *Polymers*, 11, 411, 2019.
81. Pei, L., Liu, J., Gu, X., Wang, J., Adsorption Kinetic and Mechanism of Reactive Dye on Cotton Yarns with Different Wettability in Siloxane Non-aqueous Medium. *J. Text. I.*, 111, 925–933, 2019.
82. Wang, J., Gao, Y., Zhu, L., Gu, X., Dou, H., Pei, L., Dyeing Property and Adsorption Kinetics of Reactive Dyes for Cotton Textiles in Salt-Free Non-AqueousDyeing Systems. *Polymers*, 10, 1030, 2018.
83. Luo, Y., Pei, L., Gu, X., Wang, J., Sustainable indigo dyeing and fixation mechanism of wet rubbing fastness agent on cotton fiber in silicone non-aqueous dyeing system. *AATCC J. Res.*, 7, 27–36, 2020.
84. Luo, Y., Pei, L., Wang, J., Sustainable indigo dyeing and improvement of rubbing fastness of dyed cotton fiber using different fixing agents for obtaining eco-friendly cowboy products. *J. Clean Prod.*, 251, 119728, 2020.
85. Luo, Y., Pei, L., Zhang, H., Zhong, Q., Wang, J., Improvement of the Rubbing Fastness of Cotton Fiber in Indigo/SiliconNon-Aqueous Dyeing Systems. *Polymers*, 11, 1854, 2019.
86. Alebeid, O.K., Liujun, P., Wenlong, Z., Jiping, W., Sustainable wool fibers dyeing using henna extract in non-aqueous medium. *Environ. Chem. Lett.*, 18, 489–494, 2020.
87. Alebeid, O.K., Pei, L., Suliman, H., Zhou, W., Wang, J., Study the dyeing behavior of wool fabric using Henna extract in DecamethylCyclopentasiloxane (D5) medium. *J. Text I.*, 112, 358–362, 2021.

3

Structural Coloration of Textile Materials

Showkat Ali Ganie* **and Qing Li**

State Key Laboratory of Silkworm Genome Biology, Chongqing Engineering Research Center for Biomaterial Fiber and Modern Textile, College of Sericulture, Textile Biomass Science, Southwest University, Chongqing, PR China

Abstract

Light professed by the eye and brain of a specific composition as the sensation can be defined as color. Specific color vision can be formed by visible light of a certain wavelength, since light and color are closely connected. The higher levels of chemicals consumption, water and wastewater generation are involved in conventional textile coloration since it is a wet process. There are various mechanisms like absorption, emission, diffraction, interference and photochromism by which color in textiles can be generated. Structural coloration is developing as an innovative way to yield materials for colorful textiles. The chemical coloration can be formed by changing the ecological circumstances using a single material or the periodicity of the nanostructures for the formation of numerous colors that are impossible to reproduce. A variety of textile materials are used for coloration that are derived or extracted from nature i.e. of the plant, animal, mineral and microbial origins called natural dyestuffs. These dyes have been used for the dyeing of textile materials and have become a consistent procedural of human lives from time immemorial.

Keywords: Structural coloration, textile, light, photonic crystals, thin-film interference, multilayer interference

3.1 Introduction

Physical color or structural color is a pictorial insight of selectively replicated light by a particular physical structure interacting with incident light, such as scattering, dispersion, diffraction and interference and does not requires any

**Corresponding author*: alishowkat443@gmail.com

Luqman Jameel Rather, Mohd Shabbir and Aminoddin Haji (eds.) Sustainable Practices in the Textile Industry, (75–86) © 2021 Scrivener Publishing LLC

chemical colorants [1]. Chemical color or pigmentary color is the most common way to yield colors since certain wavelengths of visible light are selectively absorbed by them to produce colors [2]. In nature as, structural colors various alive beings like butterfly wings, peacock feathers and beetles display bright colors. It is believed that the coloring characters of pigmentary and structural colors are expressively diverse from each other. Usually, structural color displays greater saturation, good illumination, minimum staining, and an lustrous effect, that are hard and difficult to attain with conventional pigmentary colors [3, 4]. Five basic optical phenomenon and their amalgamations: light scattering, multilayer interference, the diffraction grating effect, photonic crystals and thin-film interference are responsible for structural colors in nature. In an extensive diversity of study arenas biomimetic structural coloration has been enticing growing consideration. In the field of textile, photonic crystals and thin-film interference biomimetic structural colors have been typically engrossed. For obtaining biomimetic structural colors, there are numerous procedures like colloidal self-assembly, vapor deposition, ink jet printing, sputtering, electrostatic self-assembly, spin coating and dipping-adsorption. The electrostatic self-assembly and colloidal self-assembly have become the most auspicious approaches for nanocoatings on textiles to produce vivid, tunable structural colors and bright colors, because of the ease and universality of the approaches with superior nanocoatings. The nano-coatings with biomimetic structural coloration are an eco-coloration technology in comparison with the traditional coloration of textiles using chemical colorants. The amalgamation of structural and pigmentary coloration is expected to expand textile coloration approaches and produce further attractive and added-value textile goods [5].

There are four major approaches allied with structural coloration. (i) Structural color can be created via thin-film interference with an iridescent consequence where the colors change according to alterations in the angle of vision or sunlight. This integrates multilayer and monolayer thin film interference, like the surface color on the organic transparency of the wings of dragonflies and on a soap bubble. Structural color produced by multilayer thin-film interference is comparatively common, via structurally complex whereas monolayer thin-film interference is rare. (ii) Structural color can be created by a diffraction grating effect, which includes effects like iridescence. This comprises the color of music CDs, snakeskin, some crustacean hairs etc. (iii) The dispersion and scattering of light can yield a structural color without any iridescent effect, e.g. the ocean and the blue of the sky. (iv) Structural color can be a result of the occurrence of photonic crystals, opal which is a natural example, by which can also be artificially made. There are numerous objects and organisms (Figure 3.1) in nature with structural colors, which can display one or more of the above approaches/principles.

(a) (b) (c) (d) (e)

Figure 3.1 Examples of structural color by interference on a soap bubble (a), grating diffraction on a record disc (b), chromatic dispersion in the sky (c), PC in opal stones (d), multiple principles on a butterfly (e).

The most common structural coloration used on fabrics or fibers, from the cited principles is to synthesize films with structural colors that are applied as a surface coating. This includes photonic crystals self-assembly and chemical vapor deposition [6–10], liquid deposition and sol gel processes [11], physical vapor deposition [12], micro-nano structuring [13], digital inkjet printing [14] and by means of electrospinning [15]. The various tactics to structural coloration on textiles are shown in Figure 3.2.

3.2 Thin-Film Interference

Thin-film interference is a natural phenomenon in which boundaries of a thin film interfere with one another in which light waves are reflected by the lower and upper borders resulting in either reducing or enhancing the reflected light. The reflected waves from both surfaces interfere to cancel each other when the thickness of the film is an odd multiple of one quarter-wavelength of the light on it. Wave is completely transmitted because it cannot be reflected. When the thickness is a multiple of a half-wavelength of the light, the two reflected waves strengthen each other, increasing the reflection and reducing the transmission. Therefore, when white light which consists of a range of wavelengths is incident on the film, certain wavelengths in the form of colors are intensified while others are attenuated. The multiple colors seen in light reflected from soap bubbles and oil films on water are explained by the thin-film interference. It is also the approach involved in the achievement of antireflection coatings used on camera lenses and glasses. In nature structural colors are produced by thin-film interference; which can be divided into single thin-film and multilayer thin-film interference. The coloration is due to the varied structural effects, although extremely complex structures have evolved, but the coloration is due to the simplest structure, namely a thin film.

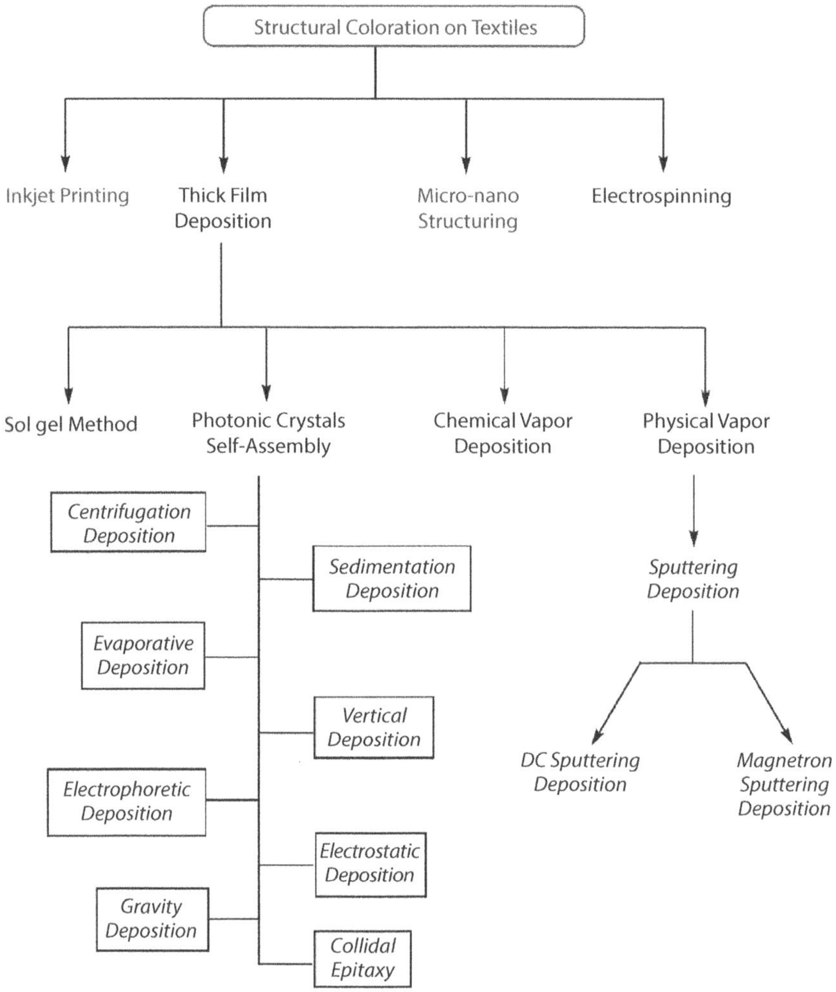

Figure 3.2 Various ways and methods of structural coloration on textiles.

3.2.1 Principle of Thin-Film Interference

Let a plane wave of light be incident on a thin film of thickness d and refractive index n with the angles of incidence and refraction as θ_1 and θ_2 as shown in Figure 3.3. Light reflected at the two surfaces interferes with each other [16].

Constructive interference equation;

$$P = 2\ nd\ cos\theta_2 = (2m+1)\frac{\lambda}{2} \tag{3.1}$$

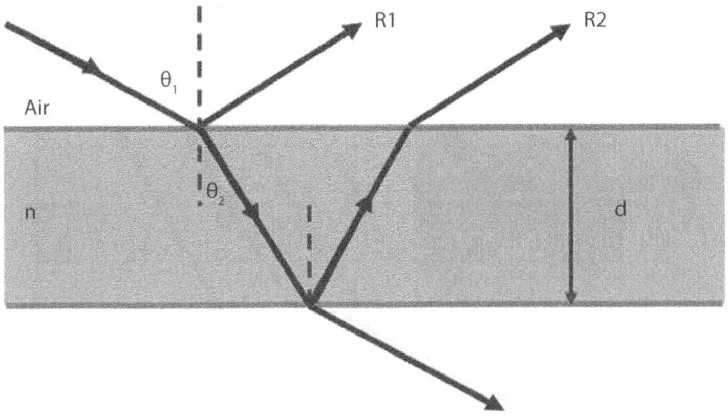

Figure 3.3 Schematic diagram of single thin-film interference.

Destructive interference equation;

$$P = 2\ nd\ cos\theta_2 = 2m\frac{\lambda}{2} \tag{3.2}$$

where n is the refractive index of thin film; d is the thickness of thin film; θ_2 is the angle of refraction; m is a nonzero integer; and λ is the reflected light wavelength.

The wavelength of constructive interference (λ) and the optical path difference (p) varied with the variation of thickness of thin film (d) under the irradiation of unidirectional polychromatic light. The refractive index (n) of thin film is also closely related to the optical path difference. The thin film can be a vivid structural color only when the product of the refractive index (n) and thickness (d) are a certain amount. Single-layer reflectors are usually found in nature, like in the wings of dragonflies and flies, in a soap bubble in the sunshine, and so on.

3.2.2 Multilayer Interference

Multilayer films are usually made of two thin films with different refractive indexes alternately; the relation of the reflection wavelength (λ) and the two kinds of thin-film refractive indexes (high refractive index n_H and low refractive n_L) and the thickness (d_H and d_L) of thin film is shown as Equation 3.3 [17]. A typical principle diagram of multilayer interference is shown in Figure 3.4.

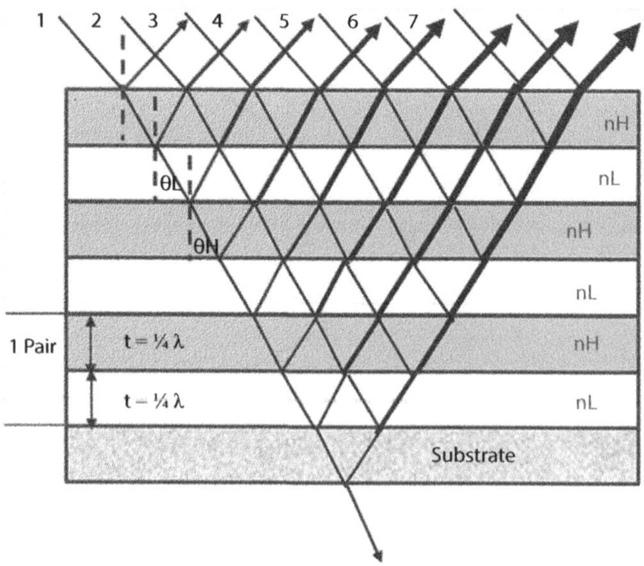

Figure 3.4 Schematic diagram of multilayer interference.

Equation 3.4 shows the relation of the refractive index of substrate (n_S) with reflection intensity (R), the number of pairs (N) and the ratio of two kinds of thin-film index (n_L/n_H) [18]. The brightness of structural color is effected by the reflection intensity (R) and the type of structural color is determined by the reflected wavelength (Equation (3.1)).

$$m\lambda = 2(n_H d_H \cos\theta_H + n_L d_L \cos\theta_L) \tag{3.3}$$

$$R = \left[\frac{\left(n_S - \frac{n_L}{n_H} \right)^{2N}}{\left(n_S + \frac{n_L}{n_H} \right)^{2N}} \right]^2 \tag{3.4}$$

In nature, multilayer interference is also commonly found, such as Hercules beetle, blue damselfish, tortoise beetle, paradise whiptail, *Paracheirodon innesi*, and so on. Regardless of thin-film interference or multilayer interference the structural colors vary with the viewing angles.

Structurally colored textile cotton fabrics made-up by the surface coating of colloidal silica nanoparticles have been successfully using a gravity sedimentation self-assembly technique. An extensive variety of spectrum colors from red to blue has been attained by simplistic control of the particle diameter of the silica nanoparticles. The structural color on the fabric was explained by the optical and morphological properties of structurally colored samples of the fabric. In addition, the structurally colored samples were prepared on glass substrates from the same silica nanoparticles batches for comparison purposes [19].

Silver/Titanium dioxide composite film placed on polyester fabrics synthesized by magnetron sputtering method was used to reproduce the single-layer film structure. The structural colors can be controlled and attuned by the thickness of titanium dioxide thin films according to film interference principle theoretical calculation. Textiles coated with such nanoparticles composite films not only comprehend the structural coloration of the surface, but also yields good anti-ultraviolet, photocatalytic, conductive and other properties, leading to decent application prospect of such approaches [20–22].

Soap-free emulsion polymerization was used to prepare colloidal photonic crystals on monodisperse nanospheres of poly (styrene-methyl methacrylate acrylic acid) and onto a chitosan-cationized woven cotton fabric deposited by electrostatic self-assembly approach. As protective and color modulating coating a second chitosan layer on the top of the photonic crystal was applied. Chitosan was selected for this work since of its positive charge that permits the creation of strong film on the surface of both chemical and natural fibers [23, 24]. The fabric was able to exhibit various iridescence at changed viewing angles altering to shorter wavelength as the viewing angle increase from 0° to 90°. The color coordinate values display the structural color changes among various coatings representing the higher quantity of stable photonics absorbed onto the cross-linked fabric surface and its greater washing durability. This cost-efficient and well-organized approach opens new tactics for the advance of dye-free coloration of textiles permitting the manufacture of garments for special applications and pioneering design with a good degree of washing and light fastness [25]. Natural coloration based on the dye-free fibers is a rising area of research due to proper consciousness of ecological sustainability. Numerous research groups have recently tried to fabricate structurally colored cylindrical fibers with a universal idea for the formation of periodic structure along the fibers by using colloidal spheres. Colors experienced in the subsequent fibers are because of the coherent scattering produced in the radial direction by the periodicity, yielding beneficially reflected light in a particular frequency. It is because of the axial symmetry of structures with cylindrical geometry the color of the periodic structured fiber is non iridescent when the direction of the incident light which is perpendicular to the fiber axis. But, if the axial direction component of the incident light is nonzero iridescence still appears [26–29]. The nanostructured fibers were immensely prepared by colloidal electrospinning with tunable structural colors. Two phenomena: Mie scattering of the colloidal spheres and reflection due to the band gap of photonic structure are responsible for the homogeneous and non-iridescent structural colors of the electrospun fibers as revealed by theoretical modeling and experimental [30] (Figure 3.5).

Pure soft poly(methylmethacrylate-butylacrylate spheres were used to produce latex films, that exhibited good elasticity in which all the spheres fused into a continuous phase, and in which no obvious structural colors was observed [31]. Photonic crystals with great structural stability attained

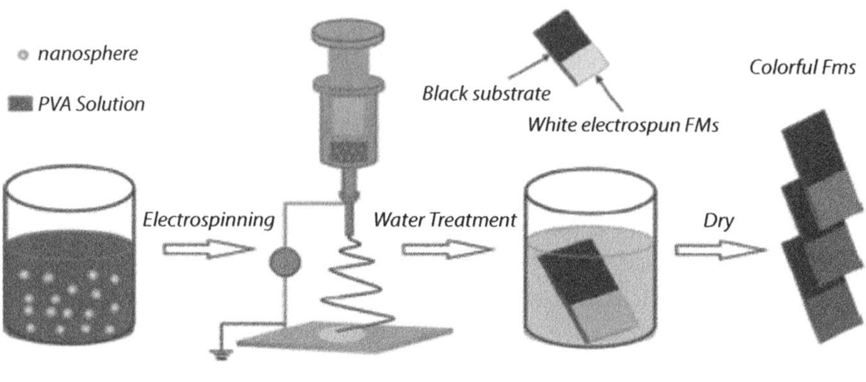

Figure 3.5 Schematic illustration of the process of colloidal electrospinning and fabrication of colorful FMs.

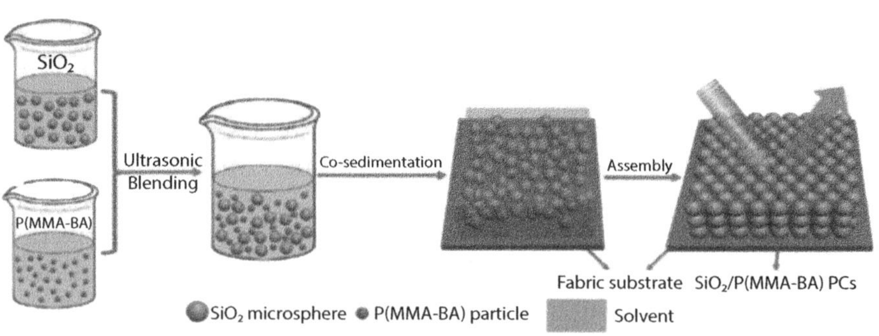

Figure 3.6 Schematic diagram of co-sedimentation self-assembly of binary colloidal dispersion on fabric substrates.

by means of the soft microspheres or core-shell microspheres with a hard core. The soft shell and with some stretchy constituents as an adhesive agents produce indistinct structural colors, even disappearance of structural colors [6] (Figure 3.6).

One of the effective ways to achieve bright and vivid structural colors is the addition of melanin with strong light absorption ability in common colloidal microsphere dispersion to create photonic crystals. Shen *et al.* [32], Jiang *et al.* [33] and Tang *et al.* [34] introduced a small quantity of carbon black, polydopamine and black dye into colloidal dispersion, after the self-assembly the conspicuous structural colors could be shown on white substrates. The monodisperse poly (styrene-butyl acrylate-methacrylic acid) dye composite microspheres with different diameters were formulated and then used to prepare on white fabric substrates by ink-jet printing self-assembly photonic crystals with structural colors (Figure 3.7). The diameters and monodispersity of dye microspheres could be controlled by regulating the poly (styrene-butyl acrylate-methacrylic acid) seed microspheres, concentrations of disperse dye and the preparation temperature. The resultant photonic crystals exhibit iridescent and vivid structural colors. Increase in diameters of microspheres resulted in the red shift of structure colors on the white fabric substrates, which was well in accordance with Bragg's law. Compared to pure photonic crystals on white substrates, the structural colors of poly (styrene-butyl acrylate-methacrylic acid) disperse dye photonic crystals are more vivid and monochromatic, and the iridescence of structural colors is more obvious because due to the remarkable absorption ability of the black dye to stray light outside the photonic band gaps [35].

Figure 3.7 Photonic crystal patterns fabricated by P(St-BA-MAA)@disperse dye microspheres via ink-jet printing.

References

1. Kinoshita, S., Structural Colors in the Realm of Nature. *World Scientific*, 1, 3, 2008. https://doi.org/10.1142/6496.
2. McPhedran, R., Nicorovici, N., McKenzie, D., Rouse, G., Botten, L., Welch, V., Parker, A., Wohlgennant, M., Vardeny, V., Structural colours through photonic crystals. *Physica B*, 338, 182–185, 2003. https://doi.org/10.1016/S0921-4526(03)00483-6.
3. Berthier, S., Boulenguez, J., Bálint, Z., Multiscaled polarization effects in *Suneve coronata* (Lepidoptera) and other insects: Application to anti-counterfeiting of banknotes. *Appl. Phys. A*, 86, 123–130, 2006. https://doi.org/10.1007/s00339-006-3723-9.
4. Land, M.F., The physics and biology of animal reflectors. *Prog. Biophys. Mol. Biol.*, 24, 75–106, 1972. https://doi.org/10.1016/0079-6107(72)90004-1.
5. Shao, J., Liu, G., Zhou, L., Biomimetic nanocoatings for structural coloration of textiles, in: *Active Coatings for Smart Textiles*, p. 269–299, Elsevier, 2016, https://doi.org/10.1016/B978-0-08-100263-6.00012-5.
6. Li, Y., Zhou, L., Liu, G., Chai, L., Fan, Q., Shao, J., Study on the fabrication of composite photonic crystals with high structural stability by co-sedimentation self-assembly on fabric substrates. *Appl. Surf. Sci.*, 444, 145–153, 2018. https://doi.org/10.1016/j.apsusc.2018.03.044.
7. Zhang, H. and Liu, X., Preparation and self-assembly of photonic crystals on polyester fabrics. *Iran. Polym. J.*, 26, 107–114, 2017. https://doi.org/10.1007/s13726-016-0501-1.
8. Zhang, Y., Zhuang, G., Jia, Y., Fan, Q., Shao, J., Structural coloration of polyester fabrics with electrostatic self-assembly of (SiO 2/PEI) n. *Text. Res. J.*, 85, 785–794, 2015. https://doi.org/10.1177/0040517514555803.
9. Zhao, Y., Zhao, X., Gu, Z., Photonic Crystals in Bioassays. *Adv. Funct. Mater.*, 20, 2970–2988, 2010. https://doi.org/10.1002/adfm.201000098.
10. Zhou, L., Li, Y., Liu, G., Fan, Q., Shao, J., Study on the correlations between the structural colors of photonic crystals and the base colors of textile fabric substrates. *Dyes Pigm.*, 133, 435–444, 2016. https://doi.org/10.1016/j.dyepig.2016.06.032.
11. Yasuda, T., Nishikawa, K., Furukawa, S., Structural colors from TiO2/SiO2 multilayer flakes prepared by sol–gel process. *Dyes Pigm.*, 92, 1122–1125, 2012. https://doi.org/10.1016/j.dyepig.2011.08.006.
12. Yuan, D., Xu, X., Huang, W., Wang, F., Wei, Q., Chen, Q., Structural colors of fabric from Ag/TiO2 composite films prepared by magnetron sputtering deposition. *Int. J. Clothing Sci. Technol.*, 29, 427–435, 2017. https://doi.org/10.1108/IJCST-04-2016-0038.
13. Liu, Y., Li, S., Niu, S., Cao, X., Han, Z., Ren, L., Bio-inspired micro-nano structured surface with structural color and anisotropic wettability on Cu substrate. *Appl. Surf. Sci.*, 379, 230–237, 2016. https://doi.org/10.1016/j.apsusc.2016.03.234.

14. Nam, H., Song, K., Ha, D., Kim, T., Inkjet Printing Based Mono-layered Photonic Crystal Patterning for Anti-counterfeiting Structural Colors. *Sci. Rep.*, 6, 30885, 2016. https://doi.org/10.1038/srep30885.
15. Yu, J. and Kan, C.-W., Review on Fabrication of Structurally Colored Fibers by Electrospinning. *Fibers*, 6, 70, 2018. https://doi.org/10.3390/fib6040070.
16. Parker, A.R. and Martini, N., Structural colour in animals–Simple to complex optics. *Opt. Laser Technol.*, 38, 315–322, 2006. https://doi.org/10.1016/j.optlastec.2005.06.037.
17. Tilley, R.J.D., *Colour and the Optical Properties of Materials*, John Wiley & Sons, Ltd, Chichester, UK, 2010, https://doi.org/10.1002/9780470974773.
18. Land, M.F., A Multilayer Interference Reflector in the Eye of the Scallop. *Pecten maximus. J. Exp. Biol.*, 45, 433–447, 1966.
19. Gao, W., Rigout, M., Owens, H., The structural coloration of textile materials using self-assembled silica nanoparticles. *J. Nanopart. Res.*, 19, 303, 2017. https://doi.org/10.1007/s11051-017-3991-7.
20. Yuan, X., Wei, Q., Chen, D., Xu, W., Electrical and optical properties of polyester fabric coated with Ag/TiO 2 composite films by magnetron sputtering. *Text. Res. J.*, 86, 887–894, 2016. https://doi.org/10.1177/0040517515595034.
21. Yuan, X., Xu, W., Huang, F., Chen, D., Wei, Q., Structural colour of polyester fabric coated with Ag/TiO 2 multilayer films. *Surf. Eng.*, 33, 231–236, 2017. https://doi.org/10.1080/02670844.2016.1216264.
22. Yuan, X., Liang, S., Ke, H., Wei, Q., Huang, Z., Chen, D., Photocatalytic property of polyester fabrics coated with Ag/TiO2 composite films by magnetron sputtering. *Vacuum*, 172, 109103, 2020. https://doi.org/10.1016/j.vacuum.2019.109103.
23. Osman, E.M., Michael, M.N., Gohar, H., The Effect of Both UV\Ozone and Chitosan on Natural Fabrics. *Int. J. Chem.*, 2, 28–39, 2010. https://doi.org/10.5539/ijc.v2n2p28.
24. Vakhitova, V.V. and Safonov, N.A., Effect of Chitosan on the Efficiency of Dyeing Textiles with Active Dyes. *Fibre Chem.*, 35, 27–28, 2003. https://doi.org/https://doi.org/10.1023/A:1023819521538.
25. Yavuz, G., Zille, A., Seventekin, N., Souto, A.P., Structural coloration of chitosan coated cellulose fabrics by electrostatic self-assembled poly (styrene-methyl methacrylate-acrylic acid) photonic crystals. *Carbohydr. Polym.*, 193, 343–352, 2018. https://doi.org/10.1016/j.carbpol.2018.03.084.
26. Liu, Z., Zhang, Q., Wang, H., Li, Y., Structural colored fiber fabricated by a facile colloid self-assembly method in micro-space. *Chem. Commun.*, 47, 12801, 2011. https://doi.org/10.1039/c1cc15588a.
27. Liu, Z., Zhang, Q., Wang, H., Li, Y., Structurally colored carbon fibers with controlled optical properties prepared by a fast and continuous electrophoretic deposition method. *Nanoscale*, 5, 6917, 2013. https://doi.org/10.1039/c3nr01766d.

28. Li, K., Zhang, Q., Wang, H., Li, Y., Red, Green, Blue (RGB) Electrochromic Fibers for the New Smart Color Change Fabrics. *ACS Appl. Mater. Interfaces*, 6, 13043–13050, 2014. https://doi.org/10.1021/am502929p.
29. Sun, X., Zhang, J., Lu, X., Fang, X., Peng, H., Mechanochromic Photonic-Crystal Fibers Based on Continuous Sheets of Aligned Carbon Nanotubes. *Angew. Chem.*, 127, 3701–3705, 2015. https://doi.org/10.1002/ange.201412475.
30. Yuan, W., Zhou, N., Shi, L., Zhang, K.-Q., Structural Coloration of Colloidal Fiber by Photonic Band Gap and Resonant Mie Scattering. *ACS Appl. Mater. Interfaces*, 7, 14064–14071, 2015. https://doi.org/10.1021/acsami.5b03289.
31. Sun, C., Yao, Y., Gu, Z., Fabrication of elastic colloidal crystal films from pure soft spheres. *Colloids Surf., A*, 402, 102–107, 2012. https://doi.org/10.1016/j.colsurfa.2012.03.030.
32. Shen, Z., Shi, L., You, B., Wu, L., Zhao, D., Large-scale fabrication of three-dimensional ordered polymer films with strong structure colors and robust mechanical properties. *J. Mater. Chem.*, 22, 8069, 2012. https://doi.org/10.1039/c2jm30546a.
33. Jiang, Y., Lan, Y., Yin, X., Yang, H., Cui, J., Zhu, T., Li, G., Polydopamine-based photonic crystal structures. *J. Mater. Chem. C*, 1, 6136, 2013. https://doi.org/10.1039/c3tc30114a.
34. Tang, B., Xu, Y., Lin, T., Zhang, S., Polymer opal with brilliant structural color under natural light and white environment. *J. Mater. Res.*, 30, 3134–3141, 2015. https://doi.org/10.1557/jmr.2015.238.
35. Liu, G., Han, P., Wu, Y., Li, H., Zhou, L., The preparation of monodisperse P(St-BA-MAA)@disperse dye microspheres and fabrication of patterned photonic crystals with brilliant structural colors on white substrates. *Opt. Mater.*, 98, 109503, 2019. https://doi.org/10.1016/j.optmat.2019.109503.

4

Enzymatic Wet Processing

Mohammad Toufiqul Hoque[1], Nur-Us-Shafa Mazumder[2] and Mohammad Tajul Islam[3*]

[1]*Niederrhein University of Applied Sciences, Mönchengladbach, Germany*
[2]*Oklahoma State University, Stillwater, USA*
[3]*Department of Textile Engineering, Ahsanullah University of Science and Technology, Dhaka, Bangladesh*

Abstract

Conventional textile wet processing is characterized by a high concentration of chemicals and very high temperatures, which can have considerable negative effects on the environment and energy consumption. Enzymes are one of the paramount interests in textile wet processing towards sustainable development and low energy consumption. Enzymes have been successfully used in many textile processes such as desizing, bio scouring, biopolishing, and bio stoning. Different studies also describe how enzymes could be used for the decolorization of dyes from textile effluents. There is evidence of improved dyeability of different fibers like cellulose and nylon when treated with enzymes before dyeing. In this chapter, the history, definition, functions, and different types of enzymes will be discussed. Enzymes used in different textile processes such as pretreatment, washing, and finishing will be described briefly. This chapter will also give insight into the enzymes used for the decolorization of dyes from the effluent, and for increasing the dyeability of cotton and nylon fibers.

***Keywords*:** Enzyme, bio scouring, biopolishing, bio finishing, enzymatic discoloration

4.1 Introduction

Enzyme finish has been regarded as one of the vital treatments for textile products and textile industries. Enzymes were first applied experimentally

**Corresponding author*: tajul.dtt@aust.edu

Luqman Jameel Rather, Mohd Shabbir and Aminoddin Haji (eds.) Sustainable Practices in the Textile Industry, (87–110) © 2021 Scrivener Publishing LLC

in the year 1914. Although the protein nature of enzymes evinced in 1926, the mass production of microbial production of enzymes was started in 1960. Enzyme treatment has gained much attention in the late 2000s and is considered as an important textile finish in wet processing technology i.e., dyeing and finishing. Basically, enzymes are referred to as a substance or a group of proteins (complex or conjugated), which can be produced by living cells and act as catalysts in many biochemical reactions. Enzymes have been considered as an important process influencer in textile finishing in recent years due to their in-build specific characteristics. All groups of enzymes are produced from renewable raw materials considered as biodegradable, which possess various characteristics like firmness, functionality, and meticulousness. Enzymes are a chemical substance, formed in plants or living creatures, considered as inherent organic catalysts [1]. The enzymes used in textile industries are prone to erasing the starch, breaking down excess hydrogen per oxide and lignin. The use of enzymes in the textile industry is rapidly increasing due to their special characteristics, as they are highly strong, coherent, non-toxic, and eco-friendly. Applications of enzymes in different denim finishing and decolorization process of textile effluents are likely to become the commercial breakthroughs. Besides, the use of enzymes in the textile industry results in many advantages, which help to increase efficiency in various wet processes for textile production.

The amendment of macrostructures to new materials and energy are generally caused by biocatalysts. Furthermore, besides for growth, maintenance, and repair of cells, enzymes are biocatalysts that cause many chemical reactions within plant or animal cells, which may contain extraordinary complex systems. Ancient Egypt is believed to be the first place where the history of enzymes and applied biocatalysts originated. Enzymes were retrieved from microbes and applied in different processes such as alcohol production, cheese making, cooking, baking, and brewing. This has been demonstrated in an old Egyptian papyrus exhibiting approaches used to

Figure 4.1 Beer manufacturing in ancient Egypt using enzymes reprinted from [2] with permission from Elsevier.

preserve alcohols and food (Figure 4.1) and similar approaches were utilized in ancient Greece as well [2, 3].

4.2 Enzymes

Enzymes can be defined as the biocatalysts in living cells, which allow for the conducting of the most complex organic reactions under the required ambient and non-threating environmental circumstances. Enzymes show good aspects in the field of green chemistry and different textile wet processing. Enzymes are also able to increase the reaction rates by 10^6–10^{12} times. Enzymes such as cellulases, catalases, laccases, pectinases, lipases, proteases, xylanases are commonly used in the textile industry [4].

4.3 Function of Enzymes

Textile processing has benefited greatly in terms of product quality and environmental impact using enzymes. Out of 7,000 known enzymes only about 75 are generally applied in textile industry processes [5]. According to *Novozymes* (4th edition, 2013), the molecular weight of enzymes range from 12,000 to over 1 million daltons. The required physical space for movement can create tiny functional groups in substrates. To increase the performance or to shorten reaction period, catalytic action is a must for specific chemical complex reactions, in which enzymes act as true catalysts. A reaction such as 'S' (substrate) → 'P' (product) takes place under given conditions; they tend to show transition state. The concentration of the transition state species will fluctuate proportionally with the rate of any chemical reaction. In enzymatic reactions, catalytic centers/'active sites' and binding groups tend to generate intermediate complexes with minimal energy contents. These complexes may encounter atomic and electronic rearrangements, after which the products are discharged, demonstrated in Figure 4.2.

4.4 Classification of Enzymes

International Union of Biochemistry and Molecular Biology (IUBMB) has recognized approximately 4,000 enzymes and categorized the nomenclature into six classifications (Figure 4.3): hydrolases, oxidoreductases, isomerases, transferases, lyases, and ligases. Between all those six categories, hydrolase-based enzymes are the mostly used in the textile industry.

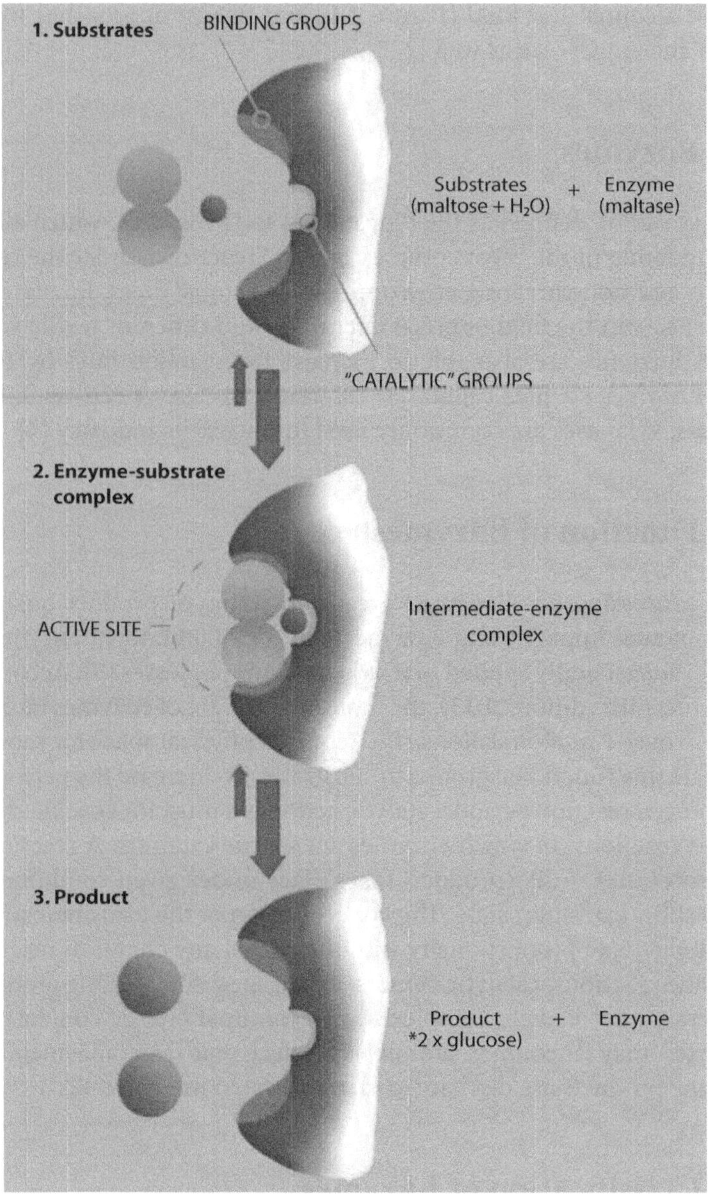

Figure 4.2 The course of an enzyme reaction [6].

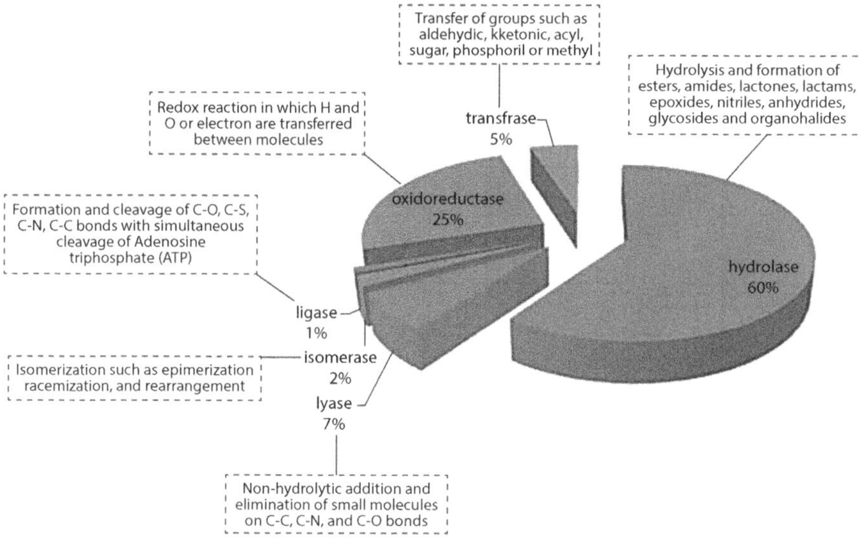

Figure 4.3 Classification of enzymes reprinted from [7] with permission from Elsevier.

Hydrolases

Enzymes of this class are capable of catalyzing the hydrolytic severance of bonds such as Carbon–Oxygen, Carbon–Nitrogen, Carbon–Carbon, and a few other bonds including phosphoric anhydride bonds. These enzymes tend to overlap in their specificities—that is why it remains difficult to generalize rules that are relevant to the enzymes of hydrolases class. The structure name normally grasps the configuration substrate X-hydrolase, where the group separated by hydrolysis is denoted by X. These enzymes are also regarded as the hydrolytic enzymes as the presence of the suffix (*–ase*) in the name of the substrate and might also be classified as transferases, since hydrolysis itself can be regarded as the acceptor as ability to transfer from a specific group to water. Table 4.1 demonstrates industrial applications of different hydrolase enzymes used in textile industry. Amylases were the only enzymes to be used predominantly in textile finishing processes almost forty years ago. And even today, these enzymes are applied to separate sizes from fabrics, which are starch-based after subsequent textile operations.

On the other hand, cellulases have been chosen later to remove hairiness or extra fibers on the surface. Additionally, cellulases have been established as stable finish assistants to the cellulose industry and denim and other garments. Textile finishing uses hydrolytic enzymes. Oxidoreductases have been considered as strong process enhancing tools. Catalases have also

Table 4.1 Industrial applications of hydrolase enzymes [8].

Enzyme	Mechanism of actions	Industrial applications
Amylase	degrading starch into elementary sugars	bio-desizing of cotton fabrics
Cellulase	degrading cellulose into soluble products	bio-scouring, bio-polishing, bio-finishing, wool carbonization
Pectinase	decomposition of pectin	bio-scouring
Xynalases	breaking down linear polysaccharidc into beta-1.4-xylan into xylose	bio-bleaching
Proteases	peptide bonds of protein are hydrolyzed into soluble amino acids and polypeptides	wool modification
Lipases	fats and oils are rotted into glycerol and fatty acids	bio-desizing, detergent formulation
Cutinases	hydrolysis of insoluble cutin	bio-scouring

been used after bleaching for removal of H_2O_2, and reduction of water consumption [9].

4.5 An-Amylase Enzyme for Desizing

Cotton warp yarns are sized before weaving to increase the strength of the yarn and reduce the hairiness. Starch based materials are mostly used as size materials, which help the weaving process by reducing end breakage during interlacing. For efficient dyeing and finishing, these size materials must be removed. The process by which size materials are removed is known as desizing. Amylases are largely used for the de-sizing of cellulose. Amylases constitute about 30% of the industrial enzyme market. Among these enzymes α-amylases are recommended as thermo-stable biocatalyst and have gained industrial interest (Table 4.2). In practice, α-amylases tend to be faster than β-amylases and due to their ability to work anywhere on the substrate molecule, they are used for textile de-sizing [4, 10]. The hydrolysis of water insoluble starch into water soluble end products are catalyzed by these enzymes and could be easily detracted from cotton in the subsequent washing steps [11, 12].

Table 4.2 Biotransformation reactions and application of α-amylase [3].

Enzyme	Temperature (°C)	Bioconversions	Application
α-amylase (bacterial)	90–100	Starch to dextrose syrups	starch hydrolysis, brewing
α-amylase (fungal)	50–60	Starch to dextrose syrups	production of maltose

The temperature requirement for de-sizing is usually at least 70 °C, but higher temperatures are also recommended [12]. Thermo-stable amylases can be used at temperatures higher than 95 °C in continuous pretreatment de-sizing [13]. The α-amylases have been isolated from bacterial sources and have achieved higher heat stability, and can show activity over wider ranges of pH, and can show excellent potential in de-sizing [12]. De-sizing duration gets longer with low temperature α-amylases (20–60 °C) than the α-amylases that work at high temperatures (95-100 °C or higher) [8]. In addition, de-sizing with high temperature α-amylases can be performed in steam chambers to endorse a fully continuous process [12]. Therefore, raising the temperature for de-sizing facilitates efficient starch removal and reduces process duration. Hao, *et al.* (2013) have experimented the effect of ultrasonic on the α-amylase on cotton fabrics sized with starch. Ultrasound irradiation and α-amylase are two important enzymatic desizing technologies in textile industry [14]. Researchers here experimented to find out the effect of ultrasound on the activity of α-amylase and its efficiency of desizing. Two different ways have been followed: i) step-wise application where ultrasound was applied first then α-amylase, ii) simultaneous application of ultrasound and α-amylase. The desizing efficiency was better in the simultaneous process compared to the step-wise method.

4.6 Pectinase Enzyme for Scouring

The non-cellulosic constituents present in Cellulosic fibers i.e. pectin (0.7–1.2%), waxes (0.4–1.0%), protein (1.1–1.9%), ashes (0.7–1.0%) and seed coat fragments (0.5–1.0%). Non-cellulosic compounds in primary cell remain inaccessible due to the presence of waxes in the thin outer wall [15, 16]. It is necessary to remove the waxes for subsequent processing and these hydrophobic non-cellulosic elements are usually eliminated by a hot scouring process (aqueous alkaline) so that the wettability of the fibers can be improved, which in turn ultimately facilitates uniform dyeing and

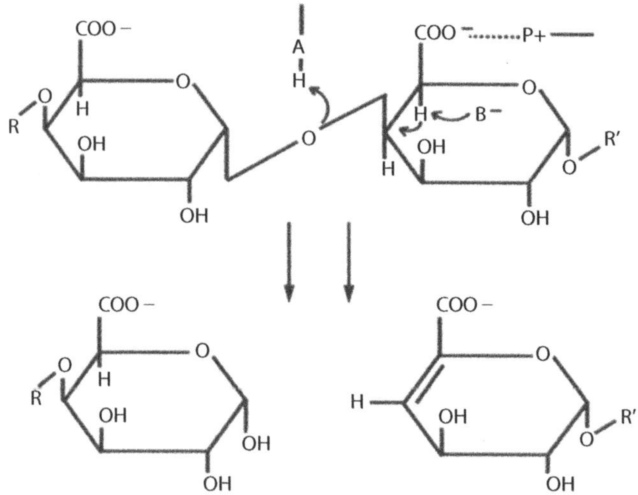

Figure 4.4 β-elimination mechanism for α-1,4-polygalacturonic acid cleavage [18].

finishing. The complex structure of pectin is disintegrated into simpler molecules by the enzymes broadly known as pectinases [12, 16].

Enzymatic bio-scouring was conducted by Joshi *et al.* [17] using pectinase isolated from marine bacteria. Pectinase scouring was carried out with a dose of 10% (o.w.f.) at pH 7, for 120 min at 60 °C temperature (Figure 4.4). Mechanical and physiochemical properties of pectinase scoured fabrics were same as that of the alkaline scoured fabrics. However, enhanced dye absorbency was observed in pectinase scouring. To make the cotton fabric finer, softer, cleaner, and brighter and to improve its characteristics such as whiteness, brightness, and yellowness dosage of 10% pectinase (o.w.f.) at 60 °C for 120 min at buffer pH 7 has been recommended. Pectinase enzyme scouring can be compelling, duplicable, and eco-friendly.

4.7 Protease Enzyme for Wool Anti-Felting

A typical property of wool is felting shrinkage while washed, and for achieving a washable wool product this property must be controlled. This irreversible shrinkage is caused by the progressive entanglement of wool fibers. The surface of the wool fiber is configured with cuticle scale, which is mainly responsible for this irreversible shrinkage (Figure 4.5).

Decreasing the abrasion between the fibers can reduce the wool shrinkage, which is necessary to make the wool product machine washable. Reducing the cuticle scales on the fiber surface can reduce the friction

Figure 4.5 Wool fiber under Scanning electron microscopy reprinted from [19] with permission from Elsevier.

between the fibers. The treatment to partly removing the scales or smoothing the edges from the overlapping scales is known as the anti-felting treatment. Fabric-felting property is improved after this treatment, which makes the wool product washable. Additive (synthetic resin layer), subtractive (oxidation, reduction) and combined process (chlorine-Hercosett) are the three anti-felting processes, which are commercially practiced.

The combined process chlorine-Hercosett is a conventionally used process as a shrink-resist treatment to wool fibers. This treatment includes a chlorination step to degrade the cuticle scale. Later, in presence of cationic polymer resin, a sulfuric acid treatment is given to cover the scales. However, this process is not eco-friendly because it causes contamination in effluent and also the natural character of wool is changed to a more synthetic handle. This processing is almost the same as using strong oxidants. Therefore, the necessity to develop more environmentally friendly processes due to the strict awareness processes and consumer awareness results to the replacement of using traditional chemicals to enzymatic ones. It has been found that the protease enzyme can hydrolyze the protein compounds and can be used to hydrolyze the cuticle scales of wool fiber [20–23]. Researchers attempted to replace the conventional chlorine process with the protease enzyme and found that this enzyme is able to smoothen the cuticle layer and helps to gain anti-shrinkage property. However, action of protease enzyme can outspread beyond the cuticle layers and attack inner fiber structure, which causes loss of weight and hence tensile strength [24]. The molecular size of the protease enzyme can be increased by covalent bonding with an entering polymer, which limits the proteolytic attack of the cuticle scales. Therefore, wool damage can be controlled. Though this process can increase shrink resistance, it can cost extra, and it can be difficult to achieve standards for machine wash ability. Smith and Shen recently applied cross-linking protein resin in order to modify the fiber surface on a wool-knitted fabric. High level of shrink resistance was demonstrated by the modified wool with machine wash ability. In addition, tensile strength and the hand feel were not compromised [22].

4.8 Cellulase Enzyme for Biopolishing and Biostoning

In general, biopolishing refers to a finishing process, which can modify fiber surface and reduces hairiness and pilling by cellulases before, during, or after dyeing [9, 13, 25]. Cellulases are regarded as the most useful enzymes applied in textile finishing, particularly cellulose-based products. The ball of fuzz, commonly referred to as a pill, can cause a major quality problem; therefore, the main goal of biopolishing with cellulase is the stoppage of hairiness and pilling. In the industrial enzyme markets, available cellulases are a combination of enzymes as shown in Figure 4.6: endogluconases (EG), exogluconases or cellobiohydrolases (CBH) and cellobiasesor or β-glucosidases [25]. Cellulases are generally categorized by the pH range such as acid cellulose (active at pH 3.8–5.8), neutral cellulose (active at pH 6.0–7.0), and alkaline cellulose (active at pH 9–10). Table 4.3 demonstrates applications of cellulases in textile industry.

The efficiency of bio-polishing can be increased by means of pre- and post-treatment processes [27]. Effective bio-polishing can be achieved through the optimization of process parameters, enzyme concentration, and the types of cellulases used [8]. Endoglucanase-based acid cellulases perform productive cellulosic bio-polishing [25, 28–31]. To increase thermal stability and ease the reuse of enzymes, diverse techniques and incapacitate matrices are used [32–34]. Moreover, improvement of stability at neutral pH can be ensured if application adds immobilization of acidic cellulases onto selective functionalized matrices [35]. In order to bring on a worn look for the most denim jeans or other denim garments, a specific finishing process, called bio-stoning, is applied (Figure 4.7). The 1980s was the period where cellulose enzymes came to play a huge role as denim washing additive to obtain a faded and worn look. Cellulases can loosen the indigo dye on the denim, thus enhancing the abrasion under the name

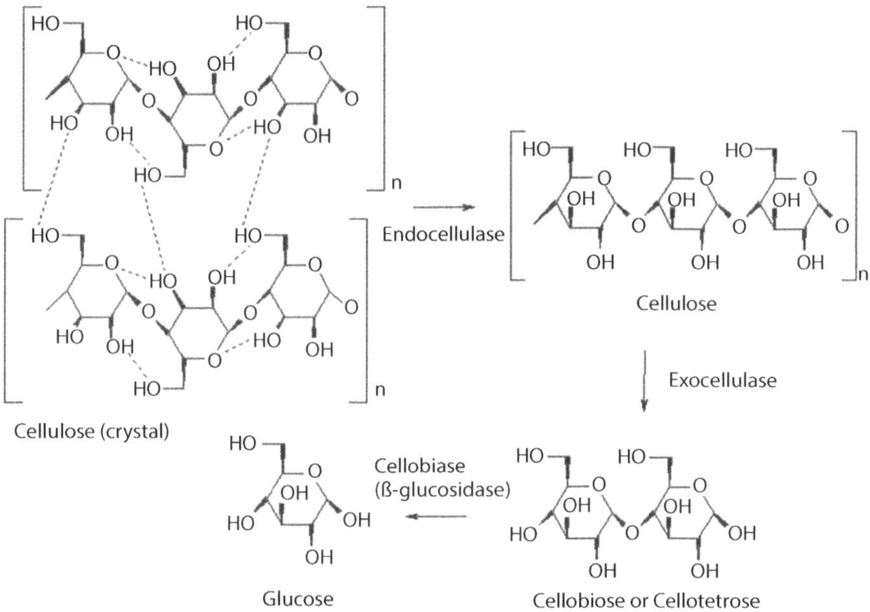

Figure 4.6 Cellulases catalyze three types of reactions: i) endocellulase, ii) exocellulase, iii) β-glucosidase [26].

Table 4.3 Application of cellulases in textile finishing [31].

Enzyme	Activity	Application
Cellulase, preferably neutral and endoglucanase rich	separate the surplus dyes from denim garments, makes cotton fabrics softer without damaging the fiber	denim fabrics bio-stoning finish, production of washing powders which are eco-friendly and have high quality
Cellulase, preferably acidic and endoglucanase rich	separation of additional microfibrils from the surface of cotton and non-denim products	bio-polishing finish of cotton and non-denim products
Cellulase, alkaline-endoglucanase rich	returning back the softness and brightness of cotton fabrics	Increase the quality of the fabrics

Figure 4.7 Left: original (raw) indigo dyed denim, right: Bio-stone washed denim reprinted from [36] with permission under Creative Commons Attribution 4.0 License.

bio-stone washing or bio-stoning [36]. Denim bio-finish has become one of the most useful and reliable enzyme-based processes in the textile finishing due to extensive research in recent times [37–39].

For denim stoning, a variety of cellulases with distinctive properties and actions are available [8]. A major problem linked with denim bio-stoning is back-staining. Back-staining is determined as the re-deposition of detached indigo onto the fabrics or garments, and ascribed to the strong-affinity interaction between indigo and cellulase enzyme [16]. Back-staining is normally caused significantly under low pH range (pH 4–6), whereas neutral pH offers less back staining. Therefore, neutral cellulases are preferred when the aim is to ensure minimal back-staining [36].

4.9 Hairiness Removal Mechanism

The removal of cellulosic fibers or hairiness from the fabric surface is required to improve the pilling or abrasive performance handle. Enzymatic hydrolysis is required for an abraded appearance. This process weakens the fiber ends which can be removed by successive mechanical action.

4.9.1 During Scouring and Bleaching in Alkaline Condition

Šimić *et al.* [40] explained in their study how a two-bath process was followed to apply enzymatic scouring and bleaching on de-sized fabrics. For this purpose, a laboratory dyeing machine was used where 1:20 liquor ratio applied in 500 m/holders. The samples weighed around 4 g and

Table 4.4 Bleaching process conditions [40].

Process conditions				
$pH_{initial}$	pH 8 with 1N Na_2CO_3	pH 8 with 1N Na_2CO_3	2 g/L	4 g/L
T (°C)	65	65	65	90

scouring was done in a buffer (pH 8) containing 2 g/l Beisol PRO and 1 g/l Felosan RG-N at 85 °C for 15 min. After that, the samples were washed out gradually at varying descending temperatures for 5 min. Afterward, four different bleaching processes were applied to the previously scoured samples. The bleaching process condition is shown in Table 4.4. Bleaching was performed for 1 h, and the samples were washed out at 60 °C, 40 °C and room temperature for 5 min after completion of bleaching and dried under given atmospheric conditions. The enzymatic scouring and bleaching with peracetic acid (PAA) method were selected, and later found to minimize fiber damage compared with conventional hydrogen peroxide (HP) bleaching process. This process was run at lower temperatures and resulted in a neutral bleaching bath. Hence, they are realized to be more environmentally friendly.

4.9.2 Applying Before Dyeing in Acidic Condition

If cellulase treatment of cotton is carried out after dyeing, then there might be a risk of shade change. Colorants can be responsible for decreasing the enzyme-effect, and thus there would need to be a higher concentration of enzymes. Before dyeing in acidic conditions, biopolishing requires the

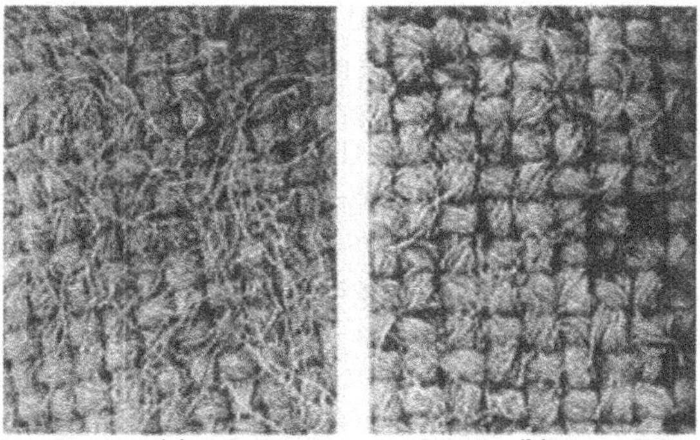

Figure 4.8 Microscopic images of: (a) cotton fabric; (b) same cotton fabric after bio-finishing with cellulase [41].

following [40]: pH 4.5–5.5, temperature ranges from 40 to 55 °C, duration 30–60 min, and concentration of enzyme 1–2% per kg of fabric. Figure 4.8 displays the disparities between before and after cellulase treated cotton fabric. Biopolishing has undoubtedly more merits over demerits, but the most common disadvantages of biopolishing which have been experienced by many researchers are loss in weight and strength.

4.10 Enzyme Decolorization of Textile Effluent

It is assumed around 1 million ton of dyes are produced each year, among which 50% are released in the effluent brooks or other dumping grounds. According to Vats and Mishra [42], the unrefined colorants in the embayment are a real threat to aquatic flora and fauna due to their enhancement for the reduction of transparency and increase chemical oxygen demand (COD), and because of this, there will be a risk of forming highly toxic and poisonous compounds. Although there are many treatments available (i.e. physical, chemical, and advanced oxidation) for the withdrawal of dyes, these are not always affordable for textile industries—mainly in southern Asia because of their cost and unfriendly attitude towards environmental health [43]. On the contrary, biological systems (cells, enzymes, plant materials) offer cheaper treatment of textile effluents, and are eco-friendly as well [44, 45].

In recent times, enzymatic approach has caught much attention in the extraction of phenolic hazardous wastes from dissolved solutions [29]. In addition, with phenolic compounds removal method, there exist other methods, under smaller proportion of biological oxygen demand (BOD) to COD, where conventional laccase or peroxidase enzyme treatment will not be enough. Therefore, a demand appears to inspect mixture of these enzymes for different wet processing steps. Hydrogen peroxide is the most utilized bleaching agent in commercial market applied at boiling temperatures. The replacement or fusion of the bleaching chemical with an enzymatic bleaching process results not only in less fiber impairment, but also is increased water saving [46]. Many resilient white-rot fungi provided a mix of enzymes, and these have capacity of breaking down varied range of tenacious and dangerous contaminated materials, including synthetic colorants [47–49]. Due to different chemistries used by the most widely known group such as lignin peroxidase (LiP), Manganese peroxidase (MnP), and laccase allow fungi to break down even with complex mixture of contaminants [50, 51]. Laccases that belong to white-rot fungi constitute metabolically miscellaneous enzyme with reduction-oxidation potential ranges between low (+375 ± 8 mV) and high (+790 ± 10 mV) [52], which boosts predominantly the ability to break down expansive phenolic and aniline compounds.

Vats and Mishra [42] mentioned enzyme-based processes offer low production cost, versatility, and availability to cater to large effluent masses produced in various wet processes, which will obviously decide the commercial success any of its processes. A study by Vats Mishra was designed to produce ligninolytic enzymes for decolorization at cheaper price [42]. *Cyathus bulleri*, previously reported [53] to generate productive laccase for decolorization a series of colorants [54], was picked for raising on agricultural wastes. Reactive orange (RO), reactive black (RB), reactive violet (RV), acid violet (AV), and Kiton/acid blue (KB) were selected as colorants. Between those dyes RO and KB belong to the massive group of economically produced dyestuffs and relatively defiant to enzyme break down [45]. Reactive orange (RO) has been examined to be decolorized by photo-oxidation technique utilizing delustrant TiO_2 [55]. There are tests and discussions [56–58] wherein the colorant was vaccinated with biomass (*Irpex lacteus* and *Bacillus* sp.), bringing about decolorization of reactive orange (RO). Only one report (Pazarbasi *et al.* [59]) was shown on decolorization of acid blue when the fungus *Trametes trogii* was refined in a dilute suspension of dye, but over long operational times, such methods are ineffective as the growth of mildew or fungi are hindered after few hours causing to oversight of the system.

4.11 Enzymes for Increasing Dyeability of Different Fibers

4.11.1 Application on Cotton

Basically, cotton is referred as a natural polymer of D-glucose chemically, in which glucose units are consecutively linked with β-1, 4-glycosidic covalent bonds [60]. Raw cotton possesses several supercilious properties compared to other cellulosic components i.e., good absorbency, outstanding biodegradability, and comfortable to wear and easy to color. To increase its appearance, hand feel, and commercial value raw cotton is customarily subjected to cellulase polishing mainly to eliminate ‘fuzz’ or ‘pilling’ from the plane. Biopolishing with cellulase has acute influences on the various properties of raw cotton such as structural, physicochemical, and coloring properties [61]. Enzyme finish on cotton can bring on systematic and continuing separation of its primary and secondary walls, which will enhance the reduction of fineness of fibers, separation of hairs or pills from the surface, cause weight loss, rupture the cotton surface and even split the microfibrils [62–64]. Gradual break down by cellulase will prompt to lose the tensile strength and molecular weight [65, 66]. Nonetheless, enzymatic polishing will obviously improve the wettability and moisture regaining of cotton, which shows enlargement of its mean pore radius and obtainable surface area [67].

The ability of dyeing cotton will not be the same considering the enzymatic treatment [68–70]. The enzymatic hydrolysis causes weight loss of cotton that influences the impregnation of anionic dyes. The untreated cotton contains high crystalline region which is why the color strength and uptake of reactive dyed cotton with cellulase are expected to be lower at similar dye concentration. In addition, cellulase treatment will produce negative charges and form aldehyde groups on cotton, consequently increasing the electrostatic repulsion between cotton and dye molecules, which can result reduced dye absorption [69]. Cellulase treatment could increase the absorption of anionic natural dyes on cotton claimed by Tsatsaroni and Liakopoulou-Kyriakides [71]. Performing enzymatic treatment can increase the affinity of anionic acid dyes to cotton was noticed by Andreaus *et al.* [72]. Anionic nanoscale pigments have been used vastly for coloration of cotton due to their faster adsorption rate, high coloring efficiency and lower energy consumption (Figure 4.9).

Hao *et al.* [61] implemented a research work utilizing a crude cellulase on to natural cotton fabrics. Firstly, the structural and physiochemical properties of cotton were probed as enzymatic modifications. Secondly,

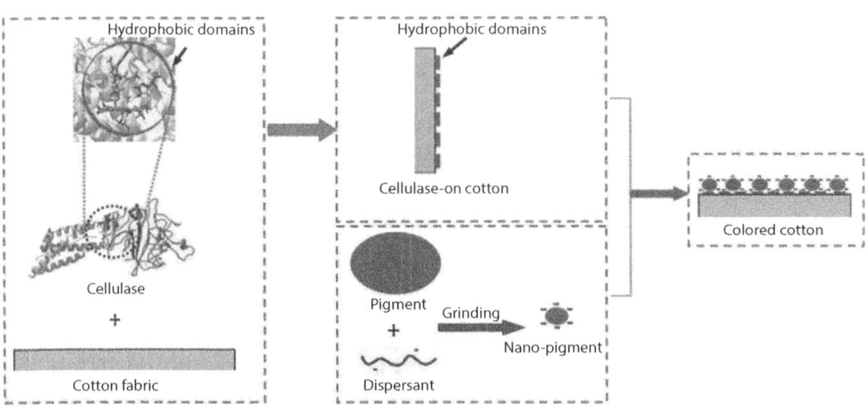

Figure 4.9 Illustrative demonstration of cellulase treatment showing intensification on pigment uptake reprinted from [61] with permission from Springer.

anionic nanoscale dispersion was prepared using copper phthalocyanine (CuPc) pigment with the help of anionic dispersants. In the end, the dyeability of cotton using the nanoscale pigment was analyzed and checked the influences of cellulase treatment. According this experiment, cotton surface still contained cellulase protein after the bio treatment—a remarkable role in launching the transfer of pigment from dye bath to cotton. The hydrophobic interaction between cellulase and pigment was the key for the improvement of dye uptake and its final color depth on cotton.

4.11.2 Application on Nylon

Polyamides (PA) are one of the most applied polymers in textile finishing or wet processing. The two most essential polyamide fibers nylon 6 and nylon 6.6 generally contain the same chemical groups in similar ratio along the chain. The only difference between these macromolecular fibers is the reversal in the sequences ($-CO-NH-CH_2-$) alternately. Therefore, there will be small differences in shape and packing. In general, synthetic fibers possess hydrophobicity, low dyeability, and insufficient washability, and to overcome such undesired properties enzymatic hydrolysis is of great interest to industrial chemists [73]. Studies proved that small monomers of nylon can be hydrolyzed by proteolytic enzymes [74, 75]. Different types of enzymes are also responsible for the generation of amino groups on the plane of nylon 6.6 and nylon 6.

One research has been published by Gashti *et al.* [76] regarding enzymatic hydrolysis of nylon 6.6 with the combination of proteolytic and lipolytic enzymes. According to the experiment, nylon 6.6 was treated with varying dosage of protease and lipase in solutions for no more than 80 min. Enzymatic catalyzation enhanced the process and caused hydrolysis of the polyamide chains and the amide functional (Figure 4.10). Enzymatic hydrolysis caused the colorants to go through the fabric more easily while the amorphous and paracrystalline regions become more obtainable. There were few notable features obtained from the mixed-enzyme treatment of nylon such as increased absorption and fixation rate (5–15% rise), lower load of dyes in effluent, shorter dyeing process due to higher dye absorption, and also reproducible dyeing was achievable. However, the only limitation of the mixed-enzyme treatment found to be the thermal properties of nylon 6.6.

Pravinzadeh *et al.* [59] used a subtilisin enzyme to improve dyeability of nylon 6 fiber [77]. For the experiment nylon 6 was treated with varying dosage of protease solutions (1%, 2%, 4%, and 6% on the weight of fabric) about for 80 min. The results showed significant effect of the enzyme

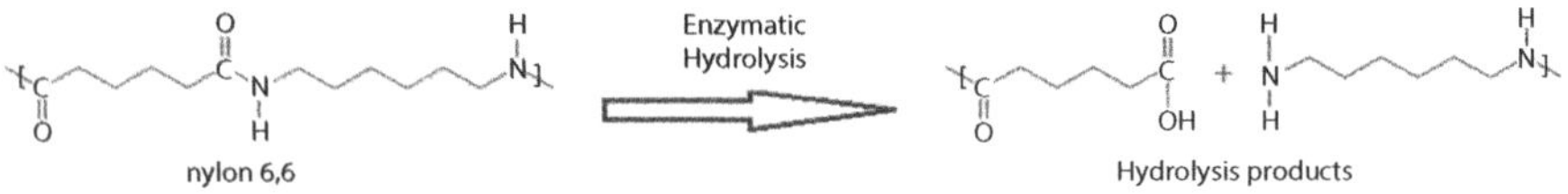

Figure 4.10 Bio hydrolysis of nylon 6.6 fibers after mixed-enzyme treatment [76].

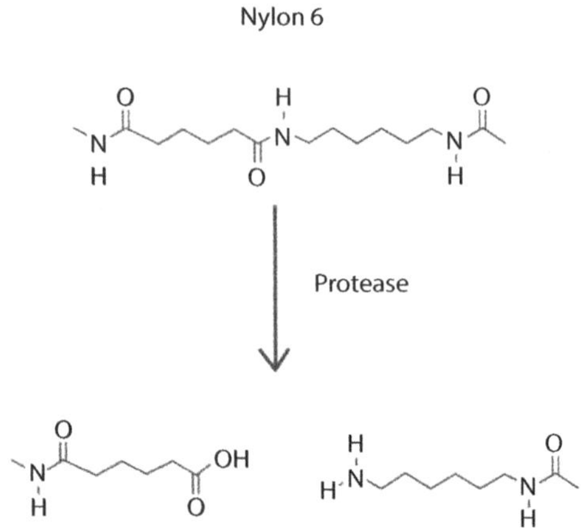

Figure 4.11 Protease treatment showing hydrolysis of nylon 6 [73].

treatment using the UV–vis spectrophotometry analysis on the remaining dye bath, which exhibited increased absorption of acid and disperse dyes into the fiber. Figure 4.11 shows the hydrolysis of nylon 6 after being treated with protease enzyme.

4.12 Conclusion

To reduce the pollution from textile industry, it is important to decrease the usage of toxic chemicals in different textile processing. Enzymes have been proven as a reliable biocatalyst, which can successfully replace the toxic chemicals. Enzymes have been used successfully in different textile commercial applications. For instance, enzymes are used in desizing, different denim finishing, dyeing of cotton and nylon to increase absorbency, etc. Enzymes can be introduced in almost every step of textile processing as an alternative to commonly used toxic chemicals; however, there are only a few textile processes where enzymes are being used successfully. Therefore, further research is required for the implementation of more commercially successful enzymatic processes. Still, there is considerable potential for the development of commercially successful new enzymatic applications in different textile processing.

References

1. Eslahi, H., Ghaffari-Moghaddam, M., Khajeh, M., Omay, D., Zakipour-Rahimabadi, E., Motalleb, G., General biography, structure and classification of enzymes. *Res. Rev. Mater. Sci. Chem.*, 3, 1, 1–83, 2014.
2. Ghanem, A., Trends in lipase-catalyzed asymmetric access to enantiomerically pure/enriched compounds. *Tetrahedron*, 63, 8, 1721–1754, 2007.
3. Haki, G.D. and Rakshit, S.K., Developments in industrially important thermostable enzymes: A review. *Bioresour. Technol.*, 89, 1, 17–34, 2003.
4. M., K., Applications of enzymes in the textile industry: A review. *2nd Int. Congr. Eng. Ecol. Mater. Process. Ind. Jahorina, Bosnia Herzegovina; Tehnol. Fak. Zvornik*, vol. March 2011, p. 230–239, 2011.
5. Quandt, C. and Kuhl, B., Operational possibilities and optimisation of enzymatic processes in the textile finishing industry. *Melliand Textilberichte*, 81, 10, 198–200, 2000.
6. Damhus, T., Kaasgaard, S., Olsen, H.S., *Novozymes Enzymes at work*, 4th ed., p. 76, Novozymes A/S, Copenhagen, Denmark, 2013.
7. Puskas, J.E., Seo, K.S., Sen, M.Y., Green polymer chemistry: Precision synthesis of novel multifunctional poly(ethylene glycol)s using enzymatic catalysis. *Eur. Polym. J.*, 47, 4, 524–534, 2011.
8. Rahman, M., Hack-Polay, D., Billah, M., Nabi, N.U., Bio-based textile processing through the application of enzymes for environmental sustainability. *Int. J. Technol. Manag. Sustain. Dev.*, 19, 1, 87–106, 2020.
9. Araújo, R., Casal, M., Cavaco-Paulo, A., Application of enzymes for textile fibres processing. *Biocatal. Biotransform.*, 26, 5, 332–349, 2008.
10. Hossain, S. and Uddin, M.K., Comparative Analysis Between Conventional Pre-treatment and Bio-Preparation. *Int. J. Eng. Technol.*, 11, 3, 14–19, 2011.
11. Opwis, K. *et al.*, Generation of methane from textile desizing liquors. *Eng. Life Sci.*, 10, 4, 293–296, 2010.
12. Shahid, M., Mohammad, F., Chen, G., Tang, R.C., Xing, T., Enzymatic processing of natural fibres: White biotechnology for sustainable development. *Green Chem.*, 18, 8, 2256–2281, 2016.
13. Choudhury, A.K.R., *Sustainable Textile Wet Processing: Applications of Enzymes*, vol. June 2014, S.S. Muthu (Ed.), p. 203–238, Springer, Singapore, 2014.
14. Hao, L., Wang, R., Fang, K., Liu, J., Ultrasonic effect on the desizing efficiency of amylase on starch-sized cotton fabrics. *Carbohydr. Polym.*, 96, 2, 474–480, 2013.

15. Hardin, I.R., Enzymatic treatment versus conventional chemical processing of cotton, in: *Advances in Textile Biotechnology*, V.A. Nierstrasz, A. Cavaco-Paulo (Eds.), p. 132–149, Woodhead Publishing Limited, Cambridge, UK, 2010.
16. Madhu, A. and Chakraborty, J.N., Developments in application of enzymes for textile processing. *J. Clean. Prod.*, 145, 114–133, 2017.
17. Joshi, M., Nerurkar, M., Badhe, P., Adivarekar, R., Scouring of cotton using marine pectinase. *J. Mol. Catal. B Enzym.*, 98, 106–113, 2013.
18. Herron, S.R., Benen, J.A.E., Scavetta, R.D., Visser, J., Jurnak, F., Structure and function of pectic enzymes: Virulence factors of plant pathogens. *Proc. Natl. Acad. Sci. U. S. A.*, 97, 16, 8762–8769, 2000.
19. Shen, J., Wool finishing and the development of novel finishes, in: *Advances in Wool Technology*, N.A.G. Johnson, I.M. Russell (Eds.), p. 147–182, Woodhead Publishing Limited, Cambridge, UK, 2008.
20. Dong, L. and Xu, L., Enzymatic Process for The Wool Fabric Anti-felting Finishing. *Mod. Appl. Sci.*, 2, 3, 91–93, 2008.
21. Shen, J., Rushforth, M., Cavaco-Paulo, A., Guebitz, G., Lenting, H., Development and industrialisation of enzymatic shrink-resist process based on modified proteases for wool machine washability. *Enzyme Microb. Technol.*, 40, 7, 1656–1661, 2007.
22. Smith, E. and Shen, J., Surface modification of wool with protease extracted polypeptides. *J. Biotechnol.*, 156, 2, 134–140, 2011.
23. Islam, M.T. and Asaduzzaman, S., Environmentally-Friendly Textile Finishing, in: *Textiles and Clothing*, M. Shabbir (Ed.), John Wiley & Sons, Inc., and Scrivener Publishing, New Jersey, USA, 2019.
24. Silva, C.J.S.M., Prabaharan, M., Gübitz, G., Cavaco-Paulo, A., Treatment of wool fibres with subtilisin and subtilisin-PEG. *Enzyme Microb. Technol.*, 36, 7, 917–922, 2005.
25. Saravanan, D., Vasanthi, N.S., Ramachandran, T., A review on influential behaviour of biopolishing on dyeability and certain physico-mechanical properties of cotton fabrics. *Carbohydr. Polym.*, 76, 1, 1–7, 2009.
26. Bhaumik, P. and Dhepe, P.L., Conversion of biomass into sugars, in: *Biomass Sugars for Non-Fuel Applications*, D. Murzin, O. Simakova (Eds.), p. 1–53, Royal Society of Chemistry, Cambridge, UK, 2016.
27. Pere, J., Puolakka, A., Nousiainen, P., Buchert, J., Action of purified Trichoderma reesei cellulases on cotton fibers and yarn. *J. Biotechnol.*, 89, 2–3, 247–255, 2001.
28. Bahtiyari, M.I., Duran, K., Körlü, A.E., Perincek, S., Usage of commercial cellulases in biopolishing of viscose fabrics. *Tekst. ve Konfeksiyon*, 20, 1, 57–64, 2010.
29. Esposito, E. and Durán, N., Potential applications of oxidative enzymes and phenoloxidase-like compounds in wastewater and soil treatment: A review. *Appl. Catal. B Environ.*, 28, 2, 83–99, 2000.

30. Bai, G., Fu, K., Jin, N., Zhu, L., Chai, H., Lu, D., Bio-polishing of cotton fabric with cellulase. *Adv. Mater. Res.*, 468–471, 46–49, 2012.
31. Mojsov, K., Biopolishing Enzymes and their applications in textiles: A review. *Tekst. Ind.*, 61, 2, 20–24, 2014.
32. Hirsh, S.L., Bilek, M.M.M., Nosworthy, N.J., Kondyurin, A., Dos Remedios, C.G., McKenzie, D.R., A Comparison of covalent immobilization and physical adsorption of a cellulase enzyme mixture. *Langmuir*, 26, 17, 14380–14388, 2010.
33. Hailei, Y., Zhong liang, S., Hongbo, S., Jinfang, C., Xiaofei, W., Haixia, Y., Immobilization of cellulase on modified mesoporous silica shows improved thermal stability and reusability. *African J. Microbiol. Res.*, 7, 25, 3248–3253, 2013.
34. Romo-Sánchez, S., Camacho, C., Ramirez, H.L., Arévalo-Villena, M., Immobilization of Commercial Cellulase and Xylanase by Different Methods Using Two Polymeric Supports. *Adv. Biosci. Biotechnol.*, 05, 06, 517–526, 2014.
35. Dinçer, A. and Telefoncu, A., Improving the stability of cellulase by immobilization on modified polyvinyl alcohol coated chitosan beads. *J. Mol. Catal. B Enzym.*, 45, 1–2, 10–14, 2007.
36. Agrawal, B.J., Bio-Stoning of Denim–An Environmental-Friendly Approach. *Curr. Trends Biomed. Eng. Biosci.*, 3, 3, 45–47, 2017.
37. Cavaco-Paulo, A., Mechanism of cellulase action in textile processes. *Carbohydr. Polym.*, 37, 3, 273–277, 1998.
38. Belghith, H., Ellouz-Chaabouni, S., Gargouri, A., Biostoning of denims by *Penicillium occitanis* (Pol6) cellulases. *J. Biotechnol.*, 89, 2–3, 257–262, 2001.
39. Yu, Y. *et al.*, Cellulase immobilization onto the reversibly soluble methacrylate copolymer for denim washing. *Carbohydr. Polym.*, 95, 2, 675–680, 2013.
40. Šimić, K., Soljačić, I., Pušić, T., Application of Cellulases in the Process of Finishing Uporaba celulaz v procesu plemenitenja. *Sci. c Rev. Znan. članek*, 58, 1, 47–56, 2015.
41. Sanap, R., Role of Enzymes in Textile Wet-Processing, https://www.textiletoday.com.bd/role-of-enzymes-in-textile-wet-processing/, 2008.
42. Vats, A. and Mishra, S., Decolorization of complex dyes and textile effluent by extracellular enzymes of *Cyathus bulleri* cultivated on agro-residues/domestic wastes and proposed pathway of degradation of Kiton blue A and reactive orange 16. *Environ. Sci. Pollut. Res.*, 24, 12, 11650–11662, 2017.
43. Vincenzo Naddeo, A.C., Wastewater Treatment by Combination of Advanced Oxidation Processes and Conventional Biological Systems. *J. Bioremediation Biodegrad.*, 04, 08, 1–8, 2013.
44. Anastasi, A., Parato, B., Spina, F., Tigini, V., Prigione, V., Varese, G.C., Decolourisation and detoxification in the fungal treatment of textile wastewaters from dyeing processes. *N. Biotechnol.*, 29, 1, 38–45, 2011.

45. Przystas, W., Zablocka-Godlewska, E., Grabinska-Sota, E., Biological removal of azo and triphenylmethane dyes and toxicity of process by-products. *Water. Air. Soil Pollut.*, 223, 4, 1581–1592, 2012.
46. Tzanov, T., Basto, C., Gübitz, G.M., Cavaco-Paulo, A., Laccases to improve the whiteness in a conventional: Bleaching of cotton. *Macromol. Mater. Eng.*, 288, 10, 807–810, 2003.
47. Novotný, *et al.*, Ligninolytic fungi in bioremediation: Extracellular enzyme production and degradation rate. *Soil Biol. Biochem.*, 36, 10, 1545–1551, 2004.
48. Rodríguez Couto, S., Dye removal by immobilised fungi. *Biotechnol. Adv.*, 27, 3, 227–235, 2009.
49. Harms, H., Schlosser, D., Wick, L.Y., Untapped potential: Exploiting fungi in bioremediation of hazardous chemicals. *Nat. Rev. Microbiol.*, 9, 3, 177–192, 2011.
50. Majeau, J.A., Brar, S.K., Tyagi, R.D., Laccases for removal of recalcitrant and emerging pollutants. *Bioresour. Technol.*, 101, 7, 2331–2350, 2010.
51. Margot, J., Maillard, J., Rossi, L., Barry, D.A., Holliger, C., Influence of treatment conditions on the oxidation of micropollutants by *Trametes versicolor* laccase. *N. Biotechnol.*, 30, 6, 803–813, 2013.
52. Mate, D.M. and Alcalde, M., Laccase engineering: From rational design to directed evolution. *Biotechnol. Adv.*, 33, 1, 25–40, 2015.
53. Salony Mishra, S. and Bisaria, V.S., Production and characterization of laccase from *Cyathus bulleri* and its use in decolourization of recalcitrant textile dyes. *Appl. Microbiol. Biotechnol.*, 71, 5, 646–653, 2006.
54. Chhabra, M., Mishra, S., Sreekrishnan, T.R., Mediator-assisted decolorization and detoxification of textile dyes/dye mixture by *Cyathus bulleri* laccase. *Appl. Biochem. Biotechnol.*, 151, 2–3, 587–598, 2008.
55. Chen, C.Y., Photocatalytic degradation of azo dye reactive orange 16 by TiO2. *Water. Air. Soil Pollut.*, 202, 1–4, 335–342, 2009.
56. Svobodová, K., Senholdt, M., Novotný, Č., Rehorek, A., Mechanism of Reactive Orange 16 degradation with the white rot fungus *Irpex lacteus*. *Process Biochem.*, 42, 9, 1279–1284, 2007.
57. Telke, A.A., Kalyani, D.C., Dawkar, V.V., Govindwar, S.P., Influence of organic and inorganic compounds on oxidoreductive decolorization of sulfonated azo dye C.I. Reactive Orange 16. *J. Hazard. Mater.*, 172, 1, 298–309, 2009.
58. Bedekar, P.A., Saratale, R.G., Saratale, G.D., Govindwar, S.P., Oxidative stress response in dye degrading bacterium Lysinibacillus sp. RGS exposed to Reactive Orange 16, degradation of RO16 and evaluation of toxicity. *Environ. Sci. Pollut. Res.*, 21, 18, 11075–11085, 2014.
59. Pazarbasİ, I., Kocyİgİt, M.B., Ozdemİr, A., Yasa, G., Karaboz, I. Decolorization of various leather dyes and leather industry effluent by Trametes trogii TEM H2. *Fresenius Environ. Bull.*, 21, 6, 1410–1416, 2012. Available: http://www.psp-parlar.de.

60. Dourado, F., Mota, M., Pala, H., Gama, F.M., Effect of cellulase adsorption on the surface and interfacial properties of cellulose. *Cellulose*, 6, 4, 265–282, 1999.
61. Hao, L., Wang, R., Wang, Fang, Liu, J., Men, Y., The influences of enzymatic processing on physico-chemical and pigment dyeing characteristics of cotton fabrics. *Cellulose*, 23, 1, 929–940, 2016.
62. Spinelli, H.J., Polymeric dispersants in ink jet technology. *Adv. Mater.*, 10, 15, 1215–1218, 1998
63. Yamada, M., Amano, Y., Horikawa, E., Nozaki, K., Kanda, T., Mode of action of cellulases on dyed cotton with a reactive dye. *Biosci. Biotechnol. Biochem.*, 69, 1, 45–50, 2005.
64. Nithya, E., Radhai, R., Rajendran, R., Shalini, S., Rajendran, V., Jayakumar, S., Synergetic effect of DC air plasma and cellulase enzyme treatment on the hydrophilicity of cotton fabric. *Carbohydr. Polym.*, 83, 4, 1652–1658, 2011.
65. Almeida, L. and Bishop, D., Effects of Agitation and Endoglucanase Pretreatment on the Hydrolysis of Cotton Fabrics by a Total Cellulase. *Text. Res. J.*, 66, 5, 287–294, 1996.
66. Rousselle, M.A. and Howley, P.S., Molecular Weight of Cotton Cellulose: Effect of Treatment with a Total Cellulase. *Text. Res. J.*, 68, 8, 606–610, 1998.
67. Csiszár, E., Urbánszki, K., Szakács, G., Biotreatment of desized cotton fabric by commercial cellulase and xylanase enzymes. *J. Mol. Catal. B Enzym.*, 11, 4–6, 1065–1072, 2001.
68. Khan, A.A., Iqbal, N., Adeel, S., Azeem, M., Batool, F., Bhatti, I.A., Extraction of natural dye from red calico leaves: Gamma ray assisted improvements in colour strength and fastness properties. *Dye. Pigment.*, 103, 50–54, 2014.
69. Zolriasatein, A.A. and Yazdanshenas, M.E., Changes in composition, appearance, physical, and dyeing properties of jute yarn after bio-pretreatment with laccase, xylanase, cellulase, and pectinase enzymes. *J. Text. Inst.*, 105, 6, 609–619, 2014.
70. Adeel, S., Usman, M., Haider, W., Saeed, M., Muneer, M., Ali, M., Dyeing of gamma irradiated cotton using Direct Yellow 12 and Direct Yellow 27: improvement in colour strength and fastness properties. *Cellulose*, 22, 3, 2095–2105, 2015.
71. Tsatsaroni, E. and Liakopoulou-Kyriakides, M., Effect of enzymatic treatment on the dyeing of cotton and wool fibres with natural dyes. *Dye. Pigment.*, 29, 3, 203–209, 1995.
72. Andreaus, J., Azevedo, H., Cavaco-Paulo, A., Effects of temperature on the cellulose binding ability of cellulase enzymes. *J. Mol. Catal. B Enzym.*, 7, 1–4, 233–239, 1999.
73. Abadulla, E., Robra, K.H., Gübitz, G.M., Silva, L.M., Cavaco-Paulo, A., Enzymatic Decolorization of Textile Dyeing Effluents. *Text. Res. J.*, 70, 5, 409–414, 2000.

74. Prijambada, I.D., Negoro, S., Yomo, T., Urabe, I., Emergence of nylon oligomer degradation enzymes in *Pseudomonas aeruginosa* PAO through experimental evolution. *Appl. Environ. Microbiol.*, 61, 5, 2020–2022, 1995.
75. Deguchi, T., Kitaoka, Y., Kakezawa, M., Nishida, T., Purification and Characterization of a Nylon-Degrading Enzyme. *Appl. Environ. Microbiol.*, 64, 4, 1366–1371, 1998.
76. Gashti, M.P., Assefipour, R., Kiumarsi, A., Gashti, M.P., Enzymatic surface hydrolysis of polyamide 6,6 with mixtures of proteolytic and lipolytic enzymes. *Prep. Biochem. Biotechnol.*, 43, 8, 798–814, 2013.
77. Parvinzadeh, M., A new approach to improve dyeability of nylon6 fibre using a subtilisin enzyme. *Color. Technol.*, 125, 4, 228–233, 2009.

Part 2

SUSTAINABLE FUNCTIONAL FINISHING OF VARIOUS TEXTILE MATERIALS

5

Coating Textiles: Towards Sustainable Processes

Imene Ghezal[1,2]

[1]Textile Engineering Laboratory, University of Monastir, Ksar-Hellal, Tunisia
[2]National Engineering School of Monastir, University of Monastir, Monastir, Tunisia

Abstract

Textile industry is a growing sector and is playing an important role in the economic stability and growth of developing countries. Nevertheless, the apparel industry is known for its long and complexed production, consuming, and recycling processes. It is also a source of environmental burden and threat on human health by cause of its excessive materials and water consumptions, toxic wastes, air emissions, and high energy consuming processes. The emerging of industrial textiles, medical apparel, and protective garments was related to the coating processes progress. In fact, coating techniques fuse polymeric coat and textile characteristics. This leads to new fabrics with interesting performances such as flame retardancy, water repellency, antimicrobial and self-cleaning properties, protection against ultra-violet and electromagnetic radiations... Coating is a widely used functional process. However, it was found that this technique has many environmental issues such as high energy and water consumption, hazardous used chemicals, and pollutant wastes. Many researches were done in order to produce environmentally friendly chemicals and to apply the non-hazardous coats by ecological processes. This chapter presents most used polymers for producing coated textiles. In a second place, it illustrates an overview of coating processes and their environment issues. Finally, it states sustainable chemistry and technologies for coated textiles manufacturing.

***Keywords*:** Coated textiles, coating processes, sustainable polymers, sustainable coating methods, environmental issues, functional substrate, eco-friendly techniques, ecological processes

---

Email: elghezalimene@hotmail.com

Luqman Jameel Rather, Mohd Shabbir and Aminoddin Haji (eds.) Sustainable Practices in the Textile Industry, (113–148) © 2021 Scrivener Publishing LLC

5.1 Introduction

Sustainability is gaining interest all over the globe since it focuses on the economic growth taking into consideration the environmental and social impacts [1]. The textile industry is one of the oldest and most profitable sectors [2, 3]. In 2024, the global textile industry growth is expected to reach USD 1.230 billion [4]. Although there's a huge interest that this sector is gaining, it still present many environmental, social, and economic issues [1]. When talking about the textile industry development environmental sustainability is the most interesting topic [1]. In fact, textile processes are long and complicated [2]. Textile finishing step is the highest consumer of water and energy and the greatest generator of toxic emissions (volatile organic compounds (VOC)) and wastewater charged with hazardous chemicals [3, 5].

Coating technology is an efficient method for finishing textiles and imparting to fabrics many properties such as water-repellency, antimicrobial property, flame retardancy, ultraviolet protection [6]. Considering the importance of this technology, there has been a continuous need to improve coating techniques and to use eco-friendly materials in order to reach sustainability [6].

This chapter presents in a first place the most used polymers for coating textiles and traditional coating techniques. It illustrates then the emerging environmentally-friendly polymers and sustainable technologies for coating textile fabrics.

5.2 Most Used Polymers for Coating Textiles

Coating is defined as the application of a polymeric layer to a material in order to modify its appearance or to ameliorate its properties. The applied continuous or discontinuous layer is generally formed of a polymeric material and additives [7]. It is dried and cured or just cooled to obtain the textile/coating system [8]. In the textile industry, polymers have been widely used to impart new functionalities to textile structures such as waterproofness, water-repellency, antimicrobial property, and flame retardancy [6].

5.2.1 Polytetrafluoroethylene (PTFE)

Polytetrafluoroethylene also known as ethene,1, 1, 2, 2-tetrafluoro-, homopolymer is the product of the polymerization reaction of tetrafluoroethylene (suspension polymerization or dispersion/emulsion polymerization) [9, 10]. The chemical structure as well as some properties of the monomer and the homopolymer are shown in Figure 5.1 [11, 12]. The non-polarizable polymeric chain of the polytetrafluoroethylene makes the interaction with water molecules impossible. This gives PTFE the water-repellent property which makes it appropriate for coating textile fabrics in order to impart to their surfaces a hydrophobic aspect. Reaching superhydrophobicity is possible for smooth surfaces treated with polytetrafluoroethylene [9].

5.2.2 Polyvinyl Acetate (PVAc)

Polyvinyl acetate (PVAc) is an amorphous polymer (T_g = 29 °C) that is synthesized by emulsion or dispersion polymerization of vinyl acetate monomer (CH_2 = $CHOOCCH_3$) [12, 14]. The monomer and polymer chemical formulas are presented in Figure 5.2 [12, 15, 16]. Vinyl acetate can also be used to produce polyvinyl acetate copolymers such as polyethylene

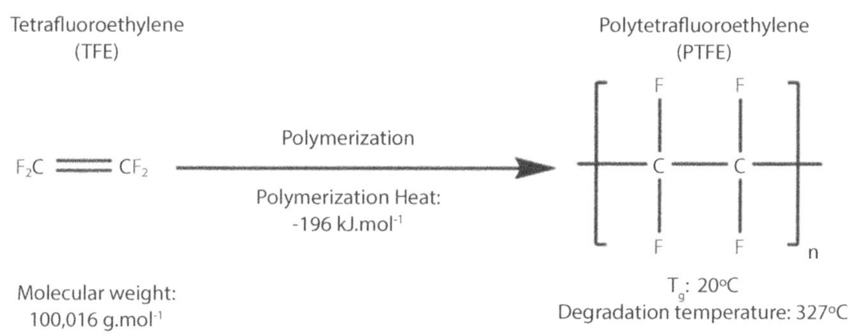

Figure 5.1 Chemical structure of tetrafluoroethylene (TFE) and polytetrafluoroethylene (PTFE) [9–13].

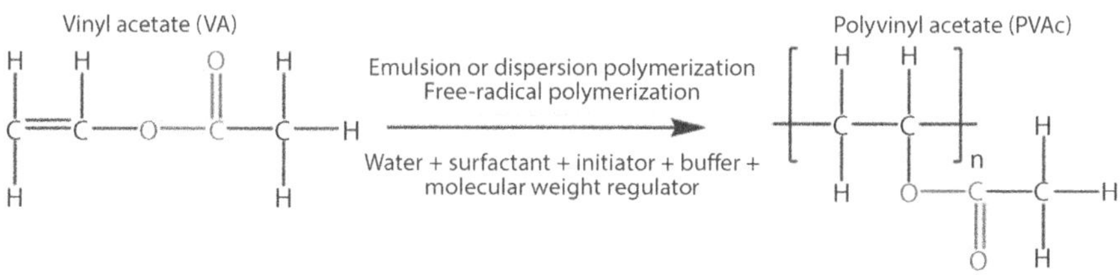

Figure 5.2 Vinyl acetate (VA) and polyvinyl acetate (PVAc) chemical formulas [12, 15, 16].

vinyl acetate [12, 16]. Polyvinyl acetate resins are used for coating upholstery. They play the role of an adhesive layer and are used to stiff textile structures [15].

5.2.3 Polyvinyl Alcohol (PVA)

Polyvinyl alcohol (PVA) can be obtained by the saponification reaction of PVAc [17]. The two-step process for PVA synthesis is represented in Figure 5.3 [17, 18]. The hydrophilic property of polyvinyl alcohol makes it appropriate for producing coated breathable substrates [19–21] and ameliorating comfort properties of some synthetic fabrics [22].

5.2.4 Polyurethanes (PUs)

Polyurethanes (PUs) are segmented copolymers and are the result of the polyaddition reaction of alcohols with isocyanates [12, 23]. Processes for PUs synthesis are presented in Figure 5.4 [12, 24–28]. Polyurethane coatings are used in the textile industry to produce waterproof breathable fabrics [29, 30] or to made synthetic leather [31, 32].

5.2.5 Polyvinyl Chloride (PVC) and Polyvinylidene Chloride (PVDC)

Polyvinyl chloride (PVC) is synthesized from vinyl chloride (VC). Polyvinylidene chloride (PVDC) is the polymerization product of vinylidene chloride (VDC) [11, 12]. Both polymers belong to chloropolymers [12]. Structural formulas as well as some characteristics of PVC and PVDC are shown in Figure 5.5 [11–13]. Vinyl polymers are used for coating textiles. Fabrics coated with vinyl resins are waterproof and can be used as gas barriers. The chlorine content of the polyvinylidene chloride makes it appropriate for producing fire retardant coatings [12, 33].

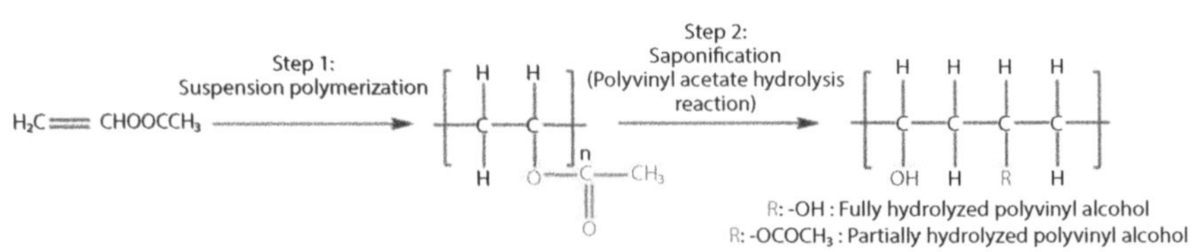

Figure 5.3 Twostep process for PVA synthesis [17, 18].

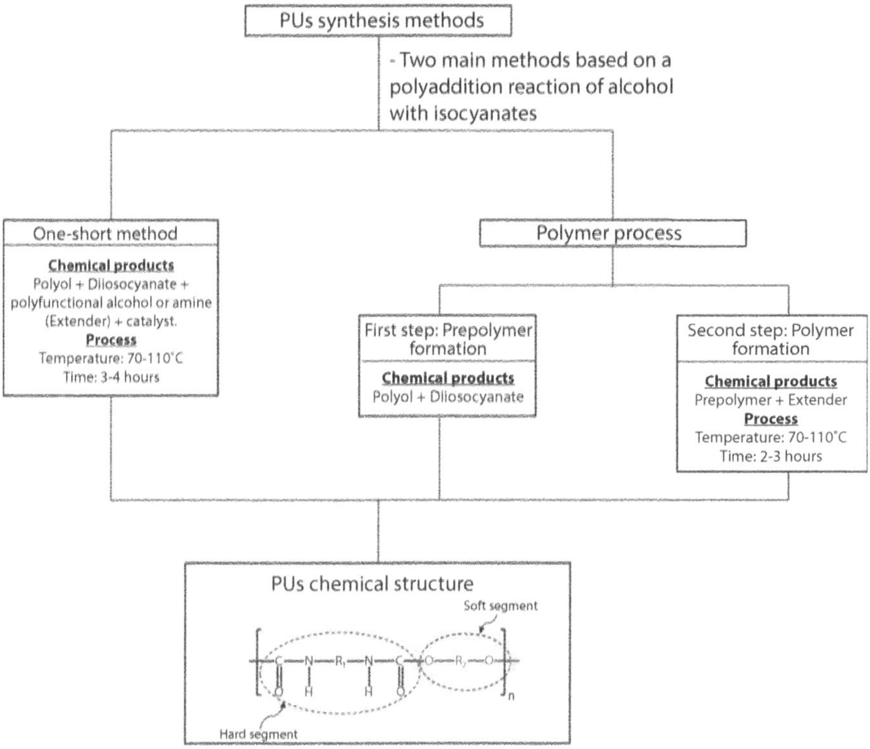

Figure 5.4 Polyurethanes (PUs) synthesis [12, 24–28].

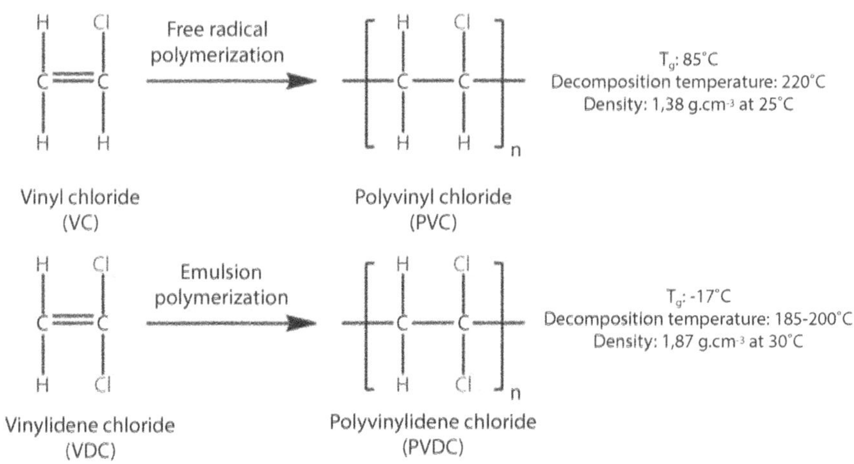

Figure 5.5 Structural formulas of vinyl chloride (VC), polyvinyl chloride (PVC), vinylidene chloride (VDC), and polyvinylidene chloride (PVDC) [11–13].

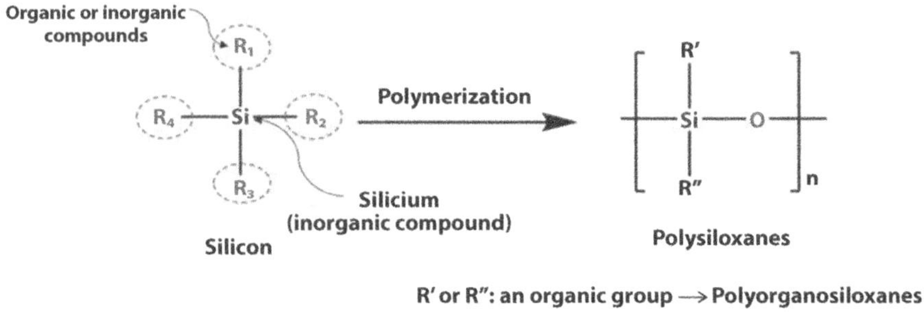

Figure 5.6 Chemical formulas of silicon and polysiloxanes [34–36].

5.2.6 Polysiloxanes

Polysiloxanes are obtained by the polymerization reaction of silanes monomers [34]. Silicone polymeric chain is formed of silicon and oxygen elements that are joined together with covalent bonds. Radicals carried by the silicium element can be organic groups (Figure 5.6) [35, 36]. In this case silicones are also denominated polyorganosiloxanes [36]. The presence of an organic radical (methyl (CH_3), ethyl (C_2H5), amino (NH_2), ….) makes the silicone an hybrid polymer regrouping both organic and inorganic elements [34, 37]. Coating textiles with silicones ameliorate their handle. It also gives treated samples water-repellent property [38, 39].

5.2.7 Acrylics

Most used acrylic polymers are acrylate and methacrylate polymers [12] (Figure 5.7). Acrylic polymers are used as back coatings or as adhesives for subsequent lamination process [40].

5.2.8 Phosphorous-Based Polymers

Phosphorous-based polymers are used to impart flame retardant property to fabrics [42]. The most known phosphorous-based polymers are polyphosphazenes and polyphosphoesters (Figure 5.8) [43, 44].

5.3 Traditional Coating Methods

There are many techniques for applying the coating to textile fabrics. Most known ones consist on the impregnation of the fabric in a polymeric

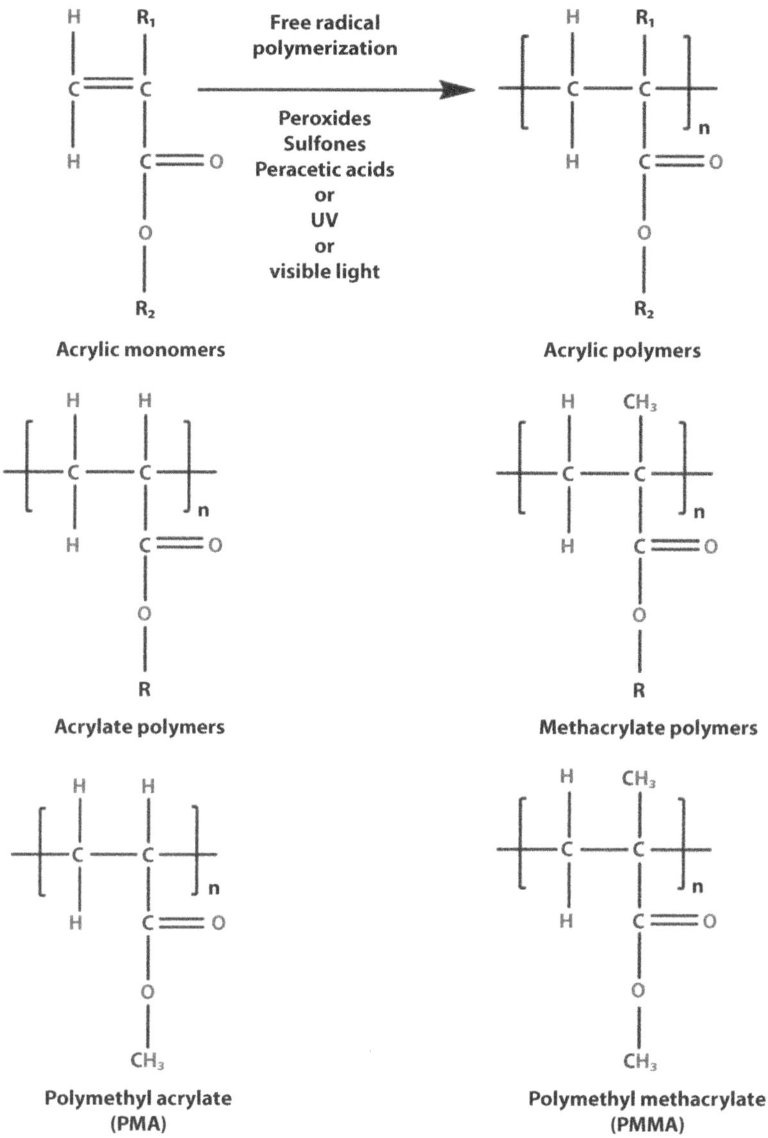

Figure 5.7 Acrylic polymers; synthesis and chemical structures [12, 40, 41].

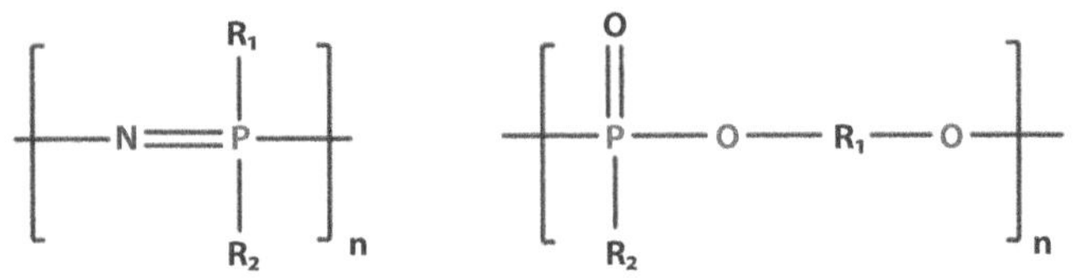

Figure 5.8 Polyphosphazenes and polyphosphoesters chemical structures [45, 46].

coating bath or on the application of a polymeric layer to one or both textile surfaces [6–8]. Here are presented most used techniques for textile coating (Figure 5.9).

a) Knife coating: The coating formulation is applied by the aim of a doctor blade (a knife) that is fixed under a transported fabric [7, 8, 47].
b) Dip coating: This technique consists on the impregnation of a textile fabric in a polymeric coating solution. The excess of the coating formulation is then eliminated by passing the substrate between rollers [13, 28].
c) Roll coating: The coat is applied to the textile surface by mean of a roll that is in contact with the polymeric coating bath [6].
d) Screen coating: By using a blade the coating paste is passed through a mesh and is applied to the textile fabric.
e) Hot-melt coating: The thermoplastic polymeric material is melted before being applied to the textile fabric [12].
f) Rotary screen coating: The coating is applied with a cylindrical perforated rotating screen that is fed with the coating paste [8].
g) Spray coating: This method consists on using gases to melt or accelerate the polymeric particles that are then sprayed on the textile surface [48].
h) Foam coating: The coat formulation is transformed to a foam that is applied to the textile surface [7].
i) Transfer coating: The coating layer is applied to the fabric by mean of a release paper on which the coating was previously spread and dried [7, 47].

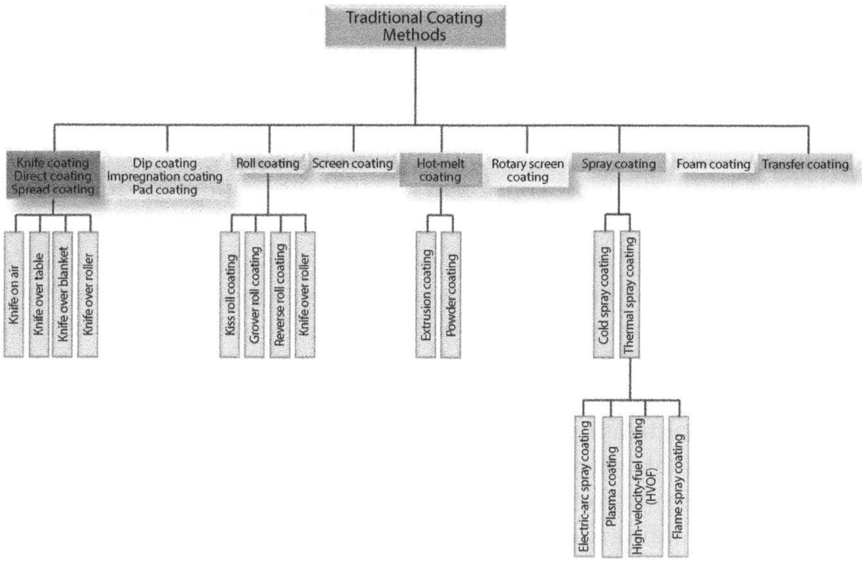

Figure 5.9 Traditional coating methods [6–8, 12, 13, 28, 47–49].

5.4 Environmental Friendly Polymers

Many commonly used polymers were classified as hazardous materials because of their monomer type or polymerization processes [50]. To avoid these toxic contaminants and to reach sustainability new environmental-friendly polymers were developed by using biotechnology [51]. They were derived from natural resources such as plants, animals, and algae [52]. This part is a review of most used environmental-friendly polymers for coating textile fabrics.

5.4.1 Cyclodextrins

In 1891, the first manuscript on cyclodextrins was reported by a French scientist named A. Villiers [53]. The first patent on cyclodextrins was then published in 1953 [53, 54]. The development period of cyclodextrins can be divided into three major stages. From 1891 to mid-1930s was the discovering period. The exploratory phase was until 1970 then closely followed the exploitation period until day [53]. Cyclodextrins are the result of starch enzymatic degradation with *Bacillus macerans* bacteria (*Cyclodextrin glycosyl transferase* or *CGTase*) [54, 55]. They are linked non-reduced cyclic oligomers of D-(+)-glucopyranose joined by α-1,4 bounds [53, 56]. α-, β-, and δ-Schardinger dextrins are formed of 6, 7, and 8 units of D-(+)-glucopyranose, respectively (Figure 5.10) [54].

Cyclohexaamylose ($C_{36}H_{60}O_{30}$), cycloheptaamylose ($C_{42}H_{70}O_{35}$), and cyclooctaamylose ($C_{48}H_{80}O_{40}$) are the most produced cyclodextrins. Still, β-cyclodextrin is the most consumed one (95%) by dint of its simple synthesis, availability, and low price if compared with other cyclodextrins [55]. When it comes to producing functional garments, chemical finishes are widely used to impart many performances to textiles such as flame retardancy, water repellency, antimicrobial property [55, 59]. With their torus shaped structures (Figure 5.11) and hydrophobic cavities cyclodextrins can be used as carriers to bio active molecules, polymers, and dyes [55]. Also, it was found that treating textiles with cyclodextrins helps in the elimination of unpleasant odors [56, 60]. Cyclodextrins are non-toxic and biodegradable which make them appropriate for producing eco-textiles [61, 62]. Notwithstanding the ability of cyclodextrins to make complexes with

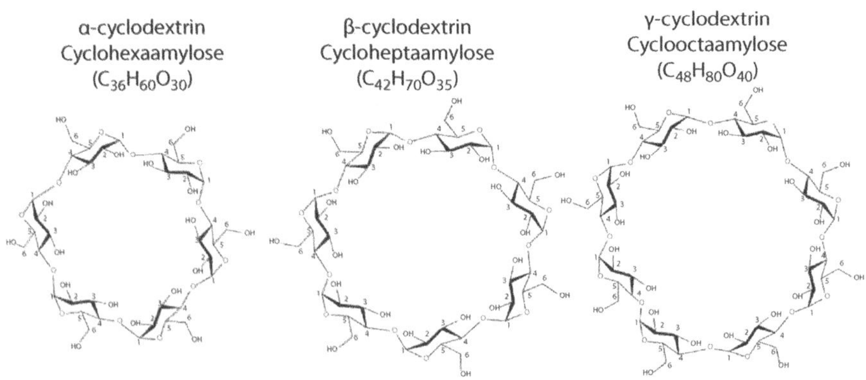

Figure 5.10 Cyclodextrins chemical structures [54–58].

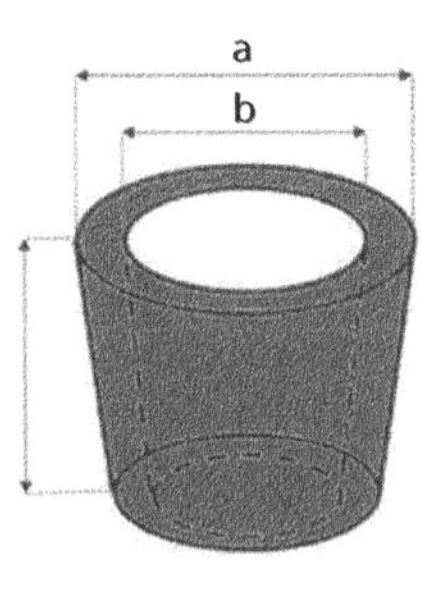

Cyclodextrins dimensions (pm)

	α-cyclodextrin	ß-cyclodextrin	γ-cyclodextrin
a	1370	1530	1690
b	500	650	850
c	780	780	780

Figure 5.11 Cyclodextrins dimensions (pm) [56, 58].

many types of molecules, they are unable to form covalent bounds with textile fabrics [63]. For this reason cyclodextrins were modified by using chemical reagents or enzymes [54].

5.4.2 Chitin and Chitosan

In 1811, Braconnot discovered fungine. By 1832, a researcher named Odier discovered the same biopolymer in exoskeleton of insects and called it chitin [64]. Chitin and chitosan chemical structures are presented in Figure 5.12. Chitin is a β-(1-4)-poly-N-acetyl-D-glucosamine. It can be derived from shrimp and crab shells, exoskeleton of mollusks and insects, and cell walls of fungi [8]. In 1859, a researcher called Rouget discovered chitosan after a partial alkaline deacetylation of chitin [65]. Chitosan is an assembly of glucosamine and N-acetyl-glucosamine linked with β-(1–4) bounds [66, 67]. Chitin derivative is called chitosan only if the acetylglucosamine monomer residues proportion in deacetylated chitin is at least equal to 60% [66, 68]. Chitosan is known for its biocompatibility and biodegradability. It also has an antimicrobial activity [69].

5.4.3 Sodium Alginate

Alginic acid is a biopolymer that can be obtained from brow algea (*Ascophyllum*, *Laminaria*, and *Marrocystic*) or bacteria (*Azolbacter* and *Pseudomonas*) [70]. The alginic acid copolymer is formed of α-L-guluronic and β-D-mannuronic acids. The two polysaccharide units are joined together with α (1 → 4) and β (1 → 4) glucosides (carboxylate) bounds [70, 73]. Alginate copolymer is obtained by chopping then treating algae with sodium carbonate). The seaweed extract dispersion is then filtered to get unrefined Na-Alginate solution. A purification is then needed to eliminate

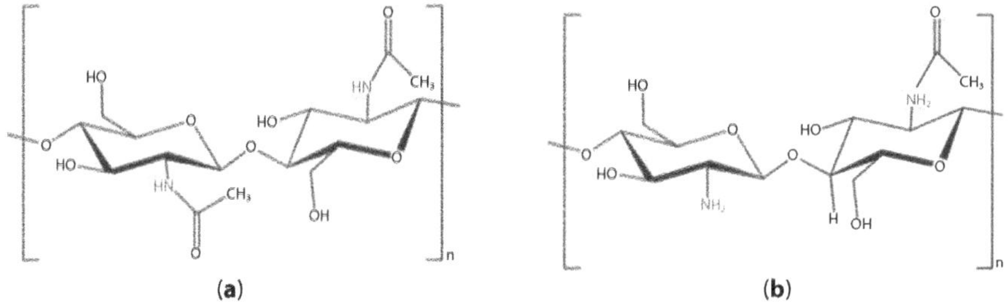

Figure 5.12 Chemical structure of chitin before and after a partial deacetylation: (a) Chitin, (b) Chitosan [64].

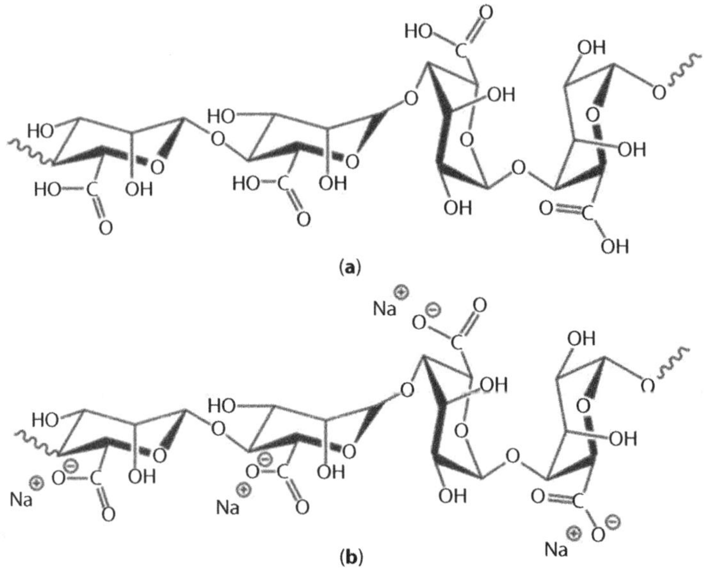

Figure 5.13 Chair structure of alginic (a) acid and (b) sodium alginate [70–72].

protein, lipopolysaccharide [70, 74]. Sodium alginate is an anionic linear polysaccharide [72, 75]. Alginic acid and sodium alginate chair structures are presented in Figure 5.13. Alginate has many properties such as biocompatibility, gelation, and swelling properties which make it widely used for coating and film formation [76].

5.4.4 Polyethylene Glycols

Polyethylene glycols (Figure 5.14) are linear polymers that are obtained by ethylene oxide and any diols polymerization [77, 78]. They are also known as Macrogols, polyethylene oxide (PEO), and polyoxyethylene (POE) [77, 79]. Molecular weights of PEGs range from two hundred to tens of thousands [81]. Depending of the molecular weight, polyethylene glycols can be liquid, soft, or hard solids. At ambient temperature,

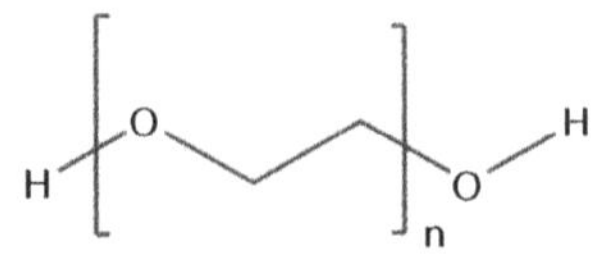

Figure 5.14 Polyethylene glycol (PEG) chemical structure [80].

polyethylene glycols with a molecular weight between 100 and 700 are in liquid form. PEG 1000 and PEG 2000 are soft solids. When the molecular weight is higher than 2,000, PEGs turn to rigid solids [77]. Polyethylene glycol belong to the list of compounds generally recognized as safe [81]. Its non-toxicity, biocompatibility, hydrophilicity, antimicrobial property, and flexibility make it of a high interest for textile coating [82, 83]. The high hydrophilicity of PEGs explains the interest that this polymer has gained in designing waterproof breathable fabrics. The incorporation of PEGs in the formulation of the coating foam ensures its stability and a good coat/textile adhesion. Also, many researches has proved that adding polyethylene glycol to the coating mixture ameliorates substrate breathability as a result of wetting, wicking and moisture management capacities enhancement [83, 84].

5.4.5 Natural Rubber

The first scientific interest in natural rubber was in 1745 by a French researcher named Charles Marie de La Condamine. In 1791, Samuel Peal wrote the first patent dealing with producing waterproof fabrics by using a turpentine-rubber solution [85]. The origin of natural rubber (NR) is plants. Nonetheless, the most important source is the *Havea brasiliensis* tree [86]. The liquid rubber form is known as latex. After a coagulation process with formic acid or acetic acid, liquid rubber becomes solid [12]. Natural rubber (Figure 5.15) is a linear amorphous polymer with unsaturated double bounds. Its chemical formula is cis-1,4-4-polyisoprene [12, 86]. In general, natural rubber is formed of 93% of hydrocarbon. It also contains inorganic salts and materials [86]. It is an eco-friendly biomaterial with many interesting properties such as flexibility, abrasion, tear, and tensile resistances, excellent adhesion to textiles, and waterproofness. All these properties make it of a great interest for textile coating [86, 88, 89].

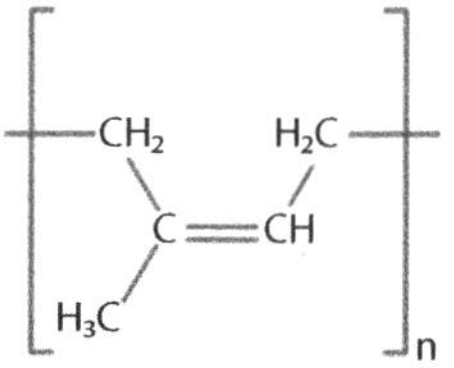

Figure 5.15 Cis-1,4-polyisoprene (Natural rubber (NR)) [87].

5.4.6 Polyvinyl Alcohol

The synthesis of polyvinyl alcohol (PVOH) began in 1924. Scientists named Herman and Haehnel have transformed polyvinyl ester to polyvinyl alcohol solution by using sodium hydroxide solution (saponification reaction) [90]. The synthetic polymer has been exploited for over half a century [91]. Commercialized polyvinyl alcohol, also commonly known as PVA or PVOH, is obtained by polyvinyl acetate polymerization. The saponification reaction occurs in aqueous sodium hydroxide environment (partial or complete hydrolysis) [92, 93]. Polyvinyl alcohol molecular weight ranges from 20,000 to 400,000 g·mol^{-1} [94]. Synthesis process of polyvinyl alcohol is shown in Figure 5.16. Polyvinyl alcohol characteristics are its easy biodegradability, biocompatibility, high water solubility, water vapor permeability, chemical attack endurance, and thermostability [20, 91–93]. Polyvinyl alcohol was widely used with polymeric blends to produce coated waterproof breathable fabrics [20].

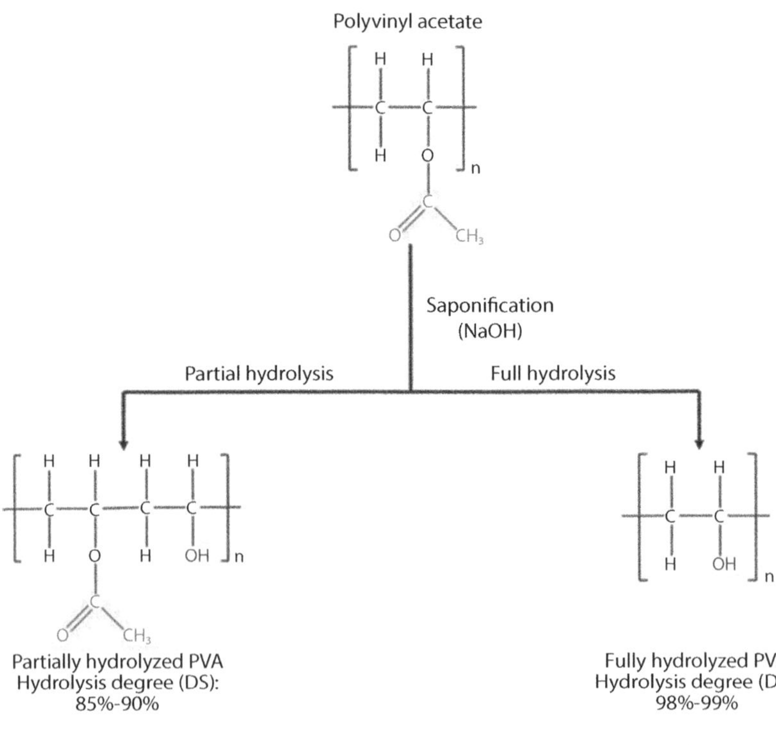

Figure 5.16 Chemical structures of polyvinyl acetate and polyvinyl alcohol [18, 92, 94, 95].

5.4.7 Dendrimers

In 1978, Vogtle discovered highly branched molecules. In 1980, Tomalia *et al.* and George R. Newkome discovered independently dendrimers [96, 97]. Dendrimers are different from traditional polymers by their molecular structure. They belong to a special polymer class [98]. Dendritic polymers are formed with a small molecule or a linear core and repeated units (branches) [96, 99, 100]. These units can be polyamino acids, polyporphyrins, polyphenols, polyethers, and polyamidoamines (PAMAM) (Figure 5.17) [96]. The architecture of dendrimers can be divided into three parts: core, interior, and end-groups [101]. Nowadays, dendrimers with fluorinated alkyl groups are widely used. Dendrimers can also be applied with fluorocarbon resins. The bionic finish impart oleophobic properties to the treated fabric with a lower amount of fluor groups then in the conventional treatment with dendrimer-free fluorocarbon resins [104–106]. The use of dendrimers with fluorine-free polymers was also found to confer water repellency to coated structures [105].

5.4.8 Sericin

Sericin is a proteinous amorphous biopolymer [107, 108]. The main source of this glycoprotein is *Bombyx mori* cocoons (15–35% by cocoon weight) [108–110]. Silk sericin (SS) protein is formed of 18 amino acids [111]. Sericin contents of amino acids are shown in Table 5.1 [112]. The high amount of hydroxyl carboxyl groups and amino groups in sericin amino acids are at the origin of the water solubility of the proteineous polymer [112]. The chemical structure of sericin is shown in Figure 5.18. Sericin was widely used to modify the textile surface chemistry [112, 114]. Polyester fabrics were coated with this eco-friendly polymer in order to ameliorate their hydrophilicity for subsequent treatments or to

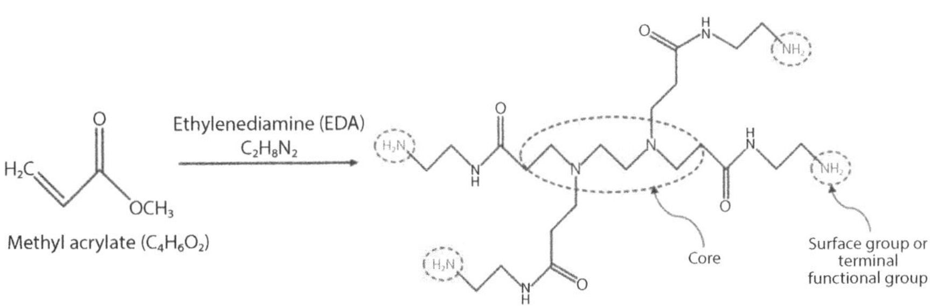

Figure 5.17 Chemical structure of polyamidoamine (PAMAM- G0) [102, 103].

Table 5.1 Sericin content of amino acids [109, 112].

Amino acids	Percentage
Serine	33.4%
Aspartic acid	16.7%
Glycine	13.5%
Other amino acids	36.4%

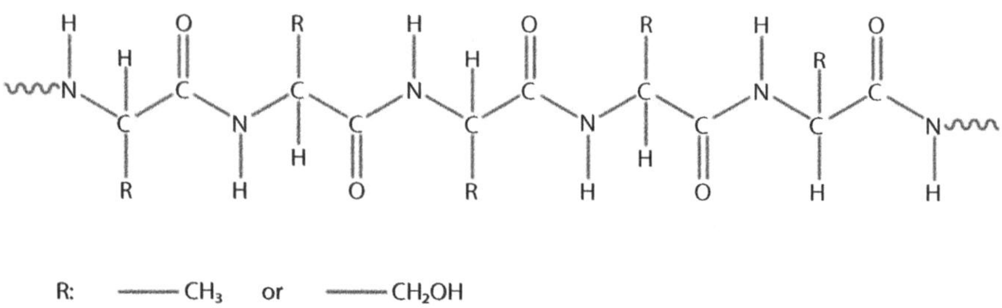

Figure 5.18 Silk sericin (SS) chemical structure [113].

improve their moisture management property [115]. Also, cotton structures were treated with sericin to gain antibacterial and UV-protection properties [110, 114, 115].

5.4.9 Polyphenols

Polyphenols are abundant in nature. They can be extracted from plants [116, 117]. The structure of these compounds is formed of phenolic units and hydroxyl groups [118]. Depending on the number of phenolic rings in polyphenols, the chemical structure of elements linking two adjacent phenols, and the chemical substituents beared by the phenolic rings, these compounds can be classified into four main classes [118, 119]: flavonoids, phenolic acids, lignians, and stilbenes. The chemical structures of polyphenols subclasses are presented in Figure 5.19. Polyphenols are emerging compounds for coating textile structures. Treated fabrics showed antibacterial property and resistance to ultraviolet radiation [116, 120, 121].

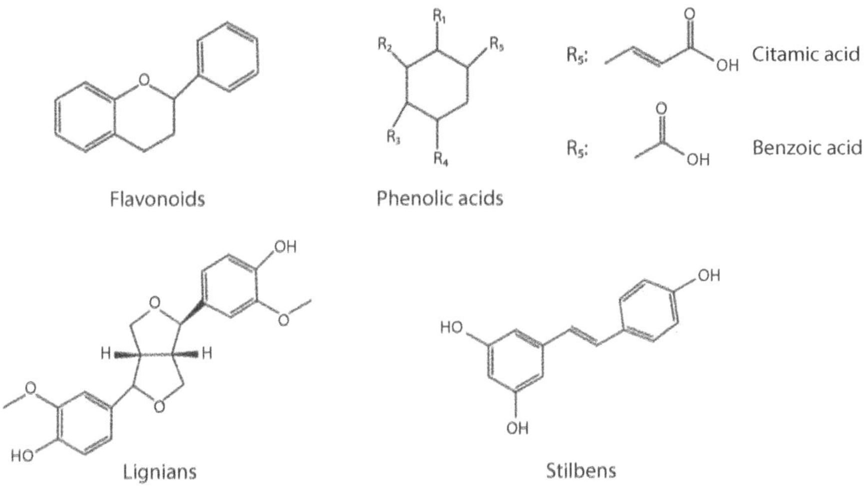

Figure 5.19 Chemical structures of polyphenols subclasses [118, 119].

5.5 Sustainable Coating Technologies

Previously presented technologies present some issues that are considered as barriers to sustainability achievement. The high water and energy consumption, the hazardous emissions and wastes, and the unhealthy working environment are the major causes for the emerging of new technologies [51].

5.5.1 Powder Coating Technique

In 1965, the technological development of powder coating technique began in Australia [122, 123]. Powder coatings were used for metal surfaces finishing [123]. In 1975, an evolution in powder coating technology occurred to respond to market needs [124]. Powder coating can be obtained by a dry process (dry mixing, melt mixing, and supercritical fluid methods) or by a wet process (evaporation, precipitation, and spray drying methods) [122, 125]. Powder particles sizes range from 30 to 60 μm [126]. Depending of the resin forming the powder coating, powders can be classified according to two categories; thermoplastic powders (TP resins) and thermosetting powders (TS resins) [127, 128]. Thermoplastic powders are formed of thermoplastic polymers with high molecular weight [125]. When heated, thermoplastic powders melt and flow [11, 123, 128]. Main used thermoplastic powder coatings are thermoplastic polyester powders, polyamide powder, polyvinylidene fluoride powders, nylon powders, polypropylene powders, polyethylene powders, polyvinyl chloride powders, thermoplastic cellulose powders, and polyurethane powders [125, 129, 130]. For thermoplastic powder coatings films thickness varies from 50 to 250 μm [125, 130]. Thermoplastic powder coatings are usually applied by using the fluidized bed dipping method [125] or by using a powder scattering device [131]. After heating the applied powder under a defined melting point (infrared of flatbed lamination), it is compacted and/or smoothed with a calendar and cooled down [11].

The high thickness of films made by thermoplastic powder coatings is one of the most important reasons for thermoset powder coatings development [125]. The most used powder coatings generally contain epoxy-based powders, acrylic powders, hydroxyl polyester powders, and carboxyl polyester powders [128]. The molecular weight of polymers used for thermoset powder coatings are lower than the molecular weight of polymers used for thermoplastic powder coatings. For this reason, thinner films (25 μm) are obtained with thermoset powder coatings than with thermoplastic ones [125]. Electrostatic spraying methods are widely used to obtain thermoset powder coatings [125]. Powder coating technology is considered as an ecofriendly method. It is a solvent free technique that does not generate volatile organic compounds. Also, the absence of water and solvents reduces the amount of hazardous wastes, eliminates drying step which helps in energy saving, and procures to workers a clean working space. Considering that over-sprayed powders can be reused, powder coating technology is an efficient method for treating surfaces [126, 131–133].

5.5.2 Sol–Gel Technology

First silica gels and alumina gels were synthesized by M. Ebelmer (1845) and Cossa (1870), respectively [134]. A sol–gel process regroups two main steps (hydrolysis and condensation reactions) for the transformation of a suspension (sol) to a three-dimensional network structure (gel) [134–136]. Precursors are used for preparing sols. They can be colloidal particles or molecular precursors. Molecular precursors can be either metal salts alkoxides or organic monomers and oligomers [134]. The most used precursors are (semi) metal alkoxides [135, 137]. Silicon alkoxides are widely used in sol–gel processes for textile coating [136]. A presentation of sol–gel process different steps is shown in Figure 5.20.

Many researches focus on imparting functionalities to textile fabrics by using the sol-gel technology. This technique was found to be efficient with lower cost and less used chemicals than the traditional finishing methods. Sol–gel technology can be used to obtain functionalized coatings

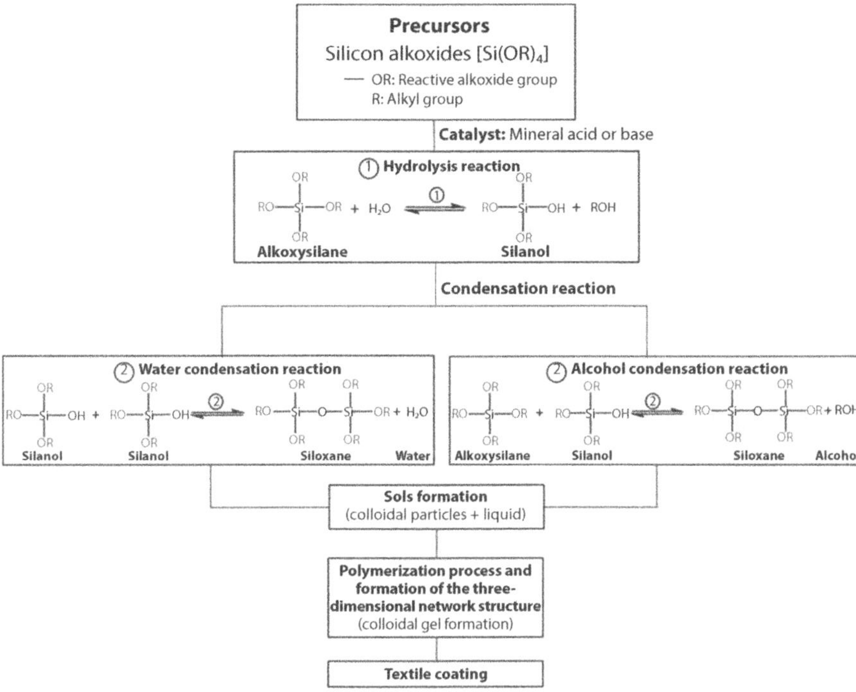

Figure 5.20 Sol–gel process with silicon alkoxides as precursors [134, 136–141].

[136, 139]. Sol–gel materials are used to confer to textile fabrics many characteristics such as flame retardancy, water-repellency, oil repellency, UV protection, antimicrobial and anti-wrinkle properties [136, 142–144].

5.5.3 Plasma Treatment

The fourth state of matter was introduced by the British chemist and physicist sir William Crookes in 1879 [145]. It results from partial or full gas ionization and dissociation [145, 146]. In 1928, Irving Langmuir called for the first time the first state of matter "plasma" [146]. Man-made plasmas are the result of gas excitation with electromagnetic power [7]. Reactive species in plasma regroup excited atoms, ions, electrons, excited molecules, radicals [145–147]. In 1980, plasma processing was used to modify chemical and topographic states of textile surfaces [7]. Plasma can be classified according to four main-classes; plasma chamber pressure, species temperature, power source, and used gas [7, 148] (Figure 5.21).

Plasma treatment can be used to modify the topography and the chemical nature of textile surface for subsequent functionalization or to impart

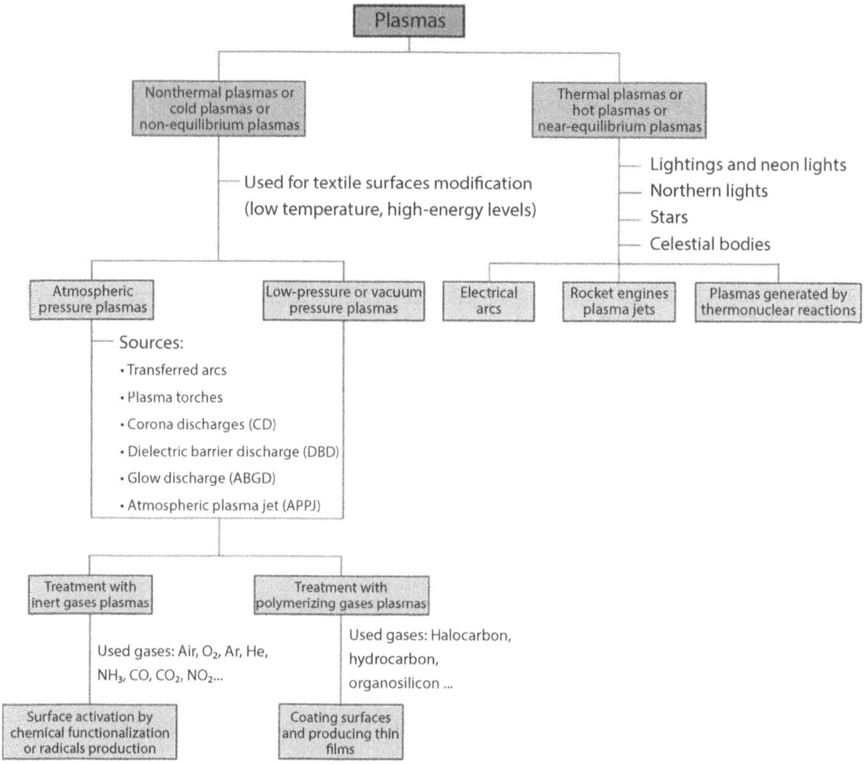

Figure 5.21 Plasmas classification [7, 145, 148–150].

many properties such as hydrophobic-oleophobic properties, antimicrobial property, flame retardant property, etc. [151]. Obtaining such properties with conventional wet processes was found to consume huge amounts of water, energy, and toxic chemicals and to generate wastes charged with hazardous compounds [145]. Plasma technology is an alternative for conventional wet processes. The dry ecofriendly treatment guarantees a clean and safe working space with low chemical products quantities and low energy and water consumption. It also eliminates hazardous wastes and VOC emissions. Plasma technology offers more efficient treatments in numerous fields with lower costs then traditional wet processes [28, 145].

5.5.4 Electro-Fluidodynamic Technology

Electrofluidodynamic technologies also known as electro-hydrodynamic technologies use high voltage to produce nanoparticles, nanofibers, nanofilaments, light weight coatings, and membranes [28, 152–154]. In 1745,

Electrostatic forces were used for the first time to produce aerosols from fluids. First patents dealing with electrospraying and electrospinning techniques were issued in 1902 and 1934, respectively [154]. Electrospraying and electrospinning techniques are both based on electrofluidodynamic technology for producing nanodroplets and nanofilaments, respectively [154–156]. Their processes are based on three principle devices: a reservoir for the polymeric solution or the nanoparticles suspension, an electric source for high voltage generation, a spinneret, and a collector [154].

Electrospraying/electrospinning principle consists of subjecting solutions, emulsions, or melts to a high voltage to attain atomization and to form an aerosol [154, 157]. The high polarization gives the fine polymeric droplet in the tip of the nozzle a conical shape (Taylor cone) [155, 158]. The jet is then collected on a grounded plate or on a substrate [155, 158]. Electrospraying/electrospinning methods are used to upgrade textile fabrics by coating their surfaces and imparting to them some functionalities such as water/oil repellency, flame retardancy, and antimicrobial property [158]. Electrospun coatings and membranes are also used to produce waterproof breathable fabrics [152, 156].

5.5.5 Supercritical Fluid Technology

A fluid is supercritical when it is used under conditions that are above its critical temperature (T_c) and critical pressure (P_c) [159, 160]. In the supercritical region, fluids combine properties of both liquids and gases [161]. They are highly mobile and have a low viscosity (critical density value: 0.47 $g{\cdot}cm^{-3}$) [162–164]. Supercritical fluids have gained a major interest in textile industry especially in fabrics coating. The most used fluid is carbon dioxide (CO_2). It is utilized in its supercritical state to replace organic solvents needed for textile finishing [165]. Carbon dioxide critical point (Figure 5.22) corresponds to a critical temperature (T_c) equal to 304 K and a critical pressure (P_c) of 7.4 MPa [163].

There are many methods for textiles coating that are based on supercritical carbon dioxide technology such as the rapid expansion of supercritical solution (RESS). This technique is based on a decompressive atomization where the coat is mixed with CO_2 then sprayed on the textile fabric [165, 167]. The polymer (solute) is dissolved in the supercritical carbon dioxide (solvent) at a high pressure. This technique consists then on decreasing drastically the applied pressure. This leads to a rapid expansion of the supercritical fluid solution and to the polymer nucleation and precipitation [165, 168]. The second technique for supercritical coatings obtention is known as the supercritical antisolvent method (SAS) [167]. In this

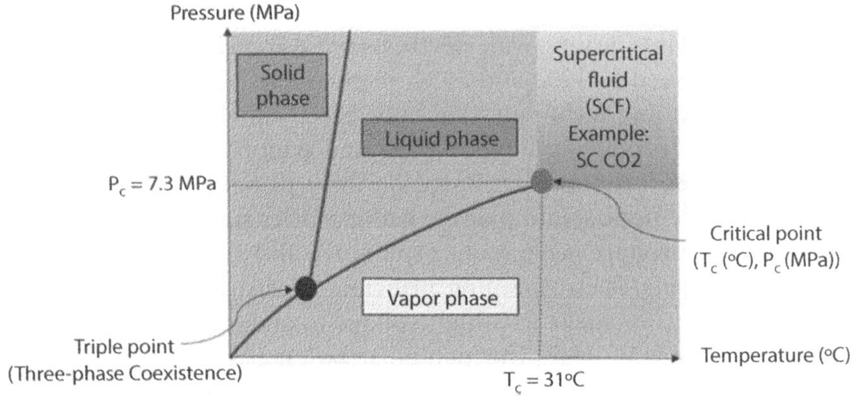

Figure 5.22 Supercritical fluid phase diagram (CO_2 as an example) [160, 163, 164, 166].

technique, the polymer is dissolved in ordinary solvents then exposed to supercritical carbon dioxide which plays the role of an antisolvent and leads to the polymer nucleation [167]. Processes based on supercritical carbon dioxide are emerging and gaining interest in the textile industry especially in fabrics coating field [28, 165]. Supercritical carbon dioxide methods are ecofriendly. They reduce the VOC emissions and water consumption and eliminate drying steps and wastes generated by conventional coating processes [165, 169]. Added to that, carbon dioxide is not expensive, non-toxic, chemically inert with the majority of polymers, and can be reused without purification [166, 170]. Coatings that are obtained with supercritical carbon dioxide methods are durable and uniform [167, 171, 172]. Major supercritical coatings are applied to impart to textile fabrics hydrophobic, omniphobic, antibacterial, flame retardant and ultraviolet protection properties [166, 167, 171–179].

5.5.6 Vapor Deposition Methods

Textile coating can be realized by using the vapor deposition method. It is a dry technique that consists on evaporating the polymeric materials into ions, atoms, and molecules [180]. Polymeric vapor is then deposed onto the textile surface to form a thin polymeric layer [181]. Vapor deposition method can be classified according to two major categories: physical vapor deposition (PVD) and chemical vapor deposition (CVD) [182]. For physical vapor deposition (PVD), transforming the polymeric material to vapor form can be achieved in a vacuum chamber by using a thermal source (thermal evaporation method) or by plasma ions bombardment (sputtering method). The vapor is then condensed through a channel and deposed on the textile substrate. For chemical vapor deposition technique (CVD), monomers and initiators are introduced in the vacuum chamber and evaporated by thermal action or by decreasing the vacuum pressure. Other gases are then introduced to react with the coat volatile compounds and to generate their deposition on the substrate [183, 184]. The chemical vapor deposition technique is characterized by its high deposition temperature and gas pressure which make it not appropriate for heat sensitive fabrics [180]. The initiated chemical vapor deposition method (iCVD) is characterized by its low temperature and is a good alternative for coating heat-sensitive materials [185, 186]. In this method a heated array or filament is used to crack initiators and to obtain free radicals. The material vapor is then deposed onto a cooled substrate [187]. PVDs and iCVDs techniques are ecofriendly, dry, free solvent, and energy saving methods [181, 188]. They are used to produce uniform nanocoatings to impart many functionalities to textile fabrics such as superhydrophobicity, conductivity, protection against electromagnetic radiations, and antimicrobial property [181, 182, 184–186, 188–190].

5.6 Conclusion

Textile coating is a growing technique that is gaining interest in the textile industry. It offers the possibility to modify textile performances and to impart new functionalities to treated fabrics. This treatment consists on applying a coating which is generally formed of a polymeric material and other additives to a textile fabric. This allows the obtention of new materials with enhanced properties. Nowadays, the word is moving towards sustainable technologies that respect both human beings and environment. This chapter reviews in the first place the most used polymers for textile coating and most used coating techniques. Based on literature, it was obvious that conventional coating processes have a negative impact on the human health and the environment. New emerging bio-based polymers and environmentally-friendly processes were presented. Green chemistry and sustainable technologies enable the manufacturing industries to produce high performance textiles with a decrease in water and energy consumption. It also eliminates volatile organic compounds emissions and hazardous wastes. The elimination of hazardous chemicals and the use of clean production processes provide a safe working environment and are the main steps for reaching environmental, social, and economic sustainability in textile coating area.

References

1. Luo, Y., Song, K., Ding, X., Wu, X., Environmental sustainability of textiles and apparel: A review of evaluation methods. *Environ. Impact Assess. Rev.*, 86, 106497, 2021.
2. Abbas, S., Chiang Hsieh, L.H., Techato, K., Taweekun, J., Sustainable production using a resource–energy–water nexus for the Pakistani textile industry. *J. Clean. Prod.*, 271, 122633, 2020.
3. Roy, M., Sen, P., Pal, P., An integrated green management model to improve environmental performance of textile industry towards sustainability. *J. Clean. Prod.*, 271, 122656, 2020.
4. Morder Intelligence, *Global Textile Industry—Growth, Trends, and Forecast (2020–2025)*, 2019.
5. Le Marechal, A.M., Krizanec, B. *et al.*, Textile finishing industry as an important source of organic pollutants, in: *Organic Pollutants Ten Years After the Stockholm Convention—Environmental and Analytical Update*, T. Puzyn and A. Mostrag-Szlichtyng (Eds.), pp. 29–54, IntechOpen, London, 2012.
6. Meirowitz, R.E., Coating processes and techniques for smart textiles, in: *Active Coatings for Smart Textiles*, I. Hu (Ed.), pp. 159–177, Woodhead Publishing, Cambridge, 2016.
7. Joshi, M. and Butola, B.S., Application technologies for coating, lamination and finishing of technical textiles, in: *Advances in the Dyeing and Finishing of Technical Textiles*, M.L. Gulrajani (Ed.), pp. 355–411, Woodhead Publishing Limited, Cambridge, 2013.
8. Pritchard, M., Sarsby, R.W., Anand, S.C. (Eds.), *Handbook of Technical Textiles*, Woodhead Publishing, Cambridge, 2000.
9. Puts, G.J., Crouse, P., Ameduri, B.M., Polytetrafluoroethylene: Synthesis and characterization of the original extreme polymer. *Chem. Rev.*, 119, 1763, 2018.
10. Radulovic, L.L. and Wojcinski, Z.W., PTFE (Polytetrafluoroethylene; Teflon®). *Encycl. Toxicol.*, 3, 1133, 2014.
11. Hall, M.E., Coating of technical textiles, in: *Handbook of Technical Textiles*, A.R. Horrocks and S.C. Anand (Eds.), pp. 173–186, Woodhead Publishing, Cambridge, 2000.
12. Akovali, G. (Ed.), *Advances in Polymer Coated Textiles*, Smithers Rapra Technology Ltd Shropshire, 2012.
13. Mohammed Reduwan Billah, S., Textile coatings, in: *Functional Polymers*, M.A.J. Mazumder, H. Sheardown, A. Al-Ahmed (Eds.), pp. 825–882, Springer, Cham, 2019.
14. Zhang, N., Wang, S., Gibril, M.E., Kong, F., The copolymer of polyvinyl acetate containing lignin-vinyl acetate monomer: Synthesis and characterization. *Eur. Polym. J.*, 123, 109411, 2020. 123.
15. Cordeiro, C.F., Vinyl acetate polymers. *Encycl. Chem. Technol.*, 12, 416, 2000.
16. Kolter, K., Dashevsky, A., Irfan, M., Bodmeier, R., Polyvinyl acetate-based film coatings. *Int. J. Pharm.*, 457, 470, 2013.

17. Peixoto, L.S., Silva, F.M., Niemeyer, M.A.L., Espinosa, G., Melo, P.A., Nele, M. *et al.*, Synthesis of poly (vinyl alcohol) and/or poly (vinyl acetate) particles with spherical morphology and core-shell structure and its use in vascular embolization. *Macromol. Symp.*, 243, 190, 2006.
18. Fan, Q., Fabric chemical testing, in: *Fabric Testing*, J. Hu (Ed.), pp. 125–147, Woodhead Publishing, Cambridge, 2008.
19. Fonseca dos Reis, E., Campos, F.S., Lage, A.P., Leite, R.C., Heneine, L.G., Vasconcelos, W.L. *et al.*, Synthesis and characterization of poly (vinyl alcohol) hydrogels and hybrids for rMPB70 protein adsorption. *Mater. Res.*, 9, 185, 2006.
20. Das, D., Chaudhuri, A., Mitra, M., Ghosh, S., Development of moisture vapour permeable waterproof cotton fabric by coating with blend of natural rubber latex and polyvinyl alcohol. *J. Text. Inst.*, 108, 1285, 2017.
21. Palanikkumaran, M., Agrawal, A.K., Jassal, M., Water-proof breathable coatings based on poly(vinyl alcohol) for cellulosic fabric. *J. Ind. Text.*, 38, 151, 2008.
22. Natarajan, S. and Moses, J., Surface modification of polyester fabric using polyvinyl alcohol in alkaline medium medium. *Indian J. Fibre Text. Res.*, 37, 287, 2015.
23. Jiang, L., Ren, Z., Zhao, W., Liu, W., Liu, H., Zhu, C., Synthesis and structure/properties characterizations of four polyurethane model hard segments. *R. Soc. Open Sci.*, 5, 1, 2018.
24. Karak, N., Biopolymers for paints and surface coatings, in: *Biopolymers and Biotech Admixtures for Eco-Efficient Construction Materials*, F. Pacheco-Torgal, V. Ivanov, N. Karak, H. Jonkers (Eds.), pp. 333–368, Woodhead Publishing, Cambridge, 2016.
25. Pachence, J.M., Bohrer, M.P., Kohn, J., Biodegradable polymers, in: *Principle of Tissue Engineering*, R. Lanza, R. Langer, J. Vacanti (Eds.), pp. 323–339, Academic Press, Cambridge, 2007.
26. Reghunadhan, A. and Thomas, S., Polyurethanes: Structure, properties, synthesis, characterization, and applications, in: *Polyurethane Polymers: Blends and Interpenetrating Polymer Networks*, S. Thomas, J. Datta, J.T. Haponiuk, A. Reghunadhan (Eds.), pp. 1–16, Elsevier, London, 2017.
27. Akindoyo, J.O., Beg, M.D.H., Ghazali, S., Islam, M.R., Jeyaratnam, N., Yuvaraj, A.R., Polyurethane types, synthesis and applications—A review. *RSC Adv.*, 6, 114453, 2016.
28. Bouasria, A., Nadi, A. *et al.*, Advances in polymer coating for functional finishing of textiles, in: *Frontiers of Textile Materials: Polymers, Nanomaterials, Enzymes, and Advanced Modification Techniques*, M. Shabbir, S. Ahmed, J.N. Sheikh (Eds.), pp. 61–86, Scrivener Publishing LLC, Beverly, 2020.
29. Gunesoglu, S., Cerci, E., Topalbekiroglu, M., The improved breathability of polyurethane coated cotton fabric via micro-cracking. *J. Text. Inst.*, 108, 1815, 2017.
30. Träubel, H., *New Materials Permeable to Water Vapor*, Springer-Verlag, 1999.

31. Michaelis, T. and Casselmann, H., Synthetic leather without solvents. *Eur. Coatings J.*, 49, 38, 2009.
32. Sur, S.H., Choi, P.J., Ko, J.W., Lee, J.Y., Lee, Y.H., Kim, H., Do, Preparation and Properties of DMF-Based Polyurethanes for Wet-Type Polyurethane Artificial Leather. *Int. J. Polym. Sci.*, 2018, 1, 2018.
33. Garnier, J., Dufils, P., Vinas, J., Vanderveken, Y., van Herk, A., Lacroix-Desmazes, P., Synthesis of poly (vinylidene chloride)-based composite latexes by emulsion polymerization from epoxy functional seeds for improved thermal stability. *Polym. Degrad. Stab.*, 97, 170, 2012.
34. Ebnesajjad, S., Adhesion promoters, in: *Surface Treatment of Materials for Adhesive Bonding*, S. Ebnesajjad (Ed.), pp. 301–329, William Andrew, Eaton, 2014.
35. Mohamed, A.L. and Hassabo, A.G., Review of silicon-based materials for cellulosic fabrics with functional applications. *J. Text. Color. Polym. Sci.*, 16, 139, 2019.
36. Noll, W., General discussion, in: *Chemistry and Technology of Silicones*, W. Noll (Ed.), pp. 1–23, Academic Press, Cambridge, 1968.
37. Eduok, U., Faye, O., Szpunar, J., Recent developments and applications of protective silicone coatings: A review of PDMS functional materials. *Prog. Org. Coatings*, 111, 124, 2017.
38. Manickam, M.M., Silicone chemistry for fabric care. *Colourage*, 56, 86, 2009.
39. Chandra, G., Review of the environmental fate and effects of silicone materials in textile applications. *Text. Chem. Color.*, 27, 21, 1995.
40. Slone, R.V., Acrylic ester polymers. *Encycl. Polym. Sci. Technol.*, 1, 96, 2010.
41. Licari, J.J. and Swanson, D.W., *Adhesives Technology for Electronic Applications Materials, Processing, Reliability*, pp. 75–141, William Andrew, 2011.
42. Zhou, X., Qiu, S., Mu, X., Zhou, M., Cai, W., Song, L. *et al.*, Polyphosphazenes-based flame retardants: A review. *Compos. Part B*, 202, 108397, 2020.
43. Henke, H., Brüggemann, O., Teasdale, I., Branched macromolecular architectures for degradable, multifunctional phosphorus-based polymers. *Macromol. Rapid Commun.*, 38, 1, 2017.
44. Teasdale, I., Stimuli-responsive phosphorus-based polymers. *Eur. J. Inorg. Chem.*, 2019, 1445, 2019.
45. Tang, X., Thankappan, S.K., Lee, P. *et al.*, Polymeric biomaterials in tissue engineering and regenerative medicine, in: *Natural and Synthetic Biomedical Polymers*, S.G. Kumbar, C.T. Laurencin, M. Deng (Ed.), pp. 351-371, Elsevier Science, Amsterdam, 2014.
46. Wang, J., Yuan, Y.Y. *et al.*, Polyphosphoesters: Controlled ring-opening polymerization and biological applications, in: *Polymer Science: A Comprehensive Reference*, K. Matyjaszewski and M. Möller (Eds.), pp. 719–747, Elsevier Science, Amsterdam, 2012.
47. Singha, K., A review on coating & lamination in textiles: Processes and applications. *Am. J. Polym. Sci.*, 2, 39, 2012.
48. Sherazi, T.A., Spray coating, in: *Encyclopedia of Membranes*, E. Drioli and L. Giorno (Eds.), Springer, Berlin, 2015.

49. Karthikeyan, J., The advantages and disadvantages of the cold spray coating process, in: *The Cold Spray Materials Deposition Process Fundamentals and Applications*, V.K. Champagne (Ed.), pp. 62–71, Woodhead Publishing, Cambridge, 2007.
50. Appendix B Common chemicals and materials, in: *Worker's Guide to Health and Safety*, vol. 461, 2015.
51. Eid, B.M. and Ibrahim, N.A., Recent developments in sustainable finishing of cellulosic textiles employing biotechnology. *J. Clean. Prod.*, 284, 124701, 2021.
52. Shrivastava, A., *Introduction to Plastics Engineering*, pp. 1–16, William Andrew, 2018.
53. Loftsson, T. and Duchêne, D., Cyclodextrins and their pharmaceutical applications. *Int. J. Pharm.*, 329, 1, 2007.
54. Maest, F., Jos, M., Firmino, H.B., Gabriella, J., Cassia, R.D., Curto, S. *et al.*, The role of β-cyclodextrin in the textile industry—Review. *Molecules*, 25, 3624, 2020.
55. Voncina, B., Vivod, V., Günay, M., Cyclodextrins in textile finishing, in: *Eco-Friendly Textile Dyeing and Finishing*, pp. 53–75, IntechOpen, London, 2013.
56. Ammayappan, L. and Moses, J.J., An overview on application of cyclodextrins in textile product enhancement. *J. Text. Assoc.*, 2009, 9, 2009.
57. Radu, C., Parteni, O., Ochiuz, L., Application of cyclodextrins in medical textiles—Review. *J. Control. Release*, 224, 146, 2016.
58. Chauhan, N., Arya, N., Yadav, N., Application of cyclodextrins in textiles. *Int. J. Chem. Stud.*, 6, 2314, 2018.
59. Luda, M.P. and Zanetti, M., Cyclodextrins and cyclodextrin derivatives as green char promoters in flame retardants formulations for polymeric materials. A review. *Polymers*, 11, 664, 2019.
60. Martin, A., Tabary, N., Leclercq, L., Junthip, J., Degoutin, S., Multilayered textile coating based on a B-cyclodextrin polyelectrolyte for the controlled release of drugs. *Carbohydr. Polym.*, 93, 718, 2013.
61. Wang, C.X. and Chen, S.L., Surface treatment of cotton using β-cyclodextrins sol–gel method. *Appl. Surf. Sci.*, 252, 6348, 2006.
62. Bhaskara-amrit, U.R., Agrawal, P.B., Warmoeskerken, M.M.C.G., Applications of β-cyclodextrins in textiles. *Autex Res. J.*, 11, 94, 2011.
63. Nazi, M., Mohammad, R., Malek, A., Kotek, R., Modification of β-cyclodextrin with itaconic acid and application of the new derivative to cotton fabrics. *Carbohydr. Polym.*, 88, 950, 2012.
64. Moattari, M., Moattari, F., Kaka, G., Hm, K., Sadraie, H., Naghdi, M., Application of Chitosan in Textiles. *Ann. Mater. Sci. Eng.*, 3, 1032, 2018.
65. Jain, A., Duvvuri, L.S., Farah, S., Beyth, N., Domb, A.J., Antimicrobial polymers. *Adv. Heal. Care Mater.*, 3, 1969, 2014.
66. Thakur, V.K. and Thakur, M.K. (Eds.), *Handbook of Sustainable Polymers Processing and Applications*, PAN Stanford Publishing, Singapore, 2016.
67. Ritthidej, G.C., Nasal delivery of peptides and proteins with chitosan and related mucoadhesive polymers, in: *Peptide and Protein Delivery*, C. Van der Walle (Ed.), pp. 47–68, Academic Press, Cambridge, 2011.

68. Sharma, S., Kumar, A., Afgan, S., Kumar, R., Stimuli-responsive Polymeric Hydrogel-Copper Nanocomposite Material for Biomedical Application and Its Alternative Application to Catalytic Field. *Chemistry SELECT*, 2, 11281, 2017.
69. Şahan, G. and Demir, A., Forms of chitosan biopolymer and their textile applications, in: *XIIIth International Izmir Textile and Apparel Symposium*, pp. 233–236, 2014.
70. Kabir, I.I., Sorrell, C.C., Mofarah, S.S., Yang, W., Yuen, C.Y., Nazir, M.T., *et al.*, Review-Alginate/polymer-based materials for fire retardancy: Synthesis, structure, properties, and applications. *Polym. Rev.*, 2020, 1, 2020.
71. Pawar, S.N. and Edgar, K.J., Alginate derivatization: A review of chemistry, properties and applications. *Biomaterials*, 33, 3279, 2012.
72. Haidara, H., Vonna, L., Vidal, L., Unrevealed self assembly and crystallization structures of Na-alginate, induced by the drying dynamics of wetting films of the aqueous polymer solution. *Macromolecules*, 43, 2421, 2010.
73. Rydzkowski, T. and Thakur, V.K., Review—Biopolymers for biomedical and pharmaceutical applications: Recent advances and overview of alginate electrospinning. *Nanomaterials*, 9, 7, 2019.
74. Sondermeijer, H.P., Witkowski, P., Woodland, D., Seki, T., Aangenendt, F.J., Van Der Laarse, A. *et al.*, Optimization of alginate purification using polyvinylidene difluoride membrane filtration: Effects on immunogenicity and biocompatibility of three-dimensional alginate scaffolds. *J. Biomaterilas Appl.*, 31, 510, 2016.
75. Gad, Y.H., Aly, R.O., Abdel-Aal, S.E., Synthesis and characterization of Na-alginate/acrylamide hydrogel and its application in dye removal. *J. Appl. Polym. Sci.*, 120, 1899, 2011.
76. Baffoun, A., Haidara, H., Dupuis, D., Viallier, P., Uha, E., Werner, A., Wetting-induced formation of quasiperiodical wrinkling patterns in alginate-based coatings. *Langmuir*, 23, 9447, 2007.
77. Anisha, A.D. and Shegokar, R., Polyethylene glycol (PEG): A versatile polymer for pharmaceutical applications. *Expert Opin. Drug Deliv.*, 13, 1257, 2016.
78. Gillett, A., Waugh, D.G. *et al.*, Laser surface modification of polymeric surfaces for microbiological applications, in: *Laser Surface Modification of Biomaterials Techniques and Applications*, R. Vilar (Ed.), pp. 197-220, Woodhead Publishing, Cambridge, 2016.
79. Alcantar, N.A., Aydil, E.S., Israelachvili, J.N., Polyethylene glycol-coated biocompatible surfaces. *J. Biomed. Mater. Res.*, 51, 343, 2000.
80. Chao, Y.-C., Su, S.-K., Lin, Y.-W., Hsu, W.-T., Huang, K.-S., Interfacial properties of polyethylene glycol/vinyltriethoxysilane (PEG/VTES) copolymers and their application to stain resistance. *J. Surfactants Deterg.*, 15, 299, 2012.
81. Chen, J., Spear, S.K., Huddleston, J.G., Rogers, R.D., Polyethylene glycol and solutions of polyethylene glycol as green reaction media. *Green Chemestry*, 7, 64, 2005.

82. Miranda, T.M.R., Santos, J., Soares, G.M.B., Soil-release behaviour of polyester fabrics after chemical modification with polyethylene glycol. *IOP Conf. Ser. Mater. Sci. Eng.*, 254, 32005, 2017.
83. Samanta, A.K., Mukhopadhyay, A., Bhagwat, M.M., Application of polyethylene glycol, cetrimide, chitosan and their mixtures on cotton muslin fabric to improve rot resistance, antimicrobial property and its salt-free reactive dyeing. *J. Text. Inst.*, 107, 1386, 2016.
84. Yen, M.-S. and Yeh, T.-C., Effect of addition of polyethylene glycol on the moisture transfer and waterproof properties of foam-finished nylon fabrics. *J. Polym. Res.*, 4, 253, 1997.
85. Hurley, P.E., History of natural rubber. *J. Macromol. Sci. Part A—Chem. Pure Appl. Chem.*, 15, 1279, 2006.
86. Jalal Uddin, A., Coatings for technical textile yarns, in: *Technical Textile Yarns*, R. Alagirusamy, A. Das (Ed.), pp. 140-184, Woodhead Publishing, Cambridge, 2010..
87. Kitjanon, J., Khuntawee, W., Sutthibutpong, T., Boonnoy, P., Phongphanee, S., Wong-ekkabut, J.W., Transferability of polymer chain properties between coarse-grained and atomistic models of natural rubber molecule validated by molecular dynamics simulations. *J. Phys. Conf. Ser.*, 901, 12096, 2017.
88. Ahmad, M.R., Yunus, W., Ahmad, W., Salleh, J., Samsuri, A., Performance of natural rubber coated fabrics under ballistic impact. *Malaysian Polym. J.*, 2, 39, 2007.
89. Pongsathit, S., Eco-friendly high-performance coating for polyester fabric. *J. Appl. Polym. Sci.*, 136, 48002, 2019.
90. Haweel, C.K. and Ammar, S.H., Preparation of poly(vinyl)alcohol from local raw material Preparation of Polyvinyl Alcohol from Local Raw Material. *Iraqi J. Chem. Pet. Eng.*, 9, 15, 2016.
91. Deshmukh, K., Ahamed, M.B. *et al.*, Biopolymer composites with high dielectric performance: Interface engineering, in: *Biopolymer Composites in Electronics*, K.K. Sadasivuni, J. Kim, M.A. AlMaadeed, D. Ponnamma, J.-J. Cabibihan (Eds.), pp. 27–128, Elsevier, London, 2017.
92. Aslam, M., Kalyar, M.A., Raza, Z.A., Polyvinyl alcohol: A review of research status and use of polyvinyl alcohol based nanocomposites. *Polym. Eng. Sci.*, 58, 2119, 2018.
93. Baker, M.I., Walsh, S.P., Schwartz, Z., Boyan, B.D., A review of polyvinyl alcohol and its uses in cartilage and orthopedic applications. *J. Biomed. Mater. Res. B. Appl. Biomater.*, 100, 1451, 2012.
94. Ben Halima, N., Poly(vinyl alcohol): Review of its promising applications and insights into biodegradation. *RSC Adv.*, 6, 39823, 2016.
95. Patra, N., Salerno, M. *et al.*, Electrospun polyvinyl alcohol/pectin composite nanofibers, in: *Electrospun Nanofibers*, M. Afshari (Ed.), pp. 599–608, Woodhead Publishing, Cambridge, 2017.
96. Abbasi, E., Aval, S.F., Akbarzadeh, A., Milani, M., Nasrabadi, H.T., Dendrimers: Synthesis, applications, and properties. *Nanoscale Res. Lett.*, 9, 1, 2014.

97. Ghosh, S., Yadav, S., Vasanthan, N., Sekosan, G., A study of antimicrobial property of textile fabric treated with modified dendrimers. *J. Appl. Polym. Sci.*, 115, 716, 2009.
98. Sadjadi, S. and Sadjadi, S., Dendritic polymers for environmental remediation, in: *New Polymer Nanocomposites for Environmental Remediation*, C. Mustansar Hussain and A. Kumar Mishra (Eds.), pp. 279–335, Elsevier, London, 2018.
99. Cheng, Y., Burda, C., Nanoparticles for photodynamic therapy, in: *Comprehensive Nanoscience and Technology*, D.L. Andrews, G.D. Scholes, G.P. Wiederrecht (Ed.), pp. 1-28, Academic Press, Cambridge, 2011.
100. Pathakoti, K., Manubolu, M. *et al.*, Nanotechnology applications for environmental industry, in: *Handbook of Nanomaterials for Industrial Applications*, C.M. Hussain (Ed.), pp. 894–907, Elsevier, London, 2020.
101. Munavali, B.B., Naik, S.R. *et al.*, Dendrimers, in: *Functional Polymers*, M. Abu Jafar Mazumder, H. Sheardown, A. Al-Ahmed (Eds.), pp. 289–345, Springer Nature, Switzerland, 2019.
102. Patil, A.A., Maiti, S., Adivarekar, R.V., The use of poly(amido) amine dendrimer in modification of cotton for improving dyeing properties of acid dye. *Int. J. Cloth. Sci. Technol.*, 31, 220, 2018.
103. Godinez, L.A., Rodriguez, F.J., Rodriguez, A., De Larrea, G.Z., Larrea, G.Z., Manriquez, J., Bustos, E., Corrigendum to "Applications of dendrimers in drug delivery agents, diagnosis, therapy, and detection". *J. Nanomater.*, 2020, 1, 2020.
104. Chowdhury, K.P., Process intensification of fluorocarbon-free and fluorocarbon-based water repellent finishes on cotton knit fabrics. *J. Textle Eng. Fash. Technol.*, 4, 199, 2018.
105. Mahltig, B., Hydrophobic and oleophobic finishes for textiles, in: *Functional Finishes for Textiles Improving Comfort, Performance and Protection*, R. Paul (Ed.), pp. 387–428, Woodhead Publishing, Cambridge, 2015.
106. Marek, J., Martinkova, L., Protective clothing, in: *Waterproof and Water Repellent Textiles and Clothing*, J. Williams (Ed.), pp. 391-445, Woodhead Publishing, Cambridge, 2018.
107. Rangi, A. and Jajpura, L., The biopolymer sericin: Extraction and applications. *J. Text. Sci. Eng.*, 5, 1, 2015.
108. Rocha, L.K.H., Favaro, L.I.L., Rios, A.C., Silva, E.C., Silva, W.F., Stigliani, T.P. *et al.*, Sericin from *Bombyx mori* cocoons. Part I: Extraction and physicochemical–biological characterization for biopharmaceutical applications. *Process Biochem.*, 61, 163, 2017.
109. Elzoghby, A.O., Elgohary, M.M. *et al.*, Implications of protein- and peptide-based nanoparticles as potential vehicles for anticancer drugs, in: *Advances in Protein Chemistry and Structural Biology*, R. Donev (Ed.), pp. 169–221, Elsevier, London, 2015.
110. Gokce, Y., Aktas, Z., Capar, G., Kutlu, E., Anis, P., Improved antibacterial property of cotton fabrics coated with waste sericin/silver nanocomposite. *Mater. Chem. Phys.*, 254, 123508, 2020.

111. Padamwar, M.N. and Pawar, A.P., Silk sericin and its applications: A review. *J. Sci. Ind. Res.*, 63, 323, 2004.
112. Rajalakshmi, M. and Koushik, C.V., Application of sericin in textile industry. *J. Interdiscipl. Cycle Res.*, XII, 1031, 2020.
113. Saha, J., Mondal, I.H., Sheikh, R.K., Habib, A., Structural and functional properties of silk sericin biopolymer from. *Bombyx mori silk cocoon waste. J. Text. Sci. Eng.*, 9, 1000390, 2019.
114. Belhaj Khalifa, I., Ladhari, N., Touay, M., Application of sericin to modify textile supports. *J. Text. Inst.*, 103, 370, 2012.
115. Gupta, D., Chaudhary, H., Gupta, C., Sericin based bioactive coating for polyester fabric. *Indian J. Fibre Text. Res.*, 40, 70, 2015.
116. Kim, S., Biopolyphenolics in textile, in: *Advances in Textile Biotechnology (Second Edition)*, A. Cavaco-Paulo, V.A. Nierstrasz, Q. Wang (Eds.), pp. 159–183, Woodhead Publishing, Cambridge, 2019.
117. Piccolella, S. and Pacifico, S., Plant-derived polyphenols: A chemopreventive and chemoprotectant worth-exploring resource in toxicology, in: *Advances in Molecular Toxicology*, J.C. Fishbein and J.M. Heilman (Eds.), pp. 161–214, Elsevier B.V., Amsterdam, 2015.
118. Souto, E.B., Sampaio, A.C. *et al.*, Polyphenols for skin cancer: Chemical properties, structure-related mechanisms of action and new delivery systems, in: *Studies in Natural Products Chemistry*, A. Rahman (Ed.), pp. 21–42, Elsevier B.V., Amsterdam, 2019.
119. Goszcz, K., Duthie, G.G., Stewart, D., Leslie, S.J., Megson, I.L., Bioactive polyphenols and cardiovascular disease: Chemical antagonists, pharmacological agents or xenobiotics that drive an adaptive response? *Br. J. Pharmacol.*, 174, 1209, 2017.
120. Fahmy, H.M., Aly, A.A., Amr, A., Sayed, S.M., Rabie, A.M., Effect of phenolic compounds and water repellents combination on performance properties of cotton/polyester blended fabric. *Egypt. J. Chem.*, 60, 69, 2017.
121. Cavaco-Paulo, A., Enzymatic coating of jute fabrics for enhancing properties via *in-situ* polymerization of polyhydric phenols. *J. Ind. Text.*, 46, 160, 2016.
122. Du, Z., Wen, S., Wang, J., Yin, C., Yu, D., Luo, J., The Review of Powder Coatings. *J. Mater. Sci. Chem. Eng.*, 4, 54, 2016.
123. Sahoo, P., Das, S.K., Davim, J., P., Surface finish coatings. *Mater. Sci. Mater. Eng.*, 3, 38, 2017.
124. Center of Globalization Governance & Competitiveness, The development and diffusion of powder coatings in the United States and Europe: A historical and value chain perspective, https://gvcc.duke.edu/cggclisting/the-development-and-diffusion-of-powder-coatings-in-the-united-states-and-europe-a-historical-and-value-chain-perspective/, 2010.
125. Farshchi, N. and Gedan-smolka, M., Polyurethane powder coatings: A review of composition and characterization. *Ind. Eng. Chem. Res.*, 59, 15121, 2020.
126. Mckeen, L.W., Fluorinated coatings: Technology, history, and applications, in: *Introduction to Fluoropolymers Materials, Technology and Applications*, S. Ebnesajjad (Ed.), pp. 231–276, William Andrew, Eaton, 2013.

127. Rosato, D.V., Rosato, R.V., Rosato, M.V., *Plastic Product Material and Process Selection Handbook*, pp. 382–393, Elsevier Science, 2004.
128. Hester, C.I. and Nicholson, R., *Powder Coatings Technology Update*, Control Technology Center, United States, 1989.
129. Takeshita, Y., Sawada, T., Handa, T., Watanuki, Y., Kudo, T., Influence of air-cooling time on physical properties of thermoplastic polyester powder coatings. *Prog. Org. Coatings*, 75, 584, 2012.
130. Bond, G.E. and Cornell, J.S., The benefits of several thermoplastic powder coatings. *Pigment Resin Technol.*, 3, 3, 1974.
131. Glawe, A., Reuscher, R., Köppe, R., Thomas, K., Hot-melt application for functional compounds on technical textiles. *J. Ind. Text.*, 33, 85, 2003.
132. Karl Fink, J., Unsaturated polyester resins, in: *Reactive Polymers Fundamentals and Applications*, J.K. Fink (Ed.), pp. 1–48, William Andrew, Eaton, 2013.
133. Shim, E., Coating and laminating processes and techniques for textiles, in: *Smart Textile Coatings and Laminates*, W.C. Smith (Ed.), pp. 11–45, Woodhead Publishing, Cambridge, 2019.
134. Pierre, A.C., *Introduction to Sol-Gel Processing*, Springer Nature, 2020.
135. Alongi, J. and Malucelli, G., State of the art and perspectives on sol–gel derived hybrid architectures for flame retardancy of textiles. *J. Mater. Chem.*, 22, 21805, 2012.
136. Norfazilah, W. and Ismail, W., Sol–gel technology for innovative fabric finishing—A Review. *J. Sol-Gel Sci. Technol.*, 78, 698, 2016.
137. Corriu, R.J.P. and Leclercq, D., Recent developments of molecular chemistry for gol–Gel processes. *Angew. Chemie Int. Ed. Eng.*, 35, 1420, 1996.
138. Malucelli, G., Hybrid organic/Inorganic coatings through dual-cure processes: State of the art and perspectives. *Coatings*, 6, 10, 2016.
139. Livage, J., Sol–gel processes. *Curr. Opin. Solid State Mater. Sci.*, 2, 132, 1997.
140. Yi, S., Sun, S., Deng, Y., Feng, S., Preparation of composite thermochromic and phase-change materials by the sol–gel method and its application in textiles. *J. Text. Inst.*, 106, 1071, 2014.
141. Kay Young, S., Sol–Gel Science for Ceramic Materials. *Mater. Matters*, 1, 8, 2006.
142. Camlibel, N.O. and Arik, B., Sol–gel applications in textile finishing processes, in: *Recent Applications in Sol–Gel Synthesis*, U. Chandra (Ed.), pp. 253–281, IntechOpen, London, 2017.
143. Alongi, J., Ciobanu, M., Tata, J., Carosio, F., Malucelli, G., Thermal stability and flame retardancy of polyester, cotton, and relative blend textile fabrics subjected to sol–gel treatments. *J. Appl. Polym. Sci.*, 119, 1961, 2011.
144. Ferrero, F. and Periolatto, M., Application of fluorinated compounds to cotton fabrics via sol–gel. *Appl. Surf. Sci.*, 275, 201, 2013.
145. Dave, H., Ledwani, L. *et al.*, Nonthermal plasma: A promising green technology to improve environmental performance of textile industries, in: *The Impact and Prospects of Green Chemistry for Textile Technology*, S.-ul-Islam,

and S.B.S. Butola (Eds.), pp. 199–249, Woodhead Publishing, Cambridge, 2019.
146. Weltmann, K., Kolb, J.F., Holub, M., Uhrlandt, D., Šimek, M., Ostrikov, K. *et al.*, The future for plasma science and technology. *Plasma Process. Polym.*, 16, 1800118, 2019.
147. -ul-Islam, S. and Mohammad, F., Emerging green technologies and environment friendly products for sustainable textiles, in: *Roadmap to Sustainable Textiles and Clothing Environmental and Social Aspects of Textiles and Clothing Supply Chain*, S. Senthilkannan Muthu (Ed.), pp. 63–82, Springer, Singapore, 2014.
148. Samanta, K., Jassal, M., Agrawal, A.K., Atmospheric pressure glow discharge plasma and its applications in textile. *Indian J. Fibre Text. Res.*, 31, 83, 2006.
149. Schütze, A., Jeong, J.Y., Babayan, S.E., Park, J., Selwyn, G.S., Hicks, R.F., The atmospheric-pressure plasma jet: A review and comparison to other plasma sources. *IEEE Trans. Plasma Sci.*, 26, 1685, 1999.
150. Zille, A., Oliveira, F.R., Souto, A.P., Plasma Treatment in Textile Industry. *Plasma Process. Polym.*, 12, 98, 2014.
151. Black, S., The role of nanotechnology in sustainable textiles, in: *Sustainable Textiles Life Cycle and Environmental Impact*, R.S. Blackburn (Ed.), pp. 302–328, Woodhead Publishing, Cambridge, 2009.
152. Heikkilä, P., Sipilä, A., Peltola, M., Harlin, A., Electrospun PA-66 coating on textile surfaces. *Text. Res. J.*, 77, 864, 2015.
153. De Falco, F., Guarino, V., Gentile, G., Cocca, M., Ambrogi, V., Ambrosio, L. *et al.*, Design of functional textile coatings via non-conventional electrofluidodynamic processes. *J. Colloid Interface Sci.*, 541, 367, 2019.
154. Chun, H.J., Park, C.H., Kwon, I.K., Khang, G. (Eds.), *Cutting-Edge Enabling Technologies for Regenerative Medicine*, Springer, Singapore, 2018.
155. Jadhav, A., Wang, L., Lawrence, C., Padhye, R., Study of electrospraying characteristics of polymer solution coating on textile substrate. *Adv. Mater. Res.*, 710, 332–334, 2011.
156. Baji, A., Agarwal, K., Oopath, S.V., Emerging Developments in the Use of Electrospun Fibers and Membranes for Protective Clothing Applications. *Polymers*, 12, 492, 2020.
157. Maya, I.C. and Guarino, V., Introduction to electrofluidodynamic techniques. Part I: process optimization, in: *Electrofluidodynamic Technologies (EFDTs) for Biomaterials and Medical Devices*, V. Guarino and L. Ambrosio (Eds.), pp. 1–17, Woodhead Publishing, Cambridge, 2018.
158. Günesoglu, C., Kut, D., Orhan, M., Performing the electrospraying process for the application of textile nano finishing particles. *Text. Res. J.*, 80, 1, 2009.
159. Cunico, L.P. and Turner, C., Supercritical fluids and gas-expanded liquids, in: *The Application of Green Solvent in Separation Processes*, F. Pena Pereira and M. Tobiszewski (Eds.), pp. 155–214, Elsevier, London, 2017.
160. Parhi, R. and Suresh, P., Supercritical Fluid Technology: A Review. *J. Adv. Pharm. Sci. Technol.*, 1, 13, 2013.

161. Braeuer, A., High pressure: Fellow and opponent of spectroscopic techniques, in: *Supercritical Fluid Science and Technology*, A. Braeuer (Ed.), pp. 1–40, Elsevier B.V., Amsterdam, 2015.
162. Liubimtsev, N.A., Deniz, A., Elmanovich, I.V., Gallyamov, M.O., Pich, A., Morphology and properties of flame-retardant superhydrophobic polymer coatings deposited on cotton fabrics from supercritical CO_2. *ACS Appl. Polym. Mater.*, 2, 2919, 2020.
163. Chiu, W., Chen, C., Chang, T.M., Tahara, Y., Hashimoto, T., Kurosu, H. *et al.*, Platinum coating on silk by a supercritical CO_2 promoted metallization technique for applications of wearable devices. *Surf. Coat. Technol.*, 350, 1028, 2018.
164. Cooper, A.I., Polymer synthesis and processing using supercritical carbon dioxide. *J. Mater. Chem.*, 10, 207, 2000.
165. Hay, J.N., Review Environmentally friendly coatings using carbon dioxide as the carrier medium. *J. Mater. Sci.*, 37, 4743, 2002.
166. Mohamed, A.L., Er-rafik, M., Moller, M., Supercritical carbon dioxide assisted silicon based finishing of cellulosic fabric: A novel approach. *Carbohydr. Polym.*, 2013, 98, 1095, 2013.
167. Gallyamov, M.O., Nikitin, L.N., Nikolaev, A.Y., Obraztsov, A.N., Bouznik, V.M., Khokhlov, A.R., Formation of superhydrophobic surfaces by the deposition of coatings from supercritical carbon dioxide. *Colloid J.*, 69, 411, 2007.
168. Debenedettl, P.G., Tom, J.W., Kwauk, X., Yeo, S., Rapid expansion of supercritical solutions (RESS): Fundamentals and applications. *Fluid Phase Equilib.*, 82, 311, 1993.
169. Montero, G.A., Smith, C.B., Hendrix, W.A., Butcher, D.L., Supercritical fluid technology in textile processing: An Overview. *Ind. Eng. Chem. Res.*, 39, 4806, 2000.
170. Nikitin, L.N., Gallyamov, M.O., Said-Galiev, E.E., Khokhlov, A.R., Buznik, V.M., Supercritical carbon dioxide: A reactive medium for chemical processes involving fluoropolymers. *Russ. J. Gen. Chem.*, 79, 578, 2009.
171. Zefirov, V.V., Lubimtsev, N.A., Stakhanov, A.I., Elmanovich, I.V., Kondratenko, M.S., Lokshin, B.V. *et al.*, Durable crosslinked omniphobic coatings on textiles via supercritical carbon dioxide deposition. *J. Supercrit. Fluids*, 133, 30, 2018.
172. Kolomytkin, D.O., Gallyamov, M.O., Khokhlov, A.R., Hydrophobic properties of carbon fabric with teflon AF 2400 fluoropolymer coating deposited from solutions in supercritical carbon dioxide. *Russ. J. Phys. Chem. B*, 5, 1106, 2011.
173. Aslanidou, D., Karapanagiotis, I., Panayiotou, C., Superhydrophobic, superoleophobic coatings for the protection of silk textiles. *Prog. Org. Coatings*, 97, 44, 2016.
174. Kazaryan, P.S., Tyutyunov, A.A., Kondratenko, M.S., Elmanovich, I.V., Stakhanov, A.I., Zefirov, V.V. *et al.*, Superhydrophobic coatings on textiles based on novel poly(perfluoro- tert-hexylbutyl methacrylate-co-

hydroxyethyl methacrylate) copolymer deposited from solutions in supercritical carbon dioxide. *J. Supercrit. Fluids*, 149, 34, 2019.
175. Kazaryan, P.S., Zefirov, V.V., El' manovich, I.V., Stakhanov, A.I., Kondratenko, M.S., Khokhlov, A.R., Omniphobic coatings based on vinyl pivalate–perfluorohexylethyl methacrylate copolymers formed in supercritical carbon dioxide. *Funct. Polym.*, A 61, 157, 2019.
176. Prorokova, N.P., Kumeeva, T.Y., Khorev, A.V., Buznik, V.M., Nikitin, L.N., Ensuring a high degree of water repellency of polyester textile materials by treating them with supercritical carbon dioxide. *Fibre Chem.*, 42, 31, 2010.
177. Kumeeva, T.Y. and Prorokova, N.P., Ultrathin hydrophobic coatings obtained on polyethylene terephthalate materials in supercritical carbon dioxide with co-solvents. *Russ. J. Phys. Chem.*, 92, 346, 2018.
178. Markovic, D., Milovanovic, S., Radetic, M., Jokic, B., Zizovic, I., Impregnation of corona modified polypropylene non-woven material with thymol in supercritical carbon dioxide for antimicrobial application. *J. Supercrit. Fluids*, 101, 215, 2015.
179. Peng, L., Guo, R., Lan, J., Jiang, S., Silver nanoparticles coating on silk fabric with pretreatment of 3-aminopropyltrimethoxysilane in supercritical carbon dioxide. *J. Ind. Text.*, 47, 1, 2016.
180. Deng, Y., Chen, W., Li, B., Wang, C., Kuang, T., Li, Y., Physical vapor deposition technology for coated cutting tools: A review. *Ceram. Int.*, 46, 18373, 2020.
181. Shahidi, S., Moazzenchi, B., Ghoranneviss, M., A review-application of physical vapor deposition (PVD) and related methods in the textile industry. *Eur. Phys. J. Appl. Phys.*, 71, 31302–p1, 2015.
182. Sakalak, H., Yilmaz, K., Gürsoy, M., Karaman, M., Roll-to roll initiated chemical vapor deposition of super hydrophobic thin films on large-scale flexible substrates. *Chem. Eng. Sci.*, 215, 115466, 2020.
183. Benelmekki, M. and Erbe, A., Nanostructured thin films—Background, preparation and relation to the technological revolution of the 21st century, in: *Frontiers of Nanoscience*, M. Benelmekki and A. Erbe (Eds.), pp. 1–34, Elsevier, London, 2019.
184. Alf, B.M.E., Asatekin, A., Barr, M.C., Baxamusa, S.H., Chelawat, H., Ozaydin-ince, G. *et al.*, Chemical vapor deposition of conformal, functional, and responsive polymer films. *Adv. Mater.*, 22, 1993, 2010.
185. Martin, T.P., Lau, K.K.S., Chan, K., Mao, Y., Gupta, M., O'Shaughnessy, W.S. *et al.*, Initiated chemical vapor deposition (iCVD) of polymeric nanocoatings. *Surf. Coat. Technol.*, 201, 9400, 2007.
186. Martin, T.P., Kooi, S.E., Chang, S.H., Sedransk, K.L., Gleason, K.K., Initiated chemical vapor deposition of antimicrobial polymer coatings. *Biomaterials*, 28, 909, 2007.
187. Kosgey, E.K., MSU Graduate Thesis: *Construction of an Initiated Chemical Vapor Deposition (iCVD) Reactor and Deposition of Polytetrafluoroethylene*

(Ptfe) Thin Films Using Perfluoro-1-Octanesulfonyl Fluoride as the Initiator, pp. 1-66, Missouri State University, 2015.

188. Chen, N., Kim, D.H., Kovacik, P., Sojoudi, H., Wang, M., Gleason, K.K., Polymer thin films and surface modification by chemical vapor deposition: Recent progress. *Annu. Rev. Chem. Biomol. Eng.*, 7, 373, 2016.
189. Hilt, F., Boscher, N.D., Duday, D., Desbenoit, N., Levalois-Grützmacher,. J., Choquet, P., Atmospheric pressure plasma-initiated chemical vapor deposition (AP-PiCVD) of poly(diethylallylphosphate) coating: A char-forming protective coating for cellulosic textile. *ACS Appl. Mater. Interfaces*, 6, 18418, 2014.
190. Bashir, T., Skrifvars, M., Persson, N., Production of highly conductive textile viscose yarns by chemical vapor deposition technique: A route to continuous process. *Polym. Adv. Technol.*, 22, 2214, 2011.

6

A Review on Hydrophobicity and Fabricating Hydrophobic Surfaces on the Textiles

Mohammad Khajeh Mehrizi* **and Zahra Shahi**

Textile Engineering Department, Engineering faculty, Yazd University, Yazd, Iran

Abstract

Non-wettable surfaces have received tremendous attention in recent years and the fabrication of these textiles has been an attractive subject for research within the last decade. These surfaces are known as hydrophobic surfaces with properties, such as the leaf of the lotus plant (*Nelumbo nucifera*). Hydrophobic textiles have many applications in apparel, and consumer markets, such as self-cleaning and antimicrobial coatings, resistant cables against ice, and oil–water separators. Surface roughness by change surface structure in nano/micro scales (surface texture) and low surface energy by silicon and fluorinated compounds (surface chemical composition) are two main factors for hydrophobicity of textiles. In this chapter, the properties of artificial hydrophobic surfaces, production, and surface manipulation techniques to produce hydrophobic textiles are presented.

Keywords: Hydrophobic surfaces, lotus effect, roughness, surface energy

6.1 Introduction

The word "hydrophobicity" was derived from two Greek words "hydro: water" and "Phobos: fear", that means "avoiding water" [1]. Generally, surfaces with a water contact angle (WCA) smaller than 90°, hydrophilic, the contact angle more than90°, hydrophobic, and non-wettable surfaces with a water contact angle higher than 150° are considered hydrophilic super-hydrophobic or ultra-hydrophobic [2, 3]. In other words, a surface

**Corresponding author*: mkhajeh@yazd.ac.ir

Luqman Jameel Rather, Mohd Shabbir and Aminoddin Haji (eds.) Sustainable Practices in the Textile Industry, (149–166) © 2021 Scrivener Publishing LLC

with a large contact angle and a small contact angle hysteresis below 10° shows higher hydrophobicity. The difference between the advancing and receding contact angles are the contact angles hysteresis [4–6]. The contact angle is an angle formed by a liquid at the gas-liquid-solid interface. It is an essential parameter for predicting the wetting behavior of surfaces (Figure 6.1) [7]. The famous hydrophobic surface is the leaves of *Nelumbo nucifera* or "lotus flower". A close look with electron microscopy of the lotus shows, the structural of leaves is a compound of nano and micro roughened (about 20–40 μm) and three-dimensional epicuticular waxes that are surface tension lowering agent. This hydrophobic property is known as the "lotus effect" (Figure 6.2) [3, 4, 8].

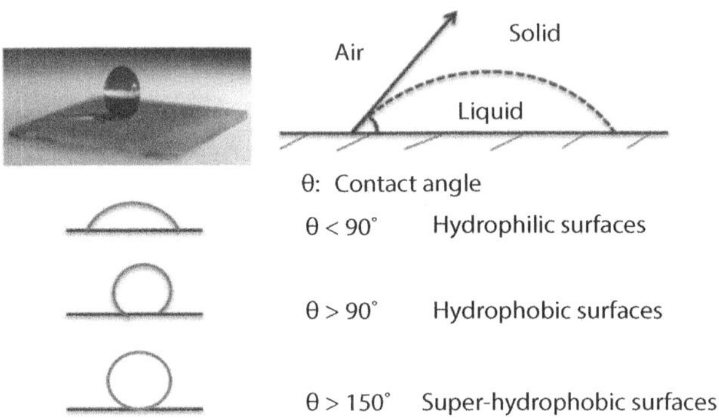

Figure 6.1 Hydrophilic, hydrophobic, and ultra-hydrophobic surfaces [2, 7].

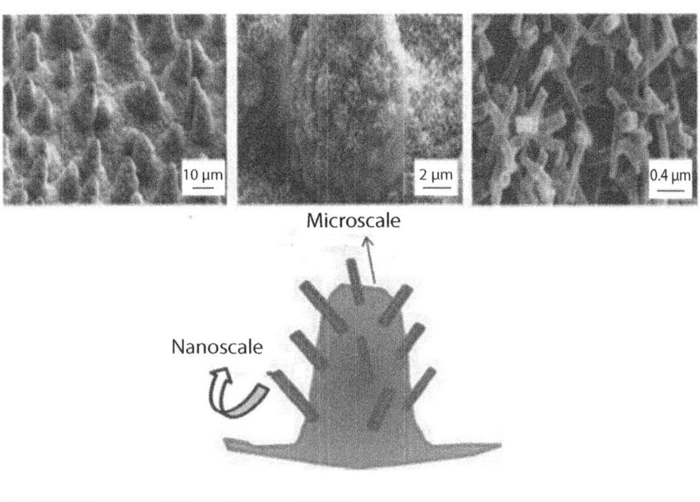

Figure 6.2 SEM images of lotus leaves [5, 9].

The air bubbles fill the valleys below of the water droplets, and the droplets comfortably sit on these nanostructures. Therefore, these leaves show significant super-hydrophobicity [5].

6.2 Self-Cleaning Surfaces

Barthlott and Ehler in 1990 discovered that lotus leaves are one of the well-known self-cleaning surfaces. The water drops roll on nano-roughening and remove contaminants and dust (Figure 6.3) [10, 11].

Also, super-hydrophobic textiles with self-cleaning performance can be used as clothing textiles with high protective, for example washing and drying of self-cleaning clothes in rainy days [11, 13]. Furthermore, self-cleaning properties reduced the number of launderings of textiles, the use of detergents, and the energy required for washing. Therefore, the performance of the textiles can be kept for long times [11]. Many surfaces exhibit hydrophobic, and self-cleaning properties, such as the butterfly's wing, duck feathers, and the leaves of cabbage, and garden nasturtium [5, 13].

6.3 Applications of Hydrophobic Surfaces

Superhydrophobic and water-repellent coatings have attracted considerable attention in recent decades [6]. Recently, super-hydrophobic textiles, due to their importance in industrial applications, have been widely studied by many researchers [14]. Primarily, medical textiles to protect infection by pathogenic fungus [15], anti-fouling clothes [16], breathable clothes [17], umbrellas, and fabrics for the surgical need to super-hydrophobicity

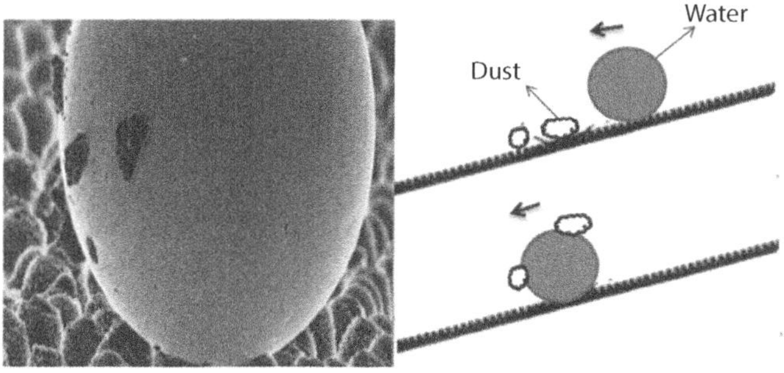

Figure 6.3 Schematic of the connection between roughening and self-cleaning [12].

[18]. Also, hydrophobic surfaces have many applications, including waterproofing of surfaces in building and civil engineering, protection of wood against water [19], filters for separating oil from water [8], anti-corrosion, anti-fogging coatings [11], ice-resistant cables [6], microfluidics domestic commodities [13].

6.4 Basic Theories: Modeling of Contact Angle

The surface wettability theories include Young's model (for smooth surfaces), Wenzel's, and Cassie–Baxter's (for roughened surfaces). In this section, relationships between contact angle with roughness and interfacial tensions are discussed [11, 20].

6.4.1 Young's Model

Young's equation with assuming smooth of surfaces describe wettability with the relationship of contact angle (θ_Y), and interfacial tension between solid (γ_{sl}) and vapor phases (γ_{lv}) (Figure 6.4) [11].

6.4.2 Wenzel Model (Homogeneous Interface)

In 1936, Wenzel considered the effect of surface roughness (without air pockets) on the wetting behaviors and modified Young's equation [2]. This model firstly introduced the concept of surface roughness. The actual surface of solid, because of roughness, will be larger of the geometric surface, that the ratio of two surfaces is called the "roughness factor" (Figure 6.5) [4]. The dependence of the roughness factor on the contact angle is shown in Figure 6.6. It shows that the hydrophobic surface becomes more

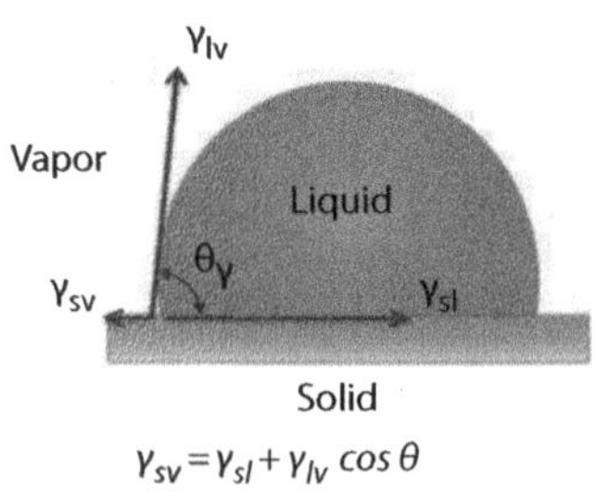

Figure 6.4 Surface tensions of solid/liquid/vapor phases in Young's model [11].

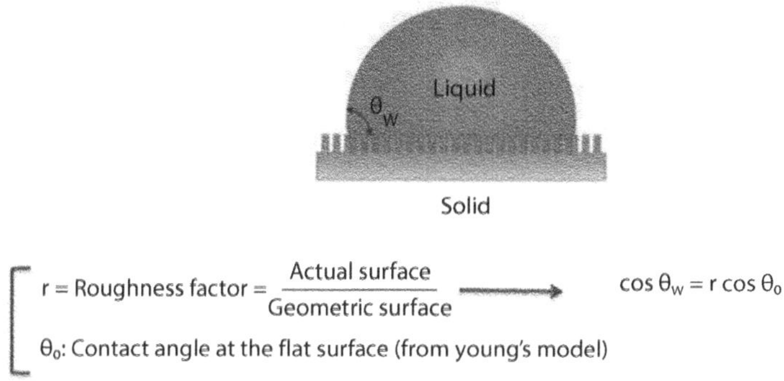

Figure 6.5 Wenzel model for static contact angle [21].

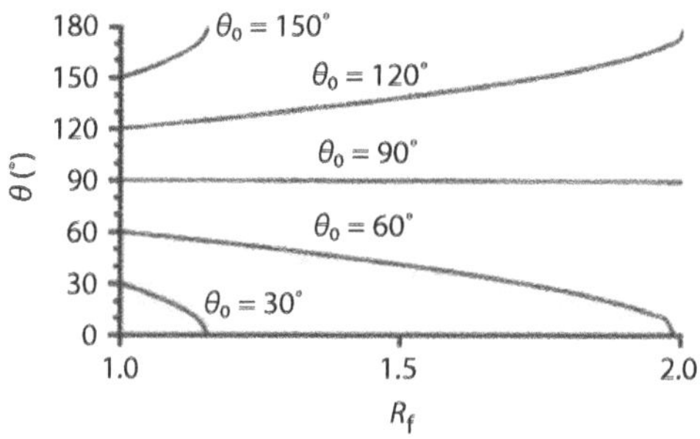

Figure 6.6 The relationship between the contact angle and roughness factor [21].

hydrophobic by increasing of roughness factor, and the hydrophilic surface becomes more hydrophilic by increasing of roughness factor [21].

6.4.3 Cassie–Baxter Model (Composite Interface)

In the rough surface, a wetting liquid will be absorbed by cavities on the surface, while a non-wetting liquid does not penetrate at cavities. Therefore, it leads to the formation of air pockets and the composite of solid–liquid–air. To calculate the contact angle, Wenzel's equation was modified by area fractions under the drop (Figure 6.7) [5]. The studies by Wenzel and Cassie–Baxter showed that factors, such as a porous and rough structure (surface texture) and low surface energy (surface chemical composition) could lead to hydrophobicity [4, 13].

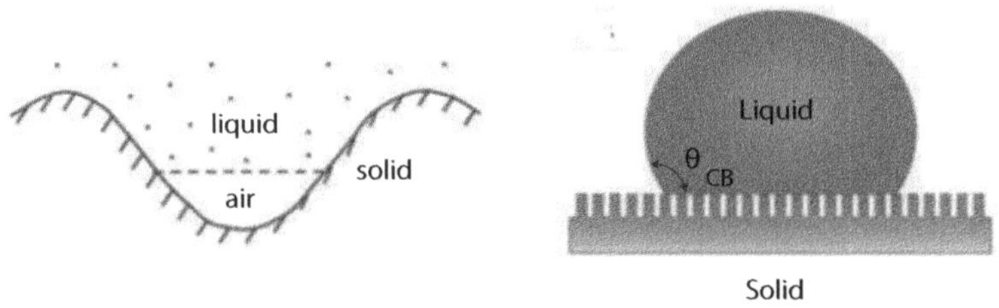

$$\cos\theta = R_f \cos\theta_0 - f_{LA}(R_f \cos\theta_0 + 1),$$

f_{LA}: The fractional flat geometric area of the liquid-air interfaces

Figure 6.7 Wenzel model for static contact angle [5].

6.5 Techniques to Make Super-Hydrophobic Surfaces

The studies by Wenzel and Cassie–Baxter showed that factors, such as a porous and rough structure (surface texture) and low surface energy (surface chemical composition) could lead to hydrophobicity [4, 13]. In other words, super-hydrophobicity can be obtained by increasing the surface roughness and decreasing the surface tension [15]. Fluorinated polymers (long fluoroalkyl chains of C8 or higher [22]) have received a lot of attention due to their very low surface energies [3, 23]. They have stability and excellent chemical resistance at high or low temperatures [24]. Also, polydimethylsiloxane (PDMS) is a famous material with low surface energy [3] that known as silicones and is biocompatible with excellent hydrophobic properties [17]. Other organic materials include polyamide, polycarbonate, alkyl ketene dimer [3], polyvinyl chloride, mercapto functional monomers and polymethyl-hydro siloxane (PMHS) [21], graphene oxide sheets [25] beeswax, paraffin waxes [26], and inorganic materials, such as ZnO and TiO_2, have been used in super-hydrophobic surfaces [3]. Natural waxes due to the high content of fatty alcohols, aldehydes, sterols, ketones, and triterpenic acids are considered hydrophobic substances [26]. Another method for creating super-hydrophobic textiles is to generate micro and nanoscale on the textile. The sol–gel chemistry, lithographic, laser etching, layer-by-layer deposition, plasma treatments, and production of nanofibers through electrospinning have been employed to produce rough surfaces [6, 11, 14, 27].

The sol–gel method can be used to create hydrophobic properties by hydrolysis of silica precursor with alcohol, at low temperatures and short times. The coating process of sol–gel solutions in textile is applied by soaking, immersion, and curing at 130 °C, and spray coating on the fabric [19]. The etching is an easy method to produce rough surfaces. Types of etching, including plasma, laser, and chemical etching, have been used to fabricate super-hydrophobic surfaces. Lithography (photolithography, X-ray lithography) is an effective technique for creating micro-/nanostructures [3]. Depositing low surface energy materials on a substrate can change the wettability of the surface [28]. Among the many techniques, plasma has been widely studied (Table 6.1) [29].

By using plasma, chemical bonds of surface are broken resulting in the formation of radical groups that leads to the physical–chemical deposition [11]. Producing different nanostructures, such as ZnO, SiO_2, TiO_2, ZrO_2, Al_2O_3 carbon nanofiber/nanotubes, and layered silicate clays with forming nana-micro scale roughness are an effective method in creating hydrophobic surfaces [8, 10, 14]. However, the use of nanoparticles is an easy way to make hydrophobic surfaces, but there are many problems. Inorganic nanoparticles can be released from the fiber and enter the skin and cause health risks [13]. The nanoparticles stacked on the textiles, and the bonding between them is very poor [22]. Also, many nanoparticles have been made from super-hard materials that are uncomfortable to wear [13].

Table 6.1 Hydrophobicity of treated fabrics with plasma [30].

Type of plasma	Textile substrate	Power (W)	Gas	Contact angle (°)
Dielectric barrier discharge	Cotton/PET Cotton	4,000 2,000	He/Ar Ar	150 140
Low pressure plasma	Cotton PET	50 100	O_2 $CH_3Si(OCH_3)_3$	146 >150
Corona	PET	5,000	$CH_3)_3SiO[(CH_3)(H)SiO]n Si(CH_3)_3$	127
Atmospheric pressure glow discharge	Cotton PET	250–400 4,000	He/Ar He	150 142

6.6 Methods of Applying Hydrophobic Coating on Textiles

There are numerous methods for applying hydrophobic coating on textiles. Dip and spray-coating methods have been extensively used to make hydrophobic surfaces [23].

6.6.1 Dip-Coating

The dip coating is one of the effective processes for the coats of material on the textiles. The process is applied in roll format before production of the garment with high throughput and cost effectiveness. However, this process is expensive, useless for small batches and unfeasible for 3D forms and limited to specific chemical products [31].

6.6.2 Spray Coating

Spray coating with pressure applied to obtain super-hydrophobic surfaces (Figure 6.8). The spray-coating is a simple method with high efficiency that can be used to preparing large-area hydrophobic coating [28].

6.7 Contact Angles (CA) Measurement

There are two types of contact angles called static and dynamic contact angles. Static contact angles are measured by putting the water drop in a fixed position on the surface, and the resulting three-phase boundary is kept. Dynamic contact angles are measured when the three-phase

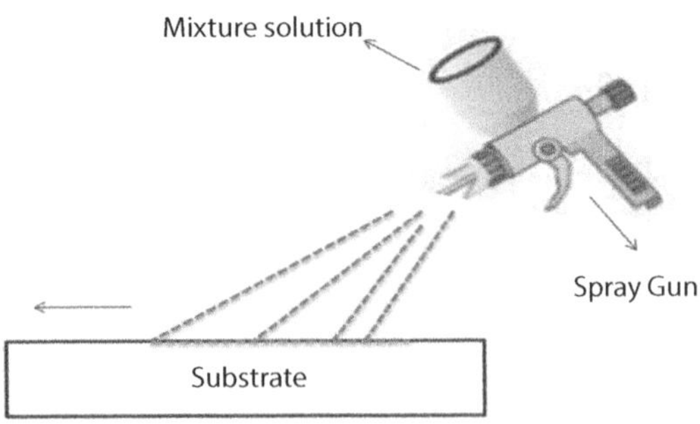

Figure 6.8 Schematic of the spray-coating of the solution [28].

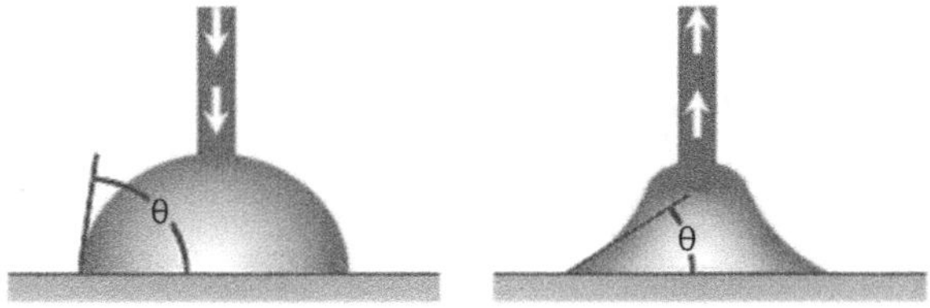

Figure 6.9 Schematic of advancing and receding angles [32].

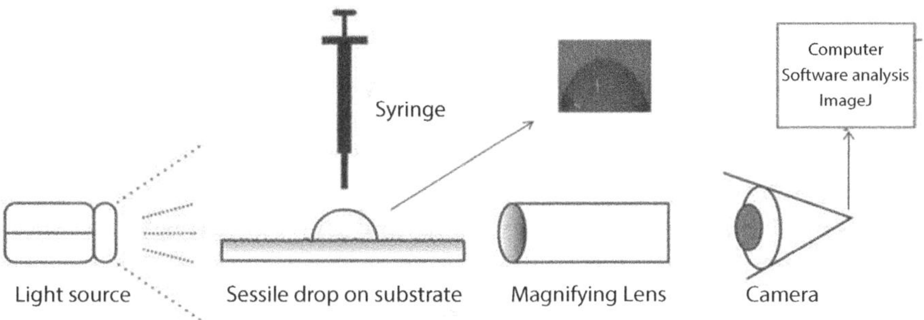

Figure 6.10 Schematic representation of the goniometer [32].

boundary is in motion. This movement is defined as advancing and receding angles (Figure 6.9) [32].

One of the common methods in the sessile drop is the goniometer telescope (Figure 6.10). The types of methods are static and dynamic sessile drops. In the static sessile drop method, the contact angles are measured by a high-resolution camera, and the droplet size remains static during the measurement. In the dynamic sessile drop, the volume of drop changes [32].

6.8 Research Records on Hydrophobic Surface Production

Xue *et al.* produced super-hydrophobic cotton fabrics with the incorporation of titanium dioxide nanoparticles by titania sol (tetrabutyl titanate) that lead to surface roughness (Figure 6.11) [33].

Ariharasudhan *et al.* produced superhydrophobic cotton nonwoven fabric (CA = 156°) by toluene-2, 4 diisocyanate (TDI) via the pad-dry-cure.

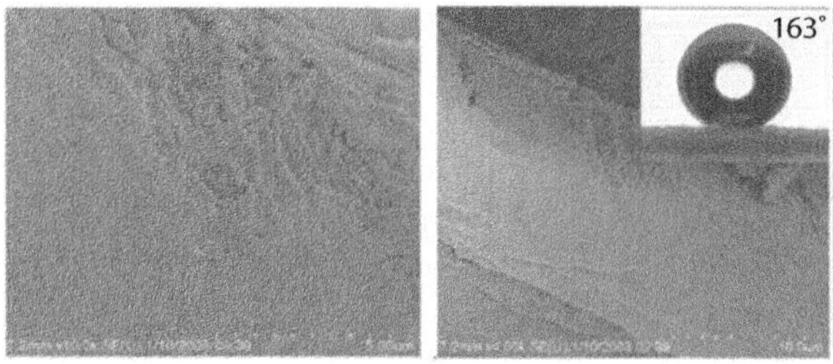

Figure 6.11 SEM images of hydrophobic cotton fabric [33].

The hydroxyl groups on cotton fiber are replaced with isocyanate groups (NCO), and it leads to roughness and hydrophobicity (Figure 6.12) [15].

Joshi *et al.* produced super-hydrophobic cotton fabric using nano-silica and nano-clay. The results showed that nano-silica because of smaller size and uniformity of roughness on the surface has performs better (CA >150°) (Figure 6.13) [10]. Yeon *et al.* produced super-hydrophobic polyester fiber using plasma (argon/hexamethyldisiloxane (HMDSO) mixtures) (Figure 6.14). As shown in Figure 6.15, the polyester fiber without plasma

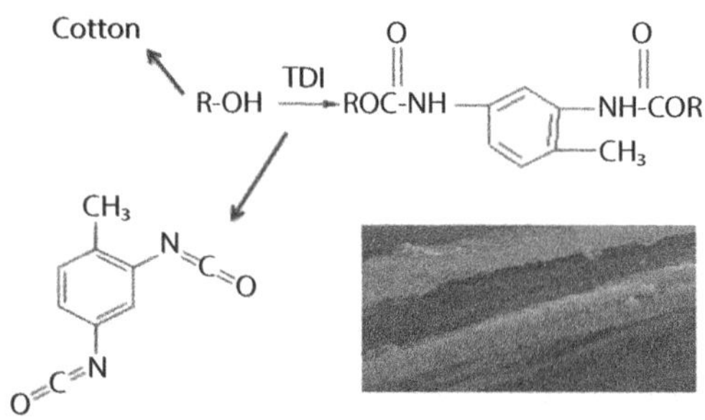

Figure 6.12 Schematic of cotton modification with TDI [34].

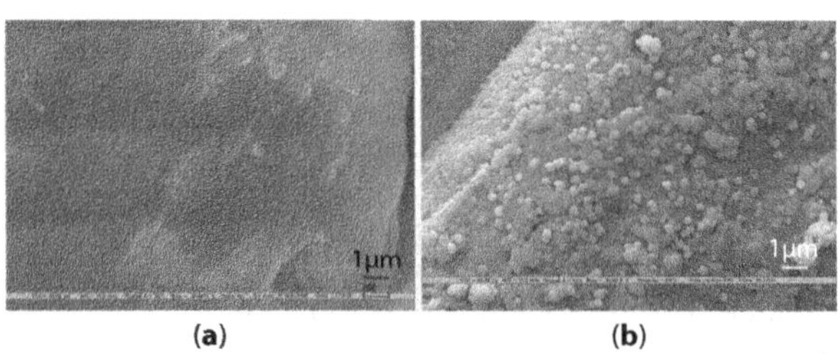

Figure 6.13 SEM images of (a) cotton fiber, (b) treated with nano-silica [10].

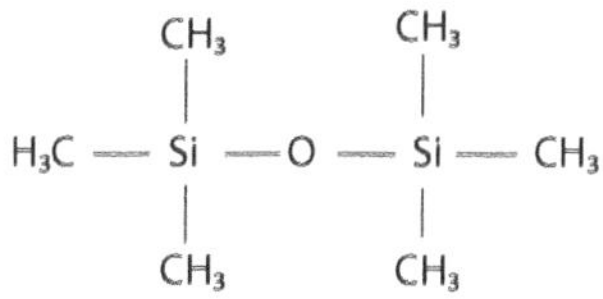

Figure 6.14 Chemical structure of hexamethyldisiloxane [29].

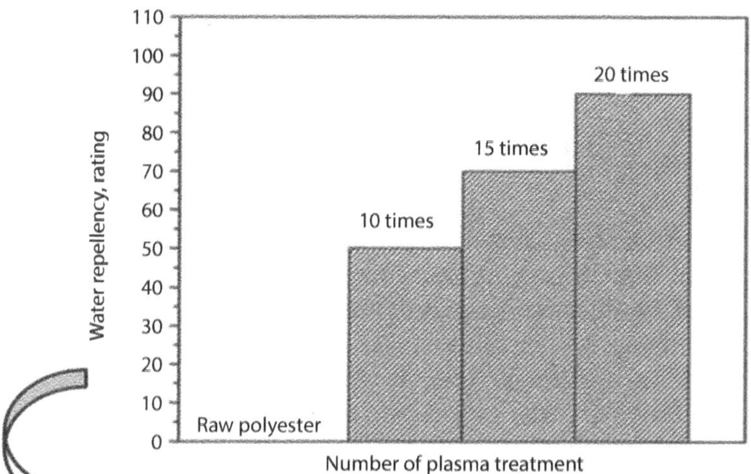

Rating	Evaluation
100	No sticking or wetting of upper surface
90	Slight random sticking or wetting of upper surface
70	Partial wetting of whole upper surface
50	Complete wetting of whole upper surface
0	Complete setting of whole upper surface

Figure 6.15 Production of hydrophobic fabrics using middle frequency (MF) plasma, Ar, HMDSO [29].

completely absorbed sprayed-water. With increasing the plasma time, silicon-oxygen bonds (Si–O–Si), and the ratings are increased [29].

Du *et al.* produced super-hydrophobic nylon fabric with tetrabutyl titanate (TBT) and octadecylamine (OA) responsible for the roughness increasing and surface energy reduction, respectively. To enhance the surface hydrophobicity was applied, 1,2,3,4-butane tetracarboxylic acid (BTCA) on the nylon and the content of the carboxylic acid group was increased (Figure 6.16) [35]. The surface roughness depends on the reaction time, and the contact angle increased with increasing reaction time

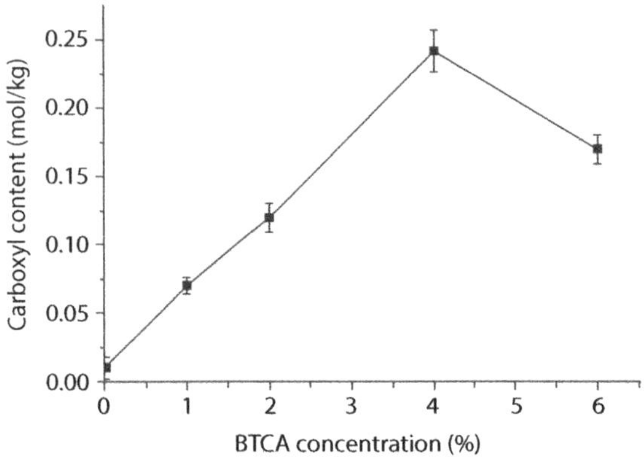

Figure 6.16 The effects of BTCA concentration on the carboxyl content [35].

(Figure 6.17). The results showed that the contact angle increased with an increase in the OA curing time and temperature (Figure 6.18).

Sasaki *et al.* produced super-hydrophobic cotton fabric using ethyl cyanoacrylate and SiO_2 particles by one-step spraying. The present fabrics are useful for medical applications and the prevention of water and blood penetration (Blood CA >150°) [16]. Liang *et al.* modified the cotton fabric with 3-(methacryloyloxy) propyltrimethoxysilane, trifluoromethyl methacrylate, and dodecafluoroheptyl methacrylate (CA >140°) (Figure 6.19) [24]. The results showed that the increasing fluorine atoms in the fluorocarbon chain caused an increase in the contact angle [24]. Gu *et al.* produced super-hydrophobic cotton fabrics through coordination assembly of

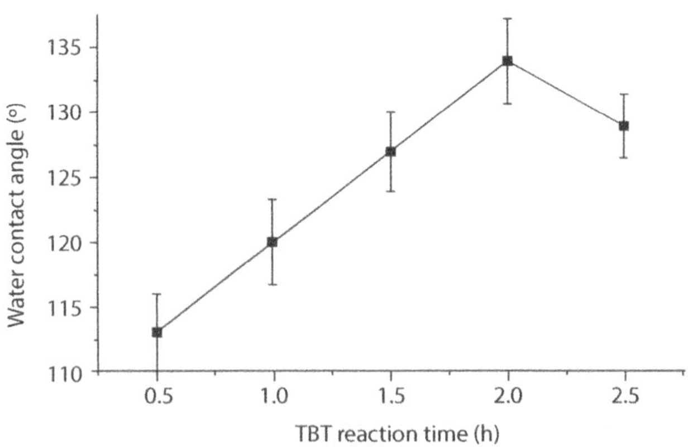

Figure 6.17 The effects of reaction time on the contact angle in BTCA–TBT–OA-treated fabric [35].

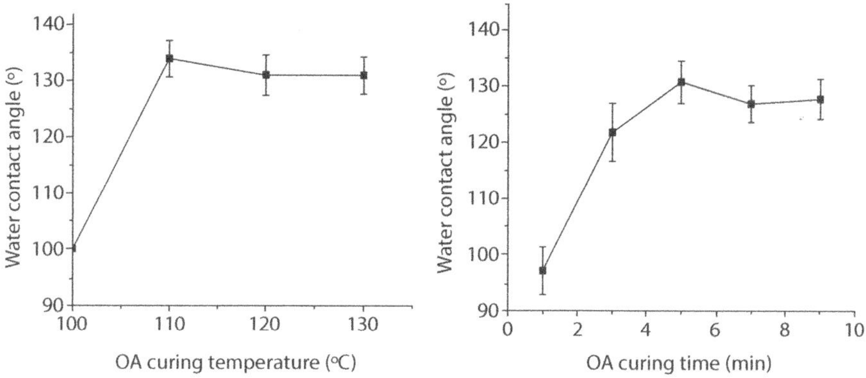

Figure 6.18 The effects of OA curing time and temperature on water contact angle [35].

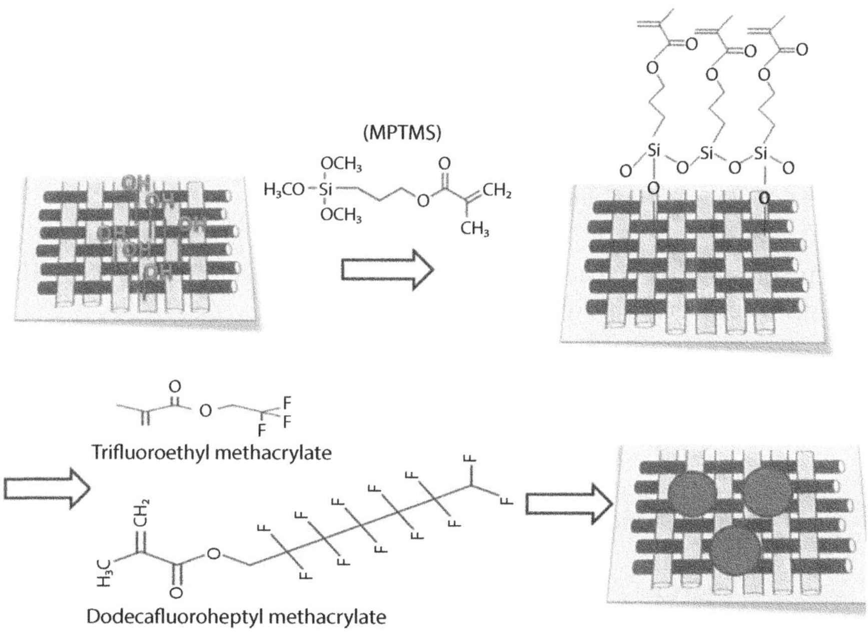

Figure 6.19 Schematic of the preparation of hydrophobic fabrics [24].

tannic acid (TA), Fe (III), and 1-octadecylamine (Fe (III)/TA/ODA). The SEM images of the cotton fabrics are shown in Figure 6.20 [27].

The hydrophobic properties of the treated fabrics were maintained after 25 laundry cycles (Figure 6.21) [27]. Soares *et al.* produced hydrophobic cotton fabrics (>150°) by the sol–gel technique and the spray of the

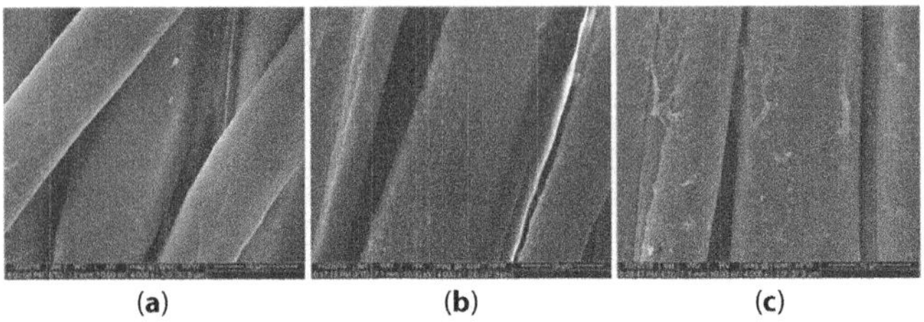

Figure 6.20 The SEM images of (a) cotton fabric, (b) Fe(III)/TA-treated fabric, and (c) Fe(III)/TA/ODA-treated [27].

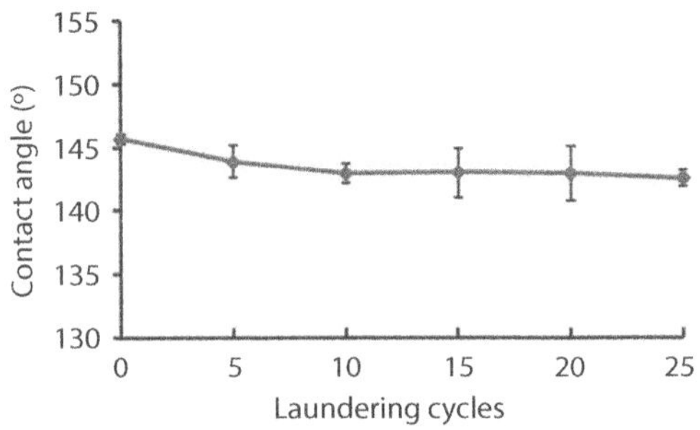

Figure 6.21 The effects of laundry cycles on contact angle [27].

precursor solution (citric acid and tetraethyl orthosilicate (TEOS)). The sol–gel reactions are as follows:

A) Hydrolysis

$$Si[OC_2H_5]_4 + 4H_2O \rightarrow Si(OH)_4 + 4C_2H_5OH \tag{6.1}$$

B) Alcohol condensation

$$Si(OH)_4 + Si[OC_2H_5]_4 \rightarrow SiOSi + 4C_2H_5OH \tag{6.2}$$

The results showed that citric acid increased the hydrophobic properties and adhesion between the silica and cotton (Figure 6.22) [33].

6.9 Conclusion

The lotus leaf surface morphology (a combination of nano and micro roughened surface) is considered an ideal super-hydrophobic. Therefore,

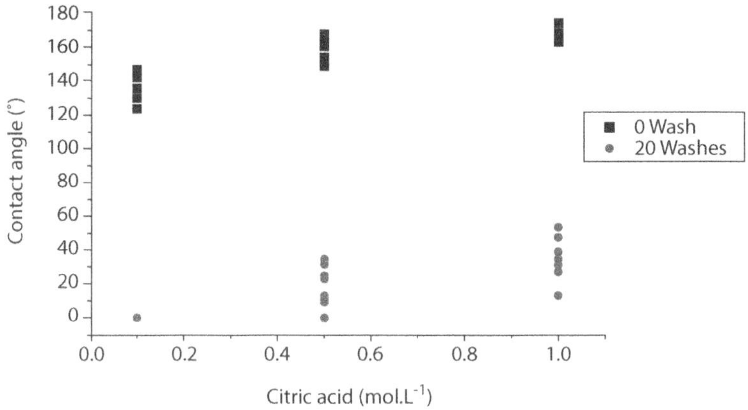

Figure 6.22 The effects of the citric acid concentration on contact angle [33].

the contact angle was increased with increasing surface roughness by micro-nano structure. Generally, hydrophobicity can be achieved by changing the surface chemistry (low surface energy) and roughness. The use of nanoparticles, such as SiO_2, TiO_2, and nanofiber due to creating roughness, is a useful method in production hydrophobic surfaces. Also, fluorinated polymers, polydimethylsiloxane, polyamide, polycarbonate have been a growing interest due to their low surface energies.

References

1. Ehrmann, A. and Blachowicz, T., *Examination of Textiles with Mathematical and Physical Methods*, pp. 125–139, Springer, New York, 2017.
2. Dodiuk, H., Rios, P.F., Dotan, A., Kenig, S., Hydrophobic, and self-cleaning coatings. *Polym. Adv. Technol.*, 18, 9, 2007.
3. Ma, M. and Hill, R.M., Superhydrophobic surfaces. *Curr. Opin. Colloid. Int. Sci.*, 11, 4, 2006.
4. Shim, M.H., Kim, J., Park, G.H., The effects of surface energy and roughness on the hydrophobicity of woven fabrics. *Text. Res. J.*, 84, 12, 2014.
5. Bhushan, B., Jung, Y.C., Koch., K., Adhesion superhydrophobicity, self-cleaning and low Micro-, nano- and hierarchical structures for. *Philos. Trans. A Math. Phys. Eng. Sci.*, 367, 1894, 2014.
6. Shi, Z., Wyman, I. *et al.*, Preparation of water-repellent cotton fabrics from fluorinated diblock copolymers and evaluation of their durability. *Polymer*, 54, 23, 2013.
7. Barashkov, N., Sakhno, T. *et al.*, Water-resistant fluorine-free technologies for synthetic fabrics. *Sci. Revs. Chem. Commun.*, 6, 4, 2016.

8. Han, S.W., Kim, H., G. *et al.*, Superhydrophobic Fabric Resistant to Aqueous Surfactant Solution as Well as Pure Water for Selective Removal of Spill-Oil. *ACS Appl. Nano Mater.*, 1, 9, 2018.
9. Chieng, B.W., Ibrahim, N.A. *et al.*, *Synthesis, Technology and Applications of Carbon Nanomaterials*, pp. 177–203, Elsevier Inc, Nederland, 2019.
10. Joshi, M., Bhattacharyya, A. *et al.*, Nanostructured coatings for super hydrophobic textiles. *Bull. Mater. Sci.*, 35, 6, 2012.
11. Park, S., Kim, J., Park, C.H., Superhydrophobic Textiles: Review of Theoretical Definitions, Fabrication, and Functional Evaluation. *J. Eng. Fibers. Fabrics*, 10, 4, 2015.
12. Barthlott, W. and Neinhuis, C., Purity of the sacred lotus, or escape from contamination in biological surfaces. *Planta.*, 202, 1997.
13. Liu, Y., Chen, X., Xin, J.H., Hydrophobic duck feathers and their simulation on textile substrates for water repellent treatment. *Bioinsp. Biomim.*, 3, 4, 2008.
14. Onar, N. and Mete, G., Development of water repellent cotton fabric with application of ZnO, Al_2O_3, TiO_2, and ZrO_2 nanoparticles modified with ormosils. *Tekstil ve Konfeksiyon.*, 26, 3, 2016.
15. Ariharasudhan, S. and Dhurai, B., Development of hydrophobic nonwoven fabric for oil spill and medical applications. *Int. Res. J. Pharm.*, 8, 8, 2017.
16. Sasaki, K., Tenjimbayashi, M. *et al.*, Asymmetric Superhydrophobic/Superhydrophilic Cotton Fabrics Designed by Spraying Polymer and Nanoparticles. *ACS Appl. Mater. Interfaces*, 8, 1, 2016.
17. Mazzon, G. and Zahid, M., Hydrophobic treatment of woven cotton fabrics with polyurethane modified amino silicone emulsions. *Appl. Surface. Sci.*, 490, 2019.
18. Ivanova, N.A. and Zaretskaya, A.K., Simple treatment of cotton textile to impart high water repellent properties. *Appl. Surface. Sci.*, 257, 5, 2010.
19. Espanhol-Soares, M., Costa, L. *et al.*, Super-hydrophobic coatings on cotton fabrics using sol–gel technique by spray. *J. Sol-Gel. Sci. Techno.*, 95, 2020.
20. Melki, S., Biguenet, F., Dupuis, D., Hydrophobic properties of textile materials: Robustness of hydrophobicity. *J. Tex. Ins.*, 110, 8, 2019.
21. Barashkov, N., Sakhno, T. *et al.*, Water-resistant fluorine-free technologies for synthetic fabrics. *Sci. Revs. Chem. Commun.*, 6, 4, 2016.
22. Lin, H., Hu, Q. *et al.*, Highly Hydrophobic Cotton Fabrics Modified by Poly (methyl hydrogen) siloxane and Fluorinated Olefin: Characterization and Applications. *Polymers*, 12, 833, 2020.
23. Cheng, Q.Y., Liu, M.C. *et al.*, Biobased super-hydrophobic coating on cotton fabric fabricated by spray-coating for efficient oil/water separation. *Poly. Test.*, 66, 2018.
24. Liang, L., Arias, M.J.L. *et al.*, Preparation of hydrophobic fabrics and effect of fluorine monomers on surface properties. *J. Eng. Fibers. Fabrics*, 14, 2019.

25. Tissera, N.D., Wijesena, R.N. *et al.*, Hydrophobic cotton textile surfaces using an amphiphilic grapheneoxide (GO) coating. *Appl. Surface. Sci.*, 324, 2015.
26. El-Bisi, M.K., Ibrahim, H.M. *et al.*, Super hydrophobic cotton fabrics via green techniques. *Der. Pharma. Chemica.*, 8, 19, 2016.
27. Gu, S.H., Yang, L. *et al.*, Fabrication of hydrophobic cotton fabrics inspired by polyphenol chemistry. *Cellulose*, 24, 2016.
28. Weng, R., Zhang, H., Liu, X., Spray-coating process in preparing PTFE-PPS composite super-hydrophobic coating. *AIP Adv.*, 4, 3, 2014.
29. Ji, Y.Y., Hong., Y.C. *et al.*, Formation of super-hydrophobic and water-repellency surface with hexamethyldisiloxane (HMDSO) coating on polyethyleneteraphtalate fiber by atmospheric pressure plasma polymerization. *Surface. Coat. Techno.*, 202, 22, 2008.
30. Zille, A., Oliveira, F.R., Souto, A.P., Plasma Treatment in Textile Industry. *Plasma Process. Polym.*, 12, 2, 2015.
31. Coulson, S., *Plasma Technologies for Textiles*, pp. 183–201, Woodhead Publishing, Netherlands, 2007.
32. Jayadev, D., Jayan, J.S. *et al.*, *Superhydrophobic Polymer Coatings*, pp. 91–121, Elsevier, Nederland, 2019.
33. Xue, C.H., H., Jia, S.T. *et al.*, Superhydrophobic cotton fabrics prepared by sol–gel coating of TiO_2 and surface hydrophobization. *Sci. Technol. Adv. Mater.*, 9, 3, 2008.
34. Chen, C.H., Wang, L., Huang, Y., Crosslinking of the electrospun polyethylene glycol/cellulose acetate composite fibers as shape-stabilized phase change materials. *Mater. Lett.*, 63, 5, 2009.
35. Du, J., Zhang, L. *et al.*, Preparation of Hydrophobic Nylon Fabric. *J. Eng. Fibers. Fabrics*, 11, 1, 2016.

7

UV Protection: Historical Perspectives and State-of-the-Art Achievements

Narcisa Vrinceanu* **and Diana Coman**

Faculty of Engineering, Department of Industrial Machines and Equipments, "Lucian Blaga" University of Sibiu, Sibiu, Romania

Abstract

The chapter presents a general review of up-to-date achievements in the UV absorbers design and fabrication for the most eco-friendly textile finishing. Nowadays, progressive and severe effects of UV radiation on human skin induced by the reduction of the ozone in the earth's atmosphere have been recorded. Consequently, lifelong vulnerability to UV irradiation leads to a list of severe skin health aftermaths: the accelerated aging, photodermatitis, skin reddening, and even malignancy. The progress of polymeric supports with UV barrier was extensively studied up to now. This historical perspective gathered in a new manner/concept the synergic effect UV radiation–polymeric surfaces by stressing the novel studies on the topic. The connection bounded by UV light and attributes of polymeric substrates, like: structural, physical, and chemical was discussed in this overview. In addition, both standard and innovative chemical substances and methodologies of UV performance for polymeric like UV shelters, nanoparticles, and layer-by-layer self-assembly (LbL) were approached in this meta-data perspective.

Keywords: UV protection, polymers, UV shields, nanostructures, UV stabilizers, dyes, UPF

7.1 Introduction

Today, the preserving of the human derma against UV light represents an extremely severe issue. As a consequence, the lower the density of

**Corresponding author*: vrinceanu.narcisai@ulbsibiu.ro

Luqman Jameel Rather, Mohd Shabbir and Aminoddin Haji (eds.) Sustainable Practices in the Textile Industry, (167–206) © 2021 Scrivener Publishing LLC

ozone coat, the more UV radiation reaches the surface of the earth. The key-concept of electromagnetic emission can be outlined by the relevant wave and the quantum theory [1, 2]. As indicated by quantum hypothesis, higher recurrence ray has an increased vitality and a diminished wavelength. Ultraviolet radiation is defined as light with energetically high and short wavelength. The world's climate is the interface between the sunlight and our bodies. Its range reaches out from 300 to 3200 nm. The radiation emitting somewhere between 280 and 420 nm is alluded to as bright light [3, 4]. There are three types of ultraviolet radiation (UVR): UV-A (320–400 nm), UV-B (280–325 nm), and UV-C (120–300 nm). The UV-C domain is consumed by the ozone coating, notwithstanding, the other two ranges: UV-A and UV-B arriving at the earth surface and causing genuine medical issues, for example, skin malignant growth, burn from the sun, and photo-maturing [5]. A and B ranges of UV beam destroy skin collagen, inducing the formation of free radicals, repress derma recovery components. The results are: spots, wrinkles and, above all, derma malignancy [6]. When UVB reaches the derma, it influences its external layer, with sun burns. This sort of sun rays is not allowed to pass the glass, being consumed by the glass and doesn't go into the room. Another sort of UVR, for example UVA, infiltrates into more profound regions of the skin and can make more serious harm the skin, and as it goes through the glass. Area C of the ultraviolet rays does not pass the atmosphere and does not reach the Earth [7].

Notwithstanding, the case of the exhaustion and diminishing of ozone coat in expanding areas, increasingly added both UVB and UVC light to the outward of the Earth. Hence, UV protecting of all of UV regions, meaning UVA, UVB and UVC turn out similarly significant [8, 9]. Past investigates illustrated amazing UV protecting impact of short-wave UV (UVC) from TiO_2 [10]. Earth is reliably illuminated by sun rays: 56% from them represents infrared (IR) radiation (760–3,000 nm wavelength), 39% noticeable light (VIS, 420–880 nm wavelength), and 5% bright (UV) radiation (240–450 nm wavelength). Bright range is reported to be isolated in a few areas. Of essential intrigue are UV areas: UVC (200–280 nm), UVB (280–315 nm), and UVA (315–400 nm). The measure of sunlight based UVB and UVA arriving at the world's surface relies upon scope, elevation, season, time, climate (shadiness), and ozone thickness. Averaged-frequency UVB light is naturally dynamic. Contrasted with UVB radiation domain, UVA rays can penetrate further the derma regardless of whether the subject is placed in the shadow [11]. There are countries like, Southern, New Zealand, and Australia wherein the absolute most significant levels of sun-oriented UV radiation on the planet [12] have been recorded.

7.2 Fundamentals Regarding UV Protection of Textile Fabrics

Textile supports are regarded as a main layer between human skin and the environment. An outstanding UV shielded polymeric support means UPF between 40 and 50+. For example, according to these criteria, a textile support fulfilling a UPF of 10 is considered to have a weak screen for outdoor users [13, 14]. There is a strong correlation between the level of UVR depletion and the fiber composition, moisture content, dyes and pigments class and concentration, whitening agents, UV-absorbing surface treatments of textiles. These parameters should be quantified in order to assess if a textile support is able to partially block solar radiation. This conclusion should not be drawn neither from the first/arbitrary evaluation nor determined from intrinsic depictions of the support's chemical structure [15]. There are numerous researches reporting the enhancement of UV and comfort performance of textile materials by using dyes, whitening agents or organic UV absorbers, according to the environmental regulations. An outline mechanism is proposed for the UV protection, explained by the surface compatibly and the interfacial bonding occurred in the matrix. For these reasons, using of anti-UV materials is of great importance to resolve the above problem issue from sunlight. Based on the Australian/New Zealand standard classification [16], the higher UPF value causes the greater protection grade (Figure 7.1) [17].

Direct and diffuse ultraviolet transference into a polymeric surface can be the urgent element deciding its UV performance [18, 19]. The UV light transmitted by a textile comprises of the micro-waves passing unaltered the pores of the fabric [20]. The ultraviolet protection factor is really the proportion of UV beam (UVA and UVB) impeded by the textile support. There is a direct relationship between a higher UPF value and a higher

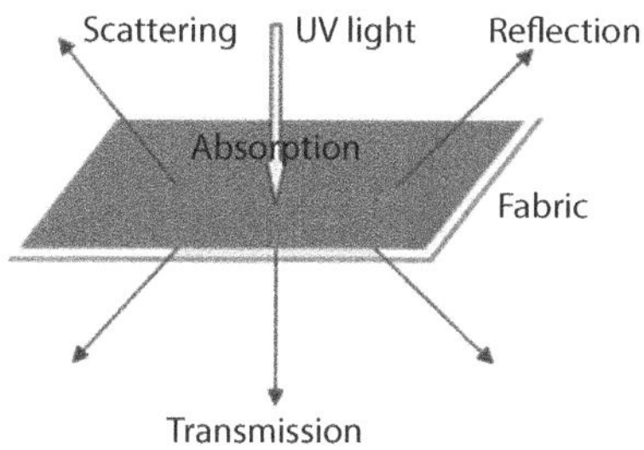

Figure 7.1 UV-transmittance behavior of a textile material.

amount of blocked ultraviolet light [15]. Shielded garments ought to possess as consistent UV beam both reflecting and potentially retaining characteristics as could be expected under the circumstances envisaging the prevention of ultraviolet light from reaching the skin and compromising the humans. The key attribute for evaluation the quality of clothing protection relies on its transmittance. The ultraviolet beam transmittance by passing textiles is characterized by means of a quantity of the sum of incidental ultraviolet light occurred within a specified wavelength limit to the total of transmitted UV incident rays arriving at the epiderma [21].

A few distinct impacts are present when there is a contact between the polymeric support and ultraviolet rays, making them split down into numerous segments. Some portion of the light is reflected at the limits of the material interface; the other one is swallowed when come into contact with the specimen. In other words, an alternative energy structure is identified in this situation. One more component of the light penetrates the polymeric support, thus touching the derma; this component is alluded to as the 'transmission' [22]. Photoprotection managed by textile supports against the harming impacts of solar UVR can be very extensive, despite the fact that the material materials are themselves vulnerable to restricted corruption brought about by UVR [23]. There are some intrinsic influences that severely affect the performance of textile materials to make a shield against UVR, like: fiber nature, construction support, like: porous structure, specific mass, specific area, density), shape and design, coloration, chemicals like colorants and ultraviolet radiation quenchers) and finishing processes [24, 25].

A solid relationship between UV insurance capacity and specific area and density or volume and diameter pores within the textures was accounted for [26–31]. Clothing can shield the skin from occurrence sun-oriented radiation on the grounds that the texture from which it is made can reflect, retain and dissipate sun powered wavelengths. The transmitted UV light of materials has two components: (i) a diffuse part, that is tailor-made by the support's absorption attributes, and (ii) the invariable segment, legitimately staging between the pores belonging to the texture. The decrease of UV light transmission should be accomplished by making some modification of the design parameters of polymeric supports. The ideal mixture/recipe of compactness, mass per surface unit, and density of woven, as well as yarn type (mono or multifilament) and diameter, permit creation of textiles with consistent ultraviolet performance [25]. Retaining, communicating, and optical properties of the textures are resolved basically by the design boundaries of materials [21, 32–34]. In design of woven texture with great UV radiation assurance, the accompanying models ought to be followed [15, 35].

(1) Structure of the filaments (most common strands communicate a higher quantity of ultraviolet light than engineered models).
(2) The weave tightness (in case that the woven texture is more firmly, then a lower quantity of UV radiation will be sent).
(3) Shading (dim shades of a similar textile support will retain UV radiation more unequivocally than light pastel shades, subsequently having bigger UV protection levels).
(4) Stretch (a more consistent stretch will induce a lower UPF).
(5) The quantity of intrinsic humidity (the wet polymeric structures will grant less UVR shield).
(6) Surface treatment (UV absorbers enhance UPF).

7.2.1 The Design of the Woven Support Represents a Relevant Factor That Directly Affect UPF

Woven design is adjusted by essential constructional parameters, in particular yarn fineness, weave type, and twist/weft thickness (Figure 7.2). Essential parameters of polymeric supports design are needy factors, where the decision of one boundary conducts to the impact of the others. In this manner, yarn fineness impacts polymeric supports densities over the weave type (cover factor, textile porosity, mass, thickness, and so forth) might be viewed as consistent and reliant on essential parameter.

Through the characterized determination of essential specifications of fabric structure/engineering, while the other fabric construction elements (weight, coating index, porous structure, thickness, and so on) might be viewed as steady and reliant on essential characteristics [15]. For outdoor

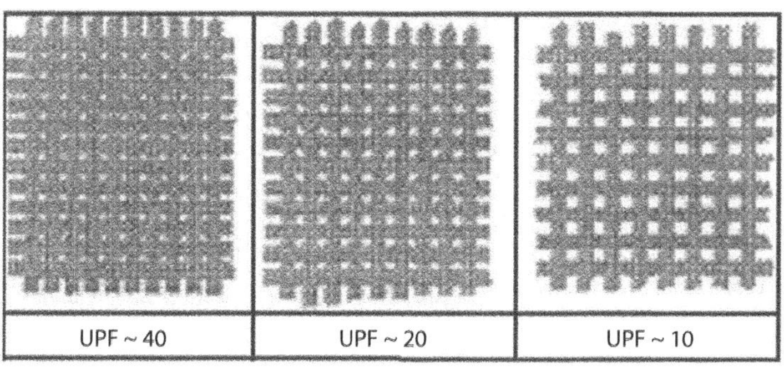

Figure 7.2 UPF for fabrics with different density [32].

textile supports, the bright (UV) light of daylight must be thought of, on the grounds that incessantly experienced UV radiation can cause skin burn from the sun with normal side effects, similar to erythema, vesicle, edema and even skin disease [36, 37].

7.2.2 The Synergism Between Structural Parameters and UV Protection of Textile Supports

It is known that fiber category can influence the transmission of ultraviolet radiation with contrasts in assimilation happening through fibrous polymers. For instance, there is a difference of ultraviolet radiation transmission between cotton and other common polymeric fibers. On the other hand, the polyester fibers absorb firmly in the UVB domain [38]. Huge conjugated aromatic polymer framework from fibers is responsible for the compelling in hindering UV radiation [18]. Water repellent fibers, like polyester possess a consistent defensive level in terms of transmittance of UVR [39]. This fact can be explained by the occurrence of benzene rings in the polymer chains, leading to an expanded absorption of UV radiation [40, 41]. With respect to cotton supports, it is similarly certain that regardless its design, greige (untreated) fabric manage the cost of unrivaled performance against UVR transmission [42].

7.2.3 Yarn Curve End up Being the Significant Determinant of the UV Security Attributes of Textile Supports

It ought to be expressed that UV shielding properties of knitting are additionally impacted by certain distinctions in line thickness brought about by yarn level introduced [43]. UPF depends on the weave's type. There is an obvious correlation between a high real twist/weft density by each weave and high texture snugness. Thus, a higher ultraviolet performance will be the result. Lower volume porosity means higher UV protection (Figure 7.2) [15].

7.2.4 The Correlation Between Fabric Porosity and Cover Factor and UV Protection

If the weave is closer, a lower quantity of ultraviolet light will be sent. In case of a UPF of 28 means that there is thorough contact between yarns and accordingly, basically, polymeric supports will allow next to zero UVR transmission [44]. The relationship between supports' weight and thickness and UPF qualities can be disclosed regarding porosity [45, 46]. Heavier texture limits UV transmission, hindering more radiation [44, 47]. Thicker supports with a larger spread index, accordingly mean a lower dynamics of UV radiation.

7.2.5 Concepts of Ultraviolet Protection Factor and Sun Protection Factor

Ultraviolet protection factor (UPF) is the scientific concept revealing the ultraviolet (UV) protection granted by the textile support to skin (Table 7.1). UPF is similar with sun protection factor (SPF). UPF is characterized as the proportion of the normal UV irradiance determined for unprotected skin to the normal UV irradiance determined for skin shielded by a textile specimen/sample (Figure 7.1) [42, 46, 48]. Both *in vivo* lab-scale integrated investigation and *in vitro* instrumental evaluation are the most common methods to test fabric to avoid sun burning. Consequently, SPF is the quantitative result of an *in vivo* assessment, and the outcome instrumentally obtained is UPF [49].

The measurement of UPF was standardized by AS/NZS 4399:1996 being actually recognized and validated by the global textile and garment industry. A broad range of skin pathologies like: sunburn, wrinkles, dryness, the inflammation of the face and sunspots can be avoided by using a valuable sunblock shield. More over the most important aspect is that the skin will be protected from any diseases due to the sun [52]. Actually, two main sunblock types are available:

a) *Organic (or chemical) sunscreen*, with the following attributes: absorption of UV radiation, conversion to heat and prevention it from destroying deep derma substrates, having a broad range of colorless and powder colors. These sunscreens have different structures, like: aminobenzoic acid, cinnamates, benzophenones, salicylates [7, 53].
b) *Non-organic (or physical) sunblock* contains mixtures of zinc oxide and titanium dioxide. UVR shielding is typically estimated with sun protection factor (SPF), the basic light important to limit erythema, contrasted to skin without sunblock [7, 54].

The wellbeing assessment of sun shielding items comprises of relevant strategies: the assessment of toxicology, the retention rate of the item after effective coating and assessment of the derma responses on subjects [7, 55].

Table 7.1 Protection series of some polymeric supports according to ASTMD6603 and AS/NZS 4399 standards [15, 50, 51].

Fabric	Listed ranking	UPF (ASTMD6603)	UPF (EN 13758-1)	Average UVA (%T)	Average UVB (%T)	Protection category
Nylon	30+	35.9	35.4	5.75	1.4	Good
Nylon	30+	35.8	35.6	5.78	2.3	Good
Nylon	30+	35	39	5.34	2.2	Good
Nylon	30+	39.4	39.5	6.26	2.14	Good
White fabric	50+	211.0	221.1	1.09	0.24	Excellent
Teal fabric	50+	67.7	78.7	2.86	0.89	Excellent
Child's Sun Hat	50+	111.5	111.7	1.6	0.68	Excellent
6-year old sun shirt swatch 1	30+	23.8	24.9	8.46	2.45	Good
6-year old sun shirt swatch 2	30+	20.6	22.8	8.7	3.87	Good
Uniform trousers (nylon	unrated	145.2	156.3	0.78	0.57	
Uniform shirt	unrated	24.5	26.6	6.23	3.24	
Uniform shirt doubled	unrated	>700	>700	0.48	0.08	

The constructive outcomes of ultraviolet radiation (UVR) on individuals are as per the following:

- the mimetic of vitamin D synthesis,
- the medication of skin pathology such as dermatitis [56].

The harmful repercussion of lengthened and reiterated sun exposure of the skin include: erythema, and chronic effects: derma photoaging, light induced disorders such as genetic, metabolic, idiopathic and allergic issues, defeat of the immune scheme, and development of derma malignancy [57]. It is mandatory to prevent the sun exposure, by using some photoprotective measures: restricting of the time risk to sunlight (peak ultraviolet radiation), using of photo shielded garments, sunglasses and sun shields [58]. There are numerous market items and processes forwarded to obtain a high leveled UPF textile supports with all types of fibres (cellulosic, wool, silk, and synthetics) [59, 60]. UV absorbers can have organic or inorganic colorless chemical structures with robust absorption within 290–360 nm [59]. Once they are entrapped inside the fibers, they will change the electronic energy into thermal energy, acting as scavengers and singlet oxygen killers. The UV absorber can be excited by the high-energy and shortwave UVR to a higher energy state, which afterwards is dissipated as longer wave radiation [60, 61]. Meanwhile, the isomerization transforms the UV absorber into a neutral absorbing sequence. UV absorbers occurred in sunscreen lotions physically hinder the ultraviolet light [59]. 2-ethyl hexyl-4-methoxy cinnamate which is the most extensively used UVB barrier, having a high refractive index (RI), grants a significant improvement to a duplicated refractive index of epiderma [60, 62]. A useful UV absorber should fulfill the following criteria: absorbance throughout the spectrum and UVR stability in order to deplete the consumed energy and to avert the photodegradation conducting to color loss [63].

Organic UV absorbers occur primarily as subordinated benzo-phenones, O-hydroxy phenyl triazines, O-hydroxy phenyl hydra-zines, hindered amines and benzoic acid esters [59, 64, 65] (Figure 7.3). The hydroxyl group is authoritative for consistent absorption in the near UV of benzophenone. These compounds are applied by coating, immersing and padding for promoting a deep UV protection [66]. Convenient sequences of UV absorbers and antioxidants can generate synergistic responses [67–75]. Despite the fact that benzophenone derivatives, benzotriazoles, and esters have levels of low energy, easy scattering and a reduced sublimation resistance they are powerful [76, 77]. That is why, additional studies were performed to enhance their functionality and boost their assessment [78–81]. On the

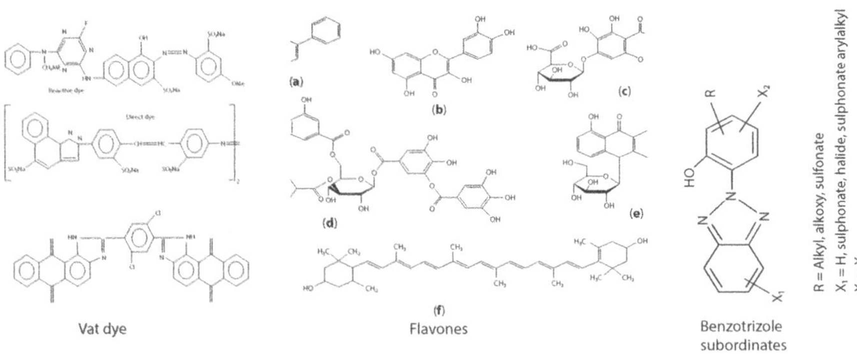

Figure 7.3 Chemical arrangements of UV shields for textile supports.

other side, thanks to their exceptional features, like: sublimation fastness, self-dispersing formulation, phenyl and triazine derivatives can be applied in high temperature dyeing and printing pastes [82]. There are also studies reporting light resistance of different colors of dyed fibers obtained from UV absorbers incorporated into the spinning dope and dye baths [76, 83]. The minimum percentage of UV absorbers should be 0.6–2.5% to grant UVR functionality of textile supports [44].

The addition with organic substance creates innovation in terms of the dispersion of fillers and its strengthened barrier effect properties [77, 84]. Moreover, the entrapping of transition metals into polymer matrices as UV shield drawn interest of researchers [78, 84]. In case of *natural fibers,* the UV absorbers are formulated onto textile support by different functionalization approaches: in (i) *mixtures with resins; (ii) by exhaustion or (iii) pad-batch.* Although practically speaking, these methods have complexity, are costly and time-consuming, some of these technologies facilitate a remarkable multifunctionality, meaning finishes with nano-oxides, still they are not available on the market [79, 80].

Natural origin cellulose (cotton, flax, hemp) and chemically modified cellulose (viscose, modal) are the most well-known fibers existent in the comfortable and thermal barrier garments, thanks to their outstanding hygienic features. Notwithstanding, these supports do not often grant an appropriate UV protection level [81]. This drawback can be diminished by engineered design variables and chemical coating approaches. There is a demand of natural life pattern, which moves the trend toward natural origin and eco-compatible goods. That is why the research studies are employing natural dyes for cellulose supports. These dyes present a minimum hazard of allergy, toxicity, being eco-friendly [82]. The increase of UV shield is mainly based on the UVR absorption features of natural dyes [83]. Diverse natural extracts such as eucalyptus leaves and tea derivatives were employed in the process of coloration, thus enhancing the UV protection of cellulose supports [14, 85]. Most of the clothing components available on the market are suitable with the dyes and various finishing chemicals are administered for the polymeric supports, by simple padding, exhaust method [59].

Inorganic UV absorbers, like titanium and zinc nano-oxides are advanced due to their exceptional properties: without toxicity, no harmfulness, chemical stability under high temperature and prolonged UVR exposure [17, 86]. These agents can be introduced during the process of spinning, inside the synthetic fibers [26, 87]. Up to now, *materials oxides (MOx)* and their included hydroxyl groups showed that UV radiation can be refracted through decomposition reactions [88] and then induced an enhancement of the UV protection [89]. On the other hand, the incorporation of *MOx* was used as elements for UV protection. *Materials oxides* have received considerable attention in several environmental applications, as it has many interesting properties that combined the mechanical, electrical, optical and biological properties.

Waste deriving from agricultural processes was successfully utilized in order to improve the UV functionality of *cotton* supports [90, 91]. A lot of interest has been concentrated onto the UV performance of cellulose based polymeric supports, both natural and regenerated cellulose, highlighting the importance of hemp fibers for the obtaining of UV protective textiles. The expanded UVR transparency of the unadulterated hemp textile support, which came about because of hemp flexibility constraints, defeated by mixing with milder and cellulose fibers (cotton, viscose) that have more elasticity [3, 92].

If polyester fibers are exposed to sunlight and UV irradiation for lengthened duration, their mechanical properties can be altered, by breaking of molecular bonds, since the energy per photon of UV radiation is higher than the energy of carbon–carbon single bonds [93, 94]. The alteration means losing of their practical value by transforming from synthetics into typical brittle materials. That is why, an effective UV shielding treatment for polyester fibers is compulsory. The UV resistance of polyesters can be enhanced by adding UV absorbers. Based on its specific attributes, like: higher molar extinction value, neutral melting point, good affinity with polymers, and low volatility, benzotriazole (BT) has hidden functions in sun shielding, corrosion blockage, and UV absorbers [85, 95–97]. The mechanism of polymer degradation is explained by the absorbing the UV light and dissipating the energy [99]. The first approach to form an amide structure was the condensation reaction between formaldehyde and benzotriazole, followed by coupling of benzotriazole with diethanolamine [99]. Then, an improved UV resistance was attained by casting hindered amine light stabilizers (HALS) [98, 100]. Nevertheless, due to their lack of reactive groups a low stability of UV resistance was noticed. Subsequently, a reactive UV absorber *2-(2-hydroxyphenyl)-2H benzotriazole* was introduced, in order to develop more stable structures with improved light fastness and UV resistance [101–107]. Thus, benzotriazole compounds were used to cover *(PET) polyethylene terephthalate,* exhibiting outstanding UV performance. There are studies reporting the synthesis of polymerizable ultraviolet-absorbing nanoparticles (NPs) with high mechanical attributes and UV blocking performance, while another study was focused onto formulated polymerizable acryloyl chloride-based UV stabilizers with enhanced UV protection [108, 109]. On the other hand, despite the interesting properties of *polyester nonwovens,* some characteristics like the inherently hydrophilic property, low mechanical resistance and insufficient sensitivity to the UV light, confine their wide applications, especially in some high-ends areas for medicine, catalysis, personal healthcare and self-cleaning [110–113].

7.3 UV Stabilizers Beginnings and Initial Development

7.3.1 UV Protection Finishing of Fabrics Using Nanoparticles

Due to attractive selectivity, cost and eco friendliness, inorganic components are regarded as novel UV absorber. The drawbacks of *conventional textile finishing methods* consist of non-permanent effects, dropping their functionality after washing or handle. Thanks to their unique large surface area and high surface energy, *nanoparticles* grant large endurance for covered supports, pointing to novel functionality [72, 114]. Thus, nano materials are acknowledged as nano fillers with huge effect onto UPF, by virtue of a higher contact point between of filling agents and PP matrices (Figure 7.4) [115]. The UV shielded nano fillers are divided in two main groups:

7.3.1.1 Inorganic Formulations With Nano-ZnO Particles

There are reports stressing the enhanced UV blocking for cotton supports by: sol–gel, wet chemical, and micro emulsion with long-term retention of ZnO nanoparticles, *dip-coating routes* especially within UVB range (280–315 nm), even after five laundering cycles [116–126]. *Some studies reported functionalization of surface,* in order to develop esterification links between nanoparticles and textile surface: mini emulsion polymerization, cold

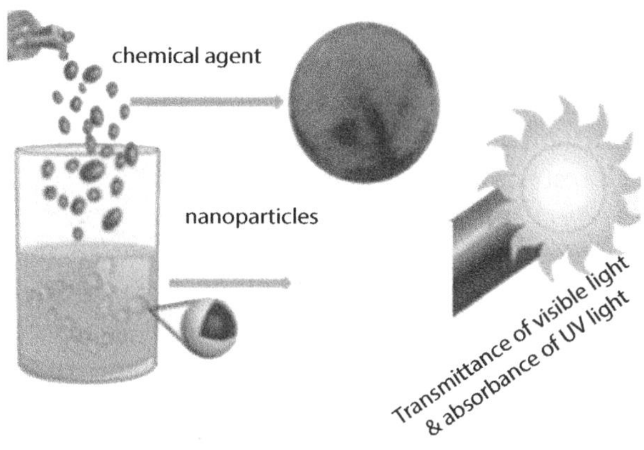

Figure 7.4 Synthesis of nanoparticles for UV shielded textiles.

plasma, micro emulsion technique and pre-treatment of cotton supports with organic hydro peroxides solution, urea (CON_2H_4), or functional coatings based on polylactic acid (PLA), created premises of ZnO nano-particles synthesis directly inside the mesopores of cotton fibers [127–137], which introduced a new non-conventional electro-fluid-dynamic technology to realize UV protection. There were identified cases when, for UVA range, ZnO shown higher UV barrier effect than TiO_2 [138].

7.3.1.2 UV Shield of Cotton Support Conferred by TiO_2 Nanoparticles

Nanosized TiO_2 finishes provide UV-protection performance to cellulosic supports due to: (1) diverging and absorbing effect of UV radiation, (2) increasing roughness deriving from surface micro/-nanoscale structures [79, 96, 139–148] and (3) high specific surface area and energy [86, 149, 150]. Abidi *et al.* identified the optimum finishing with nano-TiO_2 sols to obtain cotton fabrics with good UV resistance in the range UVB (290–315 nm), with excellent stability to home laundering [13]. *Nano-TiO_2 particles* attract interest in UV shielding due to their potential to react with different polymers like: polyethylene, polypropylene, polystyrene, polyurethane thanks to their high photocatalytic performance, stability, nontoxicity, eco compatibility and low costs [151–153]. There are known various techniques to coat *nano-TiO_2* nanoparticles onto textile supports: sol–gel, linking agent technique, thin silane films onto TiO_2 particles in polar solvents [154] octadecyltrihydrosilane on TiO_2 surfaces by ultrasonic irradiation [155–157] of siloxanes and TiO_2 in order to obtain for the first time an innovative type of UV-screening, super water repellent and powerful cotton supports [158]. In case of *pristine cotton fabrics* TiO_2 nanoparticles have some *drawbacks* such as: decreased mechanical strengths, low stain repellency, consistent flammability and weak UV-blocking ability [136, 159, 160] which restrict their extensive uses.

A *synergic effect of anatase (TiO_2) sol and polydimethylsiloxane (PDMS)* by a simple finishing method [161], reported the preparation of robust fluorine-free self-cleaning cotton textiles with high UV shield. *Aramid/Nanofibrillated cellulose (NFC)/nano-TiO_2 composite* with good UV-resistant property was possible through the ability of well dispersed TiO_2 nanoparticles [162].

7.3.1.3 Formulations Containing Nanoparticles of ZnO, Titania, Silica, Silver, Carbon-Nanotubes, Graphene and Silver Onto Cotton Textiles

In order to develop UV protective shields, facile pad-dry-cure method with ZnO, TiO_2, SiO_2, or wet chemical technique was used [163]. Recently, nano metal oxides, like TiO_2 [164], ZnO [165] and SiO_2 [166] play a magnificent role in the UV shielding and engage a lot of research interest, due to their chemical stability under UV-radiation exposure and high temperature and non-toxicity [67, 167, 203]. Many nanomaterials include *silica* [169, 170], *titania* [171], *zinc-oxide* [168, 172, 173], *silver* [174], *carbon nanotubes* [175] and *graphene* [176]. By using sol–gel methods, the modification of cotton fabric with graphene, novel UV shielded cotton fabrics were introduced [150]. *Layer-by-layer self-assembly* enticed a high UV level for cationized component of cotton of:

- fluorescent brightening agents: poly(diallyldimethylammonium-chloride) (PDDAs) [33, 38, 177–181];
- different anionic polymers with TiO_2 nanoparticles [80, 182].

The number of layers conducts to a larger UV level of protection. For instance, a film consisting of 16 multilayers of nano polyurethane/anatase (TiO_2) and TiO_2/PDDA (poly dimethyl diallyl ammonium chloride) covering the cotton fabrics, created outstanding UV protection values [183, 184]. However, there is a transposed ratio between the particles size decrease to the nanoscale and the UPF augmentation [185]. Recently, to retain UV-protective materials, a seeded sol–gel process was employed to create *silica inert shells onto TiO_2 particles* [186] or silane [187, 188], with some drawbacks in the formulation: TiO_2 colloids preparation prior to the silica coating process; excess use of organic solvent; materials will not have high transparency [187, 188].

A *combination of TiO_2 nanoparticles and natural dyes*, like Henna extract, leads to durable UV protected with cationized cotton fabrics were obtained [186–188], thanks both to the contents of TiO_2 and the dyeing temperature. Nanocomposite polyacrylonitrile *(PAN) films containing silver nanoparticles (Ag NPs)* were developed to introduce multifunctional film matrix with ultraviolet shielding. The ultraviolet blocking of films was evaluated by studying Ultraviolet Protection Factor (UPF) explained in AS/NZS4399:1996 and employing UPF computation system of UV–vis spectrophotometer under AATCC 183:2010 UVA Transmittance [139]. There are studies focused onto UV properties obtained by *embeding method of silver nanoparticles (Ag NPs) into the wool matrix* using sustainable cleaner one-pot route [189]. New enhanced UV protected functionalized *natural and synthetic fabrics including linen, silk, nylon and PET* covered with *silver nanoparticles (AgNPs)* [190, 203].

7.3.2 UV Protection of Fabrics by Dyeing of Textile Supports

Due to peculiar transmission and absorption attributes, distinct hue belonging to dyes can range markedly in terms of UV protection [14]. In other words, dyed supports will have more acceptable sun shield than bleached ones. In this way the skin will be protected, since the UV quencher will not reach it [46, 181, 191, 192]. Moreover, the UPF of unbleached cotton is more valuable than bleached cotton, probably because a low percentage of pigments remain in unbleached cotton. On the contrary, in terms of UVR transmission, there are studies reporting a double value of bleached conventional cotton than the unbleached conventional cotton [33, 38]. Making a comparison between the naturally pigmented cottons and conventional cotton (bleached or unbleached) there are undoubtedly higher UPF values of the first ones [47, 192]. There are two different facets of natural dye applications: textile dyeing and UV covering garments [193]. Increased UV protection fabrics were reported by *triazine based UV absorbers*, applied to woven cotton fabric via the *exhaust method of dyeing* [194].

By substituting the traditional mordanting with *tannin-rich natural dye* extracts [195], from *Xylocarpus granatum* (*Euphausia superba*), the dyeing of cotton imparted UV protection properties to polymeric supports. Developed textiles with shielded attributes in terms of ultraviolet light were obtained by coloring the textiles with fruit aqueous extract [196]. New UV protected *wool supports* by using eco-friendly metal mordants and then dyed with carotenoid compounds of *marigold aqueous extract* [197]. Multifunctional textiles with improved UV protections by enzymatic phosphorylation were realized by grafting of NDGA on *wool* using *laccase* [198]. There is a synergy between the pre-cure phase with natural agents, meaning mordants that induce the betterment of the dye exhaustion without any environmental hazards and the compactness of the fabric structure (cotton, flax, hemp, ramie), intensifying the UV protection ability of *cellulose* fabrics [14, 199, 200], with some drawbacks: feasibility, cost efficiency, low adherence and color fastness [201]. Many studies reported good results of UV protection, by dyeing *woolens* with *natural dyes:* madder and weld, by two methods: (i) using the mordants (CuSO4 and FeSO4); plasma sputtering of wool fabrics by Cu and Fe to increase the UPF value [202]. The novelty of the research consists in the cumulative effect of the natural dye and mordant pre-treatment in boosting the UV protection features of wool fabrics [202].

7.3.3 Other Kind of Finishes

The chemical UV shield surface treatment of textile fabrics induced sustainability facets, like: increased cost, environmental deterioration and enormous water consuming [204]. Additionally, the shortcoming relies in the unsteady washing resistance of the standard UV quenchers [13, 81]. Table 7.2 gives an updated list of UV textile finishing agents. There are reports regarding the affinity between fabric processing and ultraviolet radiation transmission, like:

1. CS(LMW)-PU finishes on *poly-cotton blended fabrics*, underlined that the surface treatment with chitosan extraordinarily increases the UV performance of PU [205].
2. *PBO/m-POSS nanocomposite fibers* with enhanced UV attributes [206].
3. UV-protective *silk fibers* obtained by functionalization with *plasma* [207], or with bioactive structure with plant origin, like *Scutellaria baicalensis* root [208].

7.4 Conclusion

This historical approach summarized the synergy between ultraviolet light and polymeric supports. The ultraviolet shield of the textiles severely changed according to some factors, like: the fabric design (thickness,

Table 7.2 Updated table with the most applied textile UV textile finishing.

Incorporated substance	**Textile material**	**Finishing method**	**UVF**	**References**
SiO_2, ZnO, TiO_2	Poly (lactic acid) nonwovens	Electrospinning	good	[209–212]
Melanin/silica core-shell particles	Cotton	Stöber method		[213]
Silicon/shape memory polyurethane (SiO_2/SMPU) hybrid sols	Aramid fibers	Sol–gel		[214]
Lavender fragrance oil-loaded cellulose/silica hybrid microcapsules	Cellulosic fabrics	One-step synthesized via emulsion solvent diffusion	159	[161, 215]
	Microcapsules with poly (urea-urethane			[216]
(MOx): TiO_2, ZnO, SiO_2	Composite-based polyethylene terephthalate (PET) fibers		80.5–113.4	[217]
TiO_2	Different polymeric supports			[218]

(*Continued*)

Table 7.2 Updated table with the most applied textile UV textile finishing. (*Continued*)

Incorporated substance	Textile material	Finishing method	UVF	References
Fe_2O_3 and TiO_2	Renewable cellulose fiber membrane			[219, 220]
Nano-TiO_2 particles assisted by nanofibrillated cellulose;	Aramid fibers	Coating		[162]
Anatase TiO_2 sol and polydimethylsiloxane (PDMS)	Cotton	Dip-coating		[161]
ZnO nanoparticles; Flower-like ZnO Nanorod; hexagonally Oriented ZnO nanorod; needled-shaped ZnO; ZnO nanoparticles—high concentration; ZnO crystallites; chitosan–ZnO nanoparticles; ZnO in chitosan nanorods by an ultrasound-assisted method	Poly (butylene succinate-co-butylene adipate) matrix; Functionalized cotton; cotton; Cellulose Nanofibers; paper (cellulose) matrix; cotton; Cellulose fibers	Cost-effective dip-coating; electrospinning and the low-temperature hydrothermal growth technique; low-temperature solvothermal and Hydrothermal method; electrospinning; microwave assisted route; microwave assisted precipitation and crystallization process; Precipitation Hydrothermal; Wet chemical method; Plasma treatment; Pad–dry–cure;	50+	[221–232] [124]

(*Continued*)

Table 7.2 Updated table with the most applied textile UV textile finishing. (*Continued*)

Incorporated substance	Textile material	Finishing method	UVF	References
Dyes from natural extracts; phthalocyanine and anthocyanins derivatives	Silk	Coating; low temperature DC glow discharge air Plasma.	36.45	[207, 232]
	Cotton	Dyeing		[195]
	Cotton	Coating		[233, 234]
	Wool fabric	Dyeing assisted by mordants: $CuSO_4$ and $FeSO_4$ Plasma sputtering		[202]
	Different natural fibers (wool, cotton and silk), and synthetic fibers (polyamide, acrylic)	Dyeing with mordants	50+	[235]
Alginate copper ions	Viscose	Functionalization	261	[16, 236, 237]
Benzotriazole ultraviolet absorbers	Polyester Fabrics	Grafting	223.74	[237]

(*Continued*)

Table 7.2 Updated table with the most applied textile UV textile finishing. (*Continued*)

Incorporated substance	Textile material	Finishing method	UVF	References
Formulations containing nanoparticles of ZnO/TiO_2	Polypropylene (PP) fibers	Ultrasonic-microwave homogeneous precipitation		[238]
	Cotton	Pad-dry-curing		[239]
Polyvinylsilsesquioxane (PVS) and nano-TiO_2	Cotton	Coating	121.5	[148]
Heterofunctional triazine	Cotton	Exhaust Dyeing	50+	[240]
Ag NPs	Natural and synthetic fabrics: woolens, linen, silk, nylon and PET	Green cleaner one-pot route; exhaustion application technique	30.4–66.1	[190, 241]
Low molecular weight chitosan (CS(LMW)) extended with polyurethane	Cotton fabrics	Finishing	18.7	[218]
Chitosan/poly (vinyl alcohol)/	Graphene Oxide Fibers	Wet-spinning route	500	[242]

(*Continued*)

Table 7.2 Updated table with the most applied textile UV textile finishing. *(Continued)*

Incorporated substance	Textile material	Finishing method	UVF	References
Modified Polyhedral Oligomeric Silsesquioxane	Poly (p-phenylene benzobisoxazol) Nanocomposite Fiber	Dry-jet wet spinning		[243]
Graphene Oxide plate	Cotton fibres	Functionalization	356.74	[244]
Naphthalene moiety	Poly (p-phenylene benzo-bisoxazole Fibers	Dry-jet wet spinning		[245, 246]

weave type, fiber diameter, density, coating factor, porous structure, specific area, dyes/pigments as well as finishing process. Standard and non-conventional UV absorbers were analyzed in order to make a barrier for textile against UV radiation. There are some restrictions nonetheless, recent studies highlighted in terms of chemical substances, poor selectivity, and unsteady washing fastness. Recently, some reports overstressed the idea of engineering of textile surfaces for UV protection. The most relevant UV stabilizers, like: natural and synthetic cellulose-based agents, natural and regenerated cellulose fibres, nanostructures like: TiO_2, ZnO and other metal nano-oxides, natural extracts as dyes and pigments were presented.

References

1. Li, S., Hu, B., Zhang, F., Polyvinylidenefluoride (PVDF) Based Membranes for Waterproofing Applications. *J. Nanoelectron. Optoelectron.*, 12, 1056, 2017.
2. McDonald, R. (Ed.), *Colour physics for industry*, pp. 1–8, Society of Dyers and Colourists, Bradford, England, 1987.
3. Lee, S., Developing UV-protective textiles based on electrospun zinc oxide nanocomposite fibers. *Fibers Polym.*, 10, 295–301, 2009. https://doi.org/10.1007/s12221-009-0295-2.
4. Zohdy, M., El-Naggar, Fathalla, Alia, Novel UV-protective formulations for cotton, PET fabrics and their blend utilizing irradiation technique. *Eur. Polym. J.*, 45, 10, 2926–2934, 2009. http://doi.org/10.1016/j.eurpolymj.2009.06.018.
5. Alebeid, O., Tao, Z., Seeahmed, A., New Approach for Dyeing and UV Protection Properties of Cotton Fabric Using Natural Dye Extracted from Henna Leaves. *Fibres Text. East Eur.*, 23, 60–65, 2015. http://doi.org/10.5604/12303666.1161758.
6. Ahmadi, F. and Jorjani, M., *Sunblock comparison SPF= 30 and SPF= 100 in advance of sunburn*, Pajoohandeh, Tehran, Iran, 2002.
7. Vosoughi Torbati, T. and Javanbakht, V., Fabrication of TiO_2/Zn2TiO4/Ag nanocomposite for synergic effects of UV radiation protection and antibacterial activity in sunscreen. *Colloid. Surf. B*, 187, 110652, 2020. https://doi.org/10.1016/j.colsurfb.2019.110652.
8. Matsumi, Y. and Kawasaki, M., Photolysis of atmospheric ozone in the ultraviolet region. *Chem. Rev.*, 103, 12, 4767–4782, 2003.
9. Magnaldo, T. and Sarasin, A., *Xeroderma pigmentosum*: From symptoms and genetics to gene-based skin therapy. *Cells Tissues Organs*, 177, 3, 189–198, 2004.
10. Horovitz, I., Avisar, D., Baker, M.A., Grilli, R., Lozzi, L., Di Camillo, D., Mamane, H., Carbamazepine degradation using a N-doped TiO_2 coated

photocatalytic membrane reactor: Influence of physical parameters. *J. Hazard. Mater.*, 310, 98–107, 2016.
11. Kullavanijaya, P. and Lim, H.W., Photoprotection. *J. Am. Acad. Dermatol.*, 52, 6, 937–958, 2005. https://doi.org/10.1016/j.jaad.2004.07.063.
12. Laperre, J. and Gamblichler, T., Sun protection offered by fabrics: on the relation between effective doses based on different action spectra. *Photodermatol. Photoimmunol. Photomed.*, 19, 1, 11–16, 2003. http://doi.org/epdf/10.1034/j.1600-0781.2003.00014.x.
13. Abidi, N., Hequet, E., Tarimala, S., Dai, L.L., Cotton fabric surface modification for improved UV radiation protection using sol–gel process. *J. Appl. Polym. Sci.*, 104, 111–117, 2007.
14. Grifoni, D., Bacci, L., Zipoli, G., Garreras, G., Baronti, S., Sabatini, F., Laboratory and outdoor assessment of UV protection offered by flax and hemp fabrics dyed with natural dyes. *Photochem. Photobiol*, 85, 313–320, 2008. https://doi.org/10.1111/j.1751-1097.2008.00439.x.
15. Dubrovski, P.D. and Golob, D., Effects of woven fabric construction and color on ultraviolet protection. *Text. Res. J.*, 79, 351–359, 2009.
16. 223AATCC Test Method 183:2004.
17. Ibrahim, N.A., El-Zairy, E.M.R., Abdalla, W.A., Khalil, H.M., Combined UV-protecting and reactive printing of cellulosic/wool blends. *Carbohydr. Polym.*, 92, 1386, 2013.
18. Xin, J.H., Daoud, W.A., Kong, Y.Y., A new approach to UV-blocking treatment for cotton fabrics. *Text. Res. J.*, 74, 97–100, 2004.
19. Ibrahim, N.A., Eid, B.M., Sharaf, S.M., *Functional Finishes for Cotton-Based Textiles: Current Situation and Future Trends*, John Wiley & Sons, Inc. and Scrivener Publishing LLC USA, 2019.
20. Abidi, N., Hequet, E.F., Abdalah, G., Cotton fabric and UV-protection. *Proc. Beltwide Cotton Conf.*, 2, 1301–1303, 2001.
21. Gabrijelčič, H., Raša, U., Franci, S., Keste, D., Influence of fabric constructional parameters and thread colour on UV radiation protection. *Fibres Text. East Eur.*, 1, 46–54, 2009.
22. Hoffmann, K., Kaspar, K., Gambrichler, T., Altmeyer, P., Change in ultraviolet (UV) transmission following the application of vaseline to non-irradiated and UVB-exposed split skin. *Ama Acad. Dermatol.*, 143, 1009–1016, 2000.
23. Andrady, A.L., Torikai, A., Redhwi, H.H., Pandey, K.K., Gies, P., Consequences of stratospheric ozone depletion and climate change on the use of materials. *Photochem. Photobiol. Sci.*, 14, 170–184, 2015. https://doi.org/10.1039/c4pp90038c.
24. Wong, W., Lam, J., Kan, C., Postle, R., Ultraviolet protection of weft-knitted fabrics. *Textil. Prog.*, 48, 1, 1–54, 2016a. https://doi.org/10.1080/00405167.2015.1126952.
25. Dimitrovski, K., Sluga, F., Urbas, R., Evaluation of the structure of monofilament PET woven fabrics and their UV protection properties. *Text. Res. J.*, 11, 10–27, 2010.

26. Algaba, I. and Riva, A., *In vitro* measurement of the ultraviolet protection factor of apparel textiles. *Color. Technol.*, 118, 2, 52–58, 2002. https://doi.org/10.1111/j.1478-4408.2002.tb00137.x.
27. Majumdar, A., Kothari, V.K., Mondal, A.K., Hatua, P., Effect of weave, structural parameters and ultraviolet absorbers on *in vitro* protection factor of bleached cotton woven fabrics. *Photodermatol. Photoimmunol. Photomed.*, 28, 2, 58–67, 2012. https://doi.org/10.1111/j.1600-0781.2011.00638.x.
28. Kocic, A., Popovic, D., Stankovic, S., Poparic, G., Influence of yarn folding on UV protection properties of hemp knitted fabrics. *Hem. Ind.*, 70, 3, 319–327, 2016. https://doi.org/10.2298/HEMIND141126036K.
29. Kan, C.W., A study on ultraviolet protection of 100% cotton knitted fabrics: effect of fabric parameters. *Sci. World J.*, 2014, 2014. https://doi.org/10.1155/2014/506049.
30. Yu, Y., Hurren, C., Milington, K., Sun, L., Wang, X., UV protection performance of textiles affected by fiber cross-sectional shape. *Textil. Res. J.*, 85, 18, 1946–1960, 2015. https://doi.org/10.1177/0040517515578335.
31. Yu, Y., Hurren, C., Milington, K., Sun, L., Wang, X., Effects of fibre parameters on the ultraviolet protection of fibre assemblies. *J. Text. Inst.*, 107, 5, 614–624, 2016. https://doi.org/10.1080/00405000.2015.1054144.
32. Algaba, I., Riva, A., Crews, P.C., Influence of fiber type and fabric porosity on the UPF of summer fabrics. *AATCC Rev.*, 4, 26–31, 2004.
33. Pailthorpe, M., Textile parameters and sun protection factors, in: *Textiles and sun protection conference proceedings*, N. Pailthorpe (Ed.), pp. 32–53, Kensington, The Society of Dyers and Colourists of Australia and New Zealand (NSW Section, 1993.
34. Zimniewska, M. and Batog, J., *Handbook of natural fibres: Processing and applications, utraviolet-blocking properties of natural fibres*, vol. 2, pp. 141–167, Woodhead Publishing Series in Textiles, New Delhi, 2012.
35. Gies, P.H., Roy, C.R., Toomey, S., McLennan, A., Protection against solar ultraviolet radiation. *Mutat. Res.*, 422, 15–22, 1998.
36. Abadie, S., Bedos, P., Rouquette, J., A human skin model to evaluate the protective effect of compounds against UVA damage. *Int. J. Cosmet. Sci.*, 41, 594–603 2019, https://doi.org/10.1111/ics.12579.
37. Makino, E., Tan, P., Mehta, R., Protective and reparative effects of a topical dual serum system against UV-induced skin damage. *J. Ama Acad. Derm.*, 81, AB110, 2019.
38. Wilson, C.A., Bevin, N.K., Laing, R.M., Niven, B.E., Solar protection—Effect of selected fabric and use characteristics on UV transmission. *Text. Res. J.*, 78, 95–104, 2008.
39. Gorensek, M. and Sluga, F., Modifying the UV blocking effect of polyester fabric. *Text. Res. J.*, 74, 469–474, 2004.
40. Pan, N. and Sun, G., *Functional textiles for improved performance, protection and health*, Elsevier, Birmingham, 2011.

41. Radetić, M., Antibacterial effect of silver nanoparticles deposited on corona-treated polyester and polyamide fabrics. *Polym. Adv. Technol.*, 19, 1816–1821, 2008. http://doi: 10.1002/pat.1205.
42. Sarkar, A.K., On the relationship between fabric processing and ultraviolet radiation transmission. *Photochem. Photobiol. Photomed.*, 23, 191–196, 2007.
43. Stankovic, S.B., Popovic, D., Poparic, G.B., Ultraviolet protection factor of gray-state plain cotton knitted fabrics. *Text. Res. J.*, 79, 1034–1042, 2009.
44. Reinert, G., Fuso, F., Hilifiker, R., Schmidt, E., UV-protecting properties of textile fabrics and their improvement. *Text. Chem. Color.*, 29, 36–43, 1997.
45. Akgun, M., Becerir, B., Alply, H.R., Ultraviolet (UV) protection of textiles. *Acta Facultatis Medicae Naissensis Rev.*, 27, 301–311, 2010.
46. Crews, P.C., Kachman, S., Beyer, A.G., Influence on UVR transmission of undyed woven fabrics. *Text. Chem. Color.*, 31, 17–26, 1999.
47. Sarkar, A.K., An evaluation of UV protection imparted by cotton fabrics dyed with natural colorants. *BMC Dermatol.*, 4, 1–8, 2004.
48. Gambichler, T., Ultraviolet protection factor of fabrics: comparison of laboratory and field-based measurements. *Photodermatol. Photoimmunol. Photomed.*, 18, 3, 135–140, 2001. http://doi:10.1034/j.1600-0781.2001.00739.x.
49. Menter, J.M., Hatch, K.L., Elsner, P., Hatch, K., Wigger-Alberti, W. (Eds.), *Textiles and the skin*, vol. 31, pp. 50–63, Karger, Basel, 2003.
50. Dubrovski, P.D., Woven fabrics and ultraviolet protection, in: *Woven fabric engineering*, P.D. Durbrovski (Ed.), pp. 273–296, Sciyo, Rijeka, 2010.
51. Algaba, I., Pepió, M., Riva, A., Modelisation of the influence of structural parameters on the ultraviolet protection factor provided by cellulosic woven fabrics. *J. Text. Inst.*, 97, 349–357, 2006.
52. Ghalamkar, F., Amir Javanbacht, A., Toosi, P., *Compare sunscreen lotus and seagull in the prevention of sunburn*, vol. 2, pp. 227–232, Pajoohandeh, Tehran, Iran, 2001.
53. Nair, M.G., Nirmala, M., Rekha, K., Anukaliani, A., Structural, optical, photo catalytic and antibacterial activity of ZnO and Co doped ZnO nanoparticles. *Mater. Lett.*, 65, 1797–1800, 2011. https://doi.org/10.1016/j.colsurfb.2019.110652.
54. Food, D., Administration, Sunscreen drug products for over-the-counter human use. *Fed. Regist.*, 43, 38206–38269, 1978.
55. Nohynek, G.J., Antignac, E., Re, T., Toutain, H., Safety assessment of personal care products/cosmetics and their ingredients. *Toxicol. Appl. Pharmacol.*, 243, 239–259, 2010.
56. Parisi, V.A. and Wilson, A.C., Pre-vitamin D3 effective ultraviolet transmission through clothing during simulated wear. *Photodermatol. Photoimmunol. Photomed.*, 21, 6, 303–310, 2005.
57. Almahroos, M. and Kurban, A.K., Sun protection for children and adolescents. *Clin. Dermatol.*, 21, 311–314, 2003. https://doi.org/10.1016/S0738-081X(03)00044-0.

58. Gies, P., van Deventer, E., Green, C.A., Sinclair, C., Tinker, R., Review of the global solar UV index 2015 workshop report. *Health Phys.*, 114, 1, 84–90, 2018. https://doi.org/10.1097/HP.0000000000000742.
59. Saravanan, D., UV Protection textile material. *AUTEX Res. J.*, 7, 53–62, 2007.
60. Holme, I., UV absorbers for protection and performance. *International Dyer*, 13, 9–10, 2003.
61. Choudhury, A.K.R., *Advances in the finishing of silk fabrics*, pp. 81–110, Elsevier BV, Kolkata, 2015, http://doi:10.1016/b978-1-78242-311-9.00005-7.
62. Sunderland, M.R., *Non-insecticidal insect-proofing of wool*, Ph.D. Thesis, Lincoln University, Christchurch, New Zealand, 2012.
63. Hustvedt, D. and Crews, P., The ultraviolet protection factor of naturally pigmented cotton. *J. Cotton Sci.*, 9, 47–55, 2005.
64. Sekar, N., UV absorbers in textiles. *Colourage, 11*, 27–28, 2000.
65. Seungsin, L., Developing UV-protective textiles based on electrospun zinc oxide nanocomposite fibers. *Fiber Polym.*, 10, 295-301, 2009.
66. Rupp, J., Bohringer, A., Yonenaga, A., Hilden, J., Textiles for protection against harmful ultraviolet radiation. *Intern Text Bull.*, 6, 8–20, 2001.
67. Choudhury, A.K.R., *Finishes for protection against microbial, insect and UV radiation*, pp. 319–382, Elsevier BV, Serampore, India, 2017, http://doi:10.1016/b978-0-08-100646-7.00011-4.
68. Decker C, Biry S., Light stabilization of polymers by radiation cured acrylic coatings. *Prog. Org. Coat.*, 29, 81-87, 1996.
69. Qi, L., Ding, Y., Dong, Q., Wen, B., Wang, F., Zhang, S., Yang, M., Photostabilization of polypropylene by surface modified rutiletype TiO_2 nanorods. *J. Appl. Polym. Sci.*, 131, 16, 2014.
70. Qi, K., Xin, J.H., Daoud, W.A., Mak, C.L., Functionalizing Polyester Fiber with a Self-Cleaning Property Using Anatase TiO_2 and Low-Temperature Plasma Treatmen. *Int. J. Appl. Ceram. Technol.*, 4, 554, 2007.
71. Bozzi, A., Yuranova, T., Kiwi, J., Self-cleaning of wool-polyamide and polyestertextiles by TiO_2-rutile modification under daylight irradiation at ambient temperature. *J. Photochem. Photobiol. A*, 172, 27–34, 2005.
72. Wong, Y.W.H., Yuen, C.W.M., Leung, M.Y.S., Ku, S.K.A., Lam, H.L.I., Selected application of nanotechnology in textiles. *AUTEX Res. J.*, 6, 1–10, 2006.
73. Hu, L., Zhang, H., Gao, A., Hou, A., Functional modification of cellulose fabrics with phthalocyanine derivatives and the UV light-induced antibacterial performance. *Carbohydr. Polym.*, 201, 382–386, 2018. https://doi.org/10.1016/j.carbpol.2018.08.087.
74. Achwal, W.B., Use of UV absorbers for minimizing photodegradation of disperse dyes as well as polyester fibres. *Colourage*, 6, 21–22, 1994.
75. Gantz, G.M. and Sumner, W.G., Stable ultraviolet light absorbers. *Text. Res. J.*, 27, 244–251, 1957.
76. Teli, M.D., Adoverelar, R., Kuma, V.R., Pardeshi, P.D., Role of chelating agents and UV absorbers in improving photo-stability of disperse dyes. *J. Text. Assoc.*, 63, 247–251, 2003.

77. Deb, H., Xiao, S., Morshed, M.N., Al Azad, S., Immobilization of cationic titanium dioxide on electrospun nanofibrous mat: Synthesis, characterization, and potential environmental application. *Fibers Polym.*, 19, 1715–172, 2018.
78. Jiang, S.D., Bai, Z.-M., Tang, G., Song, L., Stec, A.A., Hull, T.R., Hu, Y., Hu, W.-Z., Synthesis of mesoporous silica@ Co–Al layered double hydroxide spheres: layer by-layer method and their effects on the flame retardancy of epoxy resins. *ACS Appl. Mater. Interfaces*, 6, 14076–14086, 2014.
79. Riaz, S., Ashraf, M., Hussain, T., Hussain, M.T., Rehman, A., Javid, A., Iqbal, K., Basita, A., Azizb, H., Functional finishing and coloration of textiles with nanomaterials. *Color. Technol.*, 134, 5, 327–346, 2018. https://doi.org/10.1111/cote.12344.
80. Radetić, M., Functionalization of textile materials with TiO_2 nanoparticles. *J. Photochem. Photobiol. C*, 16, 62–76, 2013.
81. Alebeid, O.K. and Zhao, T., Review on: developing UV protection for cotton fabric. *J. TEXT I*, 108, 12, 2027–2039, 2017. http://doi.org/10.1080/00405000.2017.1311201.
82. Parisi, M.L., Fatarella, E., Spinelli, D., Pogni, R., Basosi, R., Environmental impact assessment of an eco-efficient production for coloured textiles. *J. Clean. Prod.*, 108, 514–524, 2015. https://doi.org/10.1016/j.clepro.2015.06.032.
83. Feng, X.X., Yhang, L.L., Chen, J.Y., Yhang, J.C., New insight into solar UV protective properties of natural dye. *J. Clean. Prod.*, 15, 4, 366e372, 2007. http://doi.org/10.1016/j.jclepro.2005.11.003.
84. Wang, J., Yuan, B., Cai, W., Qiu, S., Tai, Q., Yang, H., Hu, Y., Facile design of transition metal based organophosphorus hybrids towards the flame retardancy reinforcement and toxic effluent elimination of polystyrene. *Mater. Chem. Phys.*, 214, 209–220, 2018. http://doi:10.1016/j.matchemphys.2018.04.100.
85. Bonet-Aracil, M.A., Diaz-Garcia, P., Bou-Belda, E., Sebastis, N., Montoto, A., Rodrigo, R., UV protection from cotton fabrics dyed with different tea) and various extracts from Mediterranean flora extracts. *Dyes Pigm.*, 134, 448–452, 2016. https://doi.org/10/1016/j.dyepig.2016.07.045.
86. Yang, H., Sukang, Z., Ning, P., Studying the mechanisms of titanium dioxide as ultraviolet-blocking additive for films and fabrics by an improved scheme. *J. Appl. Polym. Sci.*, 92, 3201–3210, 2004.
87. Ding, B., Jinho, K., Kimuraand, E., Shiratori, S., Layer-by-layer structured films of TiO_2 nanoparticles and poly(acrylic acid) on electrospun nanofibres. *Nanotechnology*, 15, 913, 2004.
88. Plentz, R.S., Miotto, M., Schneider, E.E., Forte, M.M.C., Mauler, R.S., Nachtigall, S.M., Effect of a macromolecular coupling agent on the properties of aluminum hydroxide/PP composites. *J. Appl. Polym. Sci.*, 101, 1799–1805, 2006.
89. Shafei, A.E. and Abou-Okeil, A., ZnO/carboxymethyl chitosan bionano-composite to impart antibacterial and UV protection for cotton fabric. *Carbohydr. Polym.*, 83, 920–925, 2011.

90. Pandey, R., Patel, S., Pandit, P., Shanmugam, N., Jose, S., Colouration of textiles using roasted peanut skin-an agro processing residue. *J. Clean. Prod.*, 172, 1319–1326, 2018. https://doi.org/10.1016/j.jclepro.2017.10.268.
91. Zuo, D., Liang, N., Xu, J. *et al.*, UV protection from cotton fabrics finished with boron and nitrogen co-doped carbon dots. *Cellulose*, 26, 4205–4212, 2019. https://doi.org/10.1007/s10570-019-02365-5.
92. Kocic, A., Bizjak, M., Popovic, D., Poparic, G.B., Stankovic, S.B., UV protection afforded by textile fabrics made of natural and regenerated cellulose fibres. *J. Clean. Prod.*, 228, 1229–1237, 2019.
93. Day, M. and Wiles, D.M., Photochemical degradation of poly (ethylene terephthalate). III. Determination of decomposition products and reaction mechanism. *J. Appl. Polym. Sci.*, 16, 175, 1972.
94. Malik, S.K. and Arora, T., UV radiations: problems and remedies. *Man-Made Text. India.*, 46, 164, 2003.
95. Chung, K.H., Lin, Y.C., Lin, A.Y., The persistence and photostabilizing characteristics of benzotriazole and 5-methyl-1H-benzotriazole reduce the photochemical behavior of common photosensitizers and organic compounds in aqueous environments. *Environ. Sci. Pollut. Res. Int.*, 25, 5911, 2018.
96. Santos, B.A.M.C., da Silva, A.C.P., Bello, M.L., Gonçalves, A.S., Gouvêa, T.A., Rodrigues, R.F., Cabral, L.M., Rodrigues, C.R., Molecular modeling for the investigation of UV absorbers for sunscreens: Triazine and benzotriazole derivatives. *J. Photochem. Photobiol. A*, 356, 219, 2018.
97. Li, C., Ma, W., Ren, X., Synthesis and Application of Benzotriazole UV Absorbers to Improve the UV Resistance of Polyester Fabrics. *Fibers Polym.*, 20, 2289–2296, 2019. https://doi.org/10.1007/s12221-019-8722-5.
98. Sundell, P. and Sundholm, F., Polyester binders for wood containing benzotriazole and HALS light stabilizers. *J. Appl. Polym. Sci.*, 92, 3, 1413–1421, 2004. https://doi.org/10.1002/app.13672.
99. Katritzky, A.R. and Rachwal, S., Synthesis of Heterocycles Mediated by Benzotriazole. 1. Monocyclic Systems. *Chem. Rev.*, 110, 1564, 2010.
100. Hilfiker, R., Kaufmann, W., Reinert, G., Schmdt, E., Improving sun protection factors of fabrics by applying UV-absorbers. *Text. Res J.*, 66, 61–70, 1996.
101. Aultz, D.E., Development on Polymeric Benzotriazole Stabilizer. *Spec. Chem.*, 16, 71, 73–74, 1996.
102. Katritzky, A.R. and Rachwal, S., Synthesis of Heterocycles Mediated by Benzotriazole. 2. Bicyclic Systems. *Chem. Rev.*, 111, 11, 7063–7120, 2011. http://doi.org/10.1021/cr200031r.
103. Tsatsaroni, E.G. and Eleftheriadis, I.C., UV-absorbers in the dyeing of polyester with disperse dyes. *Dyes Pigm.*, 61, 2, 141–147, 2004.
104. Yang, Y.Y., Li, H.Q., He, X.D., Bing, N., Chen, Y.J., Nan, Y., Zou, Y., Synthesis and Application of New Benzotriazole Derivatives as UV Absorbers for Polyester Fabrics. *Adv. Mat. Res.*, 233–235, 219–224, 2011. Trans Tech Publications, Ltd. then space http://doi.org/10.4028/www.scientific.net/amr.233-235.219.

105. Wang, R., Xin, J., Tao, X., UV-Blocking Property of Dumbbell-Shaped ZnO Crystallites on Cotton Fabrics. *Inorg. Chem.*, 44, 3926, 2005.
106. Tsatsaroni, E.G. and Kehayoglou, A.H., Dyeing of Polyester with CI disperse yellow 42 in the presence of various UV-absorbers. Part II. *Dyes Pigm.*, 28, 2, 123–130, 1995.
107. Laifeng, L. and Bin, F., Fabrication of UV blocking nanohybrid coating via miniemulsion polymerization. *J. Colloid. Interf Sci.*, 300, 111–116, 2006.
108. Cohen, S., Haham, H., Pellach, M., Margel, S., Design of UV-Absorbing Polypropylene Films with Polymeric Benzotriaziole Based Nano-and Microparticle Coatings. *ACS Appl Mater Inter.*, 9, 1, 868–875, 2017. http://doi.org/10.1021/acsami.6b12821.
109. Shin, M.J., Shin, Y.J., Moon, Y.U., Shin, J.S., Polymerizable ultraviolet stabilizers for unsaturated polyester-based bulk molding compounds. *J. Elastomers Plast.*, 46, 569–576, 2013.
110. Pandimurugan, R. and Thambidurai, S., UV protection and antibacterial properties of seaweed capped ZnO nanoparticles coated cotton fabrics. *Int. J. Biol. Macromol.*, 105, 788–795, 2017. http://doi:10.1016/j.ijbiomac.2017.07.097.
111. Emam, H.E. and Abdelhameed, R.M., Anti-UV Radiation Textiles Designed by Embracing with Nano-MIL (Ti, In)–Metal Organic Framework. *ACS Appl. Mater. Inter.*, 9, 28034, 2017.
112. Morshed, M.N., Behary, N., Bouazizi, N., Guan, J., Chen, G., Nierstrasz, V., Stabilization of zero valent iron (Fe0) on plasma/dendrimer functionalized polyester fabrics for Fenton-like removal of hazardous water pollutants. *Chem. Eng. J.*, 374, 658–673, 2019.
113. Nabil, B., Morshed, M.N., Ahmida, E.A., Nemeshwaree, B., Christine, C., Julien, V., Olivier, T., Abdelkrim, A., Development of new multifunctional filter-based nonwovens for organics pollutants reduction and detoxification: High catalytic and antibacterial activities. *Chem. Eng. J.*, 356, 702, 2019.
114. -ul-Islam, S. and Mohammad, F., High-Energy Radiation Induced Sustainable Coloration and Functional Finishing of Textile Materials. *Ind. Eng. Chem. Res.*, 54, 15, 3727–3745, 2015. http://doi.org/doi:10.1021/acs.iecr.5b00524.
115. Garcia, M., Van Vliet, G., Jain, S., Schrauwen, B., Sarkissov, A., Van Zyl, W.E., Boukamp, B., Polypropylene/SiO_2 Nanocomposites with Improved Mechanical Properties. *Rev. Adv. Mater. Sci.*, 6, 169–175, 2004.
116. Becheri, A., Maximilian, D., Nostro, P.L., Baglioni, P., Synthesis and characterization of zinc oxide nanoparticles: Application to textiles as UV-absorbers. *J. Nanopart Res.*, 10, 679–689, 2008.
117. Burçin, A.Ç., Leyla, B., Önderm, T., Numan, H., Synthesis of ZnO nanoparticles using PS-b-PAA reverse micelle cores for UV protective, self-cleaning and antibacterial textile applications. *Colloid Surf. A*, 414, 132–139, 2012.
118. Burkinshaw, S.M., *Chemical principles of synthetic fibre dyeing*, pp. 2–7, Blacky Academic & Professional, London, 1995.

119. Paul, R., Bautista, L., Varga, M.D.L., Botet, J.M., Casals, E., Puntes, V., Marsal, F., Nano-cotton fabrics with high ultraviolet protection. *Text. Res. J.*, 80, 454, 2010.
120. Yadav, A., Virendra, R., Kathe, A.A., Functional finishing in cotton fabrics using zinc oxide nanoparticles. *Mater. Sci.*, 29, 641–645, 2006. http://doi.org/10.1007/s12034-006-0017-y.
121. Yazdanshenas, M.E. and Mohammad, S.K., Bifunctionalization of cotton textiles by ZnO nanostructures: Antimicrobial activity and ultraviolet protection. *Text. Res. J.*, 83, 993–1004, 2013.
122. Nadanathangam, V., Sampath, K., Kathe, A.A., Varadarajan, P.V., Prasad, V., Functional finishing of cotton fabrics using zinc oxide-soluble starch nanocomposites. *Nanotechnology*, 17, 5087–5095, 2006.
123. Farouk, A., Textor, T., Schollmeyer, E., Tarbuk, A., Grancarić, A.M., Sol–gel-derived inorganic–organic hybrid polymers filled with ZnO nanoparticles as an ultraviolet protection finish for textiles. *Autex Res J.*, 9, 58–63, 2010.
124. Yang, M., Liu, W., Jiang, C., Xie, Y., Shi, H., Zhang, F., Wang, Z., Facile construction of robust superhydrophobic cotton textiles for effective UV protection, self-cleaning and oil–water separation. *Colloid Surf. A Physicochem. Eng. Asp.*, 570, 172–181, 2019.
125. Mai, Z., Xiong, Z., Shu, X., Liu, X., Zhang, H., Yin, X., Zhou, Y., Liu, M., Zhang, M., Xu, W., Chen, D., Multifunctionalization of cotton fabrics with polyvinylsilsesquioxane/ZnO composite coatings. *Carbohyd. Polym.*, 199, 516–525, 2018. https://doi.org/10.1016/j.carbpol.2018.07.052.
126. Rehan, M., Nadab, A.A., Khattabd, T.A., Abdelwahede, N.A.M., Abou El-Khei, A.A., Development of multifunctional polyacrylonitrile/silver nanocomposite films: Antimicrobial activity, catalytic activity, electrical conductivity, UV protection and SERS-active sensor. *J. Mater. Res. Technol.*, 1773, 15, 2020. http://doi.org/10.1016/j.jmrt.2020.05.079.
127. El-Hady, M.M., Farouk, A., Sharaf, S., Flame retardancy and UV protection of cotton-based fabrics using nano ZnO and polycarboxylic acids. *Carbohydr Polym.*, 92, 1, 400–406, 2013. http://doi.org/10.1016/j.carbpol.2012.08.085.
128. El-Shafei, A. and Abou-Okeil, A., ZnO/carboxymethyl chitosan bionano-composite to impart antibacterial and UV protection for cotton fabric. *Carbohydr Polym.*, 83, 920–925, 2011.
129. Jesionowski, T., Kołodziejczak-Radzimska, A., Ciesielczyk, F., Sojka-Ledakowicz, J., Olczyk, J., Sielski, J., Synthesis of zinc oxide in an emulsion system and its deposition on PES nonwoven fabrics. *Fibres Text. East Eur.*, 19, 70–75, 2011.
130. Rezayi, T. and Entezari, M.H., Wettability properties vary with different morphologies of ZnO nanoparticles deposited on glass and modified by stearic acid. *New J. Chem.*, 40, 2582–2591, 2016.
131. Salač, J., Šerá, J., Jurča, M., Verney, V., Marek, A.A., Koutný, M., Photodegradation and Biodegradation of Poly (Lactic) Acid Containing Orotic Acid as a Nucleation Agent. *Materials*, 12, 3, 481, 2019. http://doi.org/10.3390/ma12030481.

132. Yan, L., Yunling, Z., Yanyan, H., Fabrication and UV-blocking property of nano-ZnO assembled cotton fibers via a two-step hydrothermal method. *Cellulose*, 18, 1643–1649, 2011.
133. Ling, C. and Guo, L., Testing the fireproof properties of nano-ZnO coated cotton fabric. *Fibres Text. East Eur.*, 4, 65–70, 2018. http://doi.org/10.5604/01.3001.0013.1820.
134. De Falco, F., Guarino, V., Gentile, G., Cocca, M., Ambrogi, V., Ambrosio, L., Avella, M., Design of functional textile coatings via non-conventional electrofluidodynamic processes. *J. Colloid. Interf. Sci.*, 541, 367–375, 2019. http://doi.org/10.1016/j.jcis.2019.01.086.
135. Wang, R.H., Xin, J.H., Tao, X.M., UV-blocking property of dumbbell-shaped ZnO crystallites on cotton fabrics. *Inorg Chem.*, 44, 3926–3930, 2005.
136. Yetisen, A.K. *et al.*, Nanotechnology in textiles. *ACS Nano.*, 10, 3042–3068, 2016. https://doi.org/10.1021/acsnano.5b08176.
137. Stoppa, M. and Chiolerio, A., Wearable Electronics and Smart Textiles: A Critical Review. *Sensors*, 14, 11957–11992, 2014.
138. Silvestre, C., Cimmino, S., Pezzuto, M., Marra, A., Ambrogi, V., Dexpert-Ghys, J., Duraccio, D., Preparation and characterization of isotactic polypropylene/zinc oxide microcomposites with antibacterial activity. *Polym. J.*, 45, 9, 938, 2013.
139. Bokhimi, X., Morales, A., Toledo-Antonio, J.A., Pedrazab, F., Local order in titania polymorphs. *Int. J. Hydrogen Energ.*, 26, 1279–1287, 2001.
140. Daoud, W.A., John, H.X., Yi-He, Z., Surface functionalization of cellulose fibers with titanium dioxide nanoparticles and their combined bactericidal activities. *Surf Sci.*, 599, 69–75, 2005.
141. Daoud, W.A. and John, H.X., Low temperature sol–gel processed photocatalytic titania coating. *J. Sol–Gel Sci. Techn.*, 29, 25–29, 2004.
142. Mahmoodi, N.M., Mokhtar, A., Nargess, Y.L., Nooshin, S.T., Kinetics of heterogeneous photocatalytic degradation of reactive dyes in an immobilized TiO_2 photocatalytic reactor. *J. Colloid. Interf. Sci.*, 295, 59–164, 2006.
143. Rios, P.F., Dodiuk, H., Keing, S., McCharthy, S., Dotan, A., Durable ultrahydrophobic surfaces for self-cleaning applications. *Polym. Adv. Technol.*, 19, 1684–1691, 2008.
144. Sawada, K., Masakatsu, S., Mitsuo, U., Chan, P.H., Hydrophilic treatment of polyester surfaces using TiO_2 photocatalytic reactions. *Text. Res J.*, 73, 819–822, 2003.
145. Bae, H.S., Lee, M.K., Kim, W.W., Rhee, C.K., Dispersion properties of TiO_2 nano-powder synthesized by homogeneous precipitation process at low temperatures. *Colloid Surf. A*, 220, 169–177, 2003.
146. Yuqing, M., Jianguo, G., Jianrong, C., Yuquing, Ion sensitive field effect transducer-based biosensors. *Biotechnol. Adv.*, 21, 2003, 527–534, 2003.
147. Liu, Y.Y., Xin, J.H., Choi, C.H., Cotton Fabrics with Single-Faced Superhydrophobicity. *Langmuir*, 28, 17426, 2012.

148. Chen, D., Mai, Z., Liu, X. *et al.*, UV-blocking, superhydrophobic and robust cotton fabrics fabricated using polyvinylsilsesquioxane and nano-TiO_2. *Cellulose*, 25, 3635–3647, 2018. https://doi.org/10.1007/s10570-018-1790-7.
149. Liu, Z., Hong, L., Guo, B., Physicochemical and electrochemical characterization of anatase titanium dioxide nanoparticles. *J. Power Sources*, 143, 231–235, 2005.
150. Zheng, C., Guoqiang, C., Zhenming, Q., Ultraviolet resistant/antiwrinkle finishing of cotton fabrics by sol–gel method. *J. Appl. Polym. Sci.*, 122, 2090–2098, 2011.
151. Basfar, A.A., Ali, K.M.I., Vaidya, M.M., Bahamdan, A.A., Alam, M.A., Improved ultraviolet (UV) radiation stability of the polypropylene (PP) films of woven jumbo bags for outdoor applications. *Polym. Plast. Technol. Eng.*, 49, 8, 841–847, 2010.
152. Kemp, T.J. and McIntyre, R.A., Influence of transition metal-doped titanium (IV) dioxide on the photodegradation of polystyrene. *Polym. Degrad. Stab.*, 91, 12, 3010–3019, 2006.
153. Chen, X.D., Wang, Z., Liao, Z.F., Mai, Y.L., Zhang, M.Q., Roles of anatase and rutile TiO_2 nanoparticles in photooxidation of polyurethane. *Polym. Test.*, 26, 2, 202–208, 2007.
154. Ukaji, E., Takeshi, F., Masahide, S., Noboru, S., The effect of surface modification with silane coupling agent on suppressing the photo-catalytic activity of fine TiO_2 particles as inorganic UV filter. *Appl. Surf. Sci.*, 254, 1563–1569, 2007.
155. Shafi Kurikka, V.P.M., Abraham, U., Xingzhong, Y., Nan-Loh, Y., Michaeland, H., Michae, G., Sonochemical preparation of silane-coated titania particles. *Langmuir.*, 17, 1726–1730, 2001.
156. Kyung, P.O., Kang, Y.S., Beong, G.J., Synthesis of TiO_2 nanoparticles coated with SiO_2 for suppression of photocatalytic activity and increased dispersion stability. *Indian Eng Chem.*, 10, 733–738, 2004.
157. Zhang, Y., Yanfei, W., Min, C., Limin, W., Fabrication method of TiO_2–SiO_2 hybrid capsules and their UV-protective property. *Colloid. Surf. A Physicochem. Eng. Asp.*, 353, 216–225, 2010.
158. Dongzhi, C., Mai, Z., Liu, X., Ye, D., Zhang, H., Yin, X., Zhou, Y., UV-blocking, superhydrophobic and robust cotton fabrics fabricated using polyvinylsilsesquioxane a158nd nano-TiO_2. *Cellulose*, 6, 3635–3647, 2018.
159. Pakdel, E., Daoud, W.A., Sun, L., Wang, X., Visible and UV functionality of TiO_2 ternary nanocomposites on cotton. *Appl. Surf. Sci.*, 321, 447–456, 2014. https://doi.org/10.1016/j.apsusc.2014.10.018.
160. Xi, G., Wang, J., Luo, G., Zhu, Y., Fan, W., Huang, M., Wang, H., Liu, X., Healable superhydrophobicity of novel cotton fabrics modified via one-pot mist copolymerization. *Cellulose*, 23, 915–927, 2016, https://doi.org/10.1007/s10570-015-0773-1.
161. Jiang, C., Liu, W., Yang, M., Liu, C., He, S., Xie, Y., Wang, Z., Facile fabrication of robust fluorine-free self-cleaning cotton textiles with superhydrophobicity

photocatalytic activity and UV durability. *Colloid. Surf. A Physicochem. Eng. Asp.*, 559, 235–242, 2018. https://doi.org/10.1016/j.colsurfa.2018.09.048.

162. Zhao, Y., Dang, W., Lu, Z., Deng, J., Hao, Y., Su, Z., Zhang, M., Fabrication of mechanically robust and UV-resistant aramid fiber-based composite paper by adding nano-TiO_2 and nanofibrillated cellulose. *Cellulose*, 25, 3913–3925, 2018. https://doi.org/10.1007/s10570-018-1818-z.
163. Khan, Muhammad Zaman, Baheti, Vijay, Ashraf, Munir, Hussain, Tanveer, Ali, Azam, Javid2, Amjed and Rehman, Abdur, Ugur, S.S., Sarıšık, M., Aktaş, H.A., Nano-TiO_2 based multilayer film deposition on cotton fabrics for UV-protection. *Fiber Polym.*, 12, 190–196, 2018.
164. Erdem, N., Erdogan, U.H., Cireli, A.A., Onar, N., Structural and ultraviolet-protective properties of nano-TiO_2-doped polypropylene filaments. *J. Appl. Polym. Sci.*, 115, 1, 152–157, 2010.
165. Salla, J., Pandey, K.K., Srinivas, K., Improvement of UV resistance of wood surfaces by using ZnO nanoparticles. *Polym. Degrad. Stab.*, 97, 4, 592–596, 2012.
166. Erdem, N., Cireli, A.A., Erdogan, U.H., Flame retardancy behaviors and structural properties of polypropylene/nano-SiO_2 composite textile filaments. *J. Appl. Polym. Sci.*, 111, 4, 2085–2091, 2009.
167. Radetić, M., Functionalization of textile materials with TiO_2 nanoparticles. *J. Photochem. Photobiol. C: Photochem. Rev.*, 16, 62, 2013.
168. El-Naggar, M.E., Hassabo, A.G., Mohamed, A.L., Shaheen, T.I., Surface modification of SiO_2 coated ZnO nanoparticles for multifunctional cotton fabrics. *J. Colloid Interf. Sci.*, 498, 413, 2017.
169. Lin, J., Chen, H., Fei, T., Liu, C., Zhang, J., Highly transparent and thermally stable superhydrophobic coatings from the deposition of silica aerogels. *Appl. Surf. Sci.*, 273, 776–786, 2013. https://doi.org/10.1016/j.apsusc.2013.02.134.
170. Zhang, M., Wang, S.L., Wang, C.Y., Li, J., A facile method to fabricate superhydrophobic cotton fabrics. *Appl. Surf. Sci.*, 261, 561–566, 2012. https://doi.org/10.1016/j.apsusc.2012.08.055.
171. Deng, Z.-Y., Wang, W., Mao, L.-H., Wang, C.-F., Chen, S., Versatile superhydrophobic and photocatalytic films generated from TiO_2–SiO_2@PDMS and their applications on fabrics. *J. Mater. Chem. A*, 2, 4178–4184, 2014. https://doi.org/10.1039/c3ta14942k.
172. Prasad, V., Arputharaj, A., Bharimalla, A.K., Patil, P.G., Vigneshwaran, N., Durable multifunctional finishing of cotton fabrics by *in situ* synthesis of nano-ZnO. *Appl. Surf. Sci.*, 390, 936–940, 2016. https://doi.org/10.1016/j.apsusc.2016.08.155.
173. Zhu, H., Wang, N., Yao, L., Chen, Q., Zhang, R., Qian, J., Hou, Y., Guo, W., Fan, S., Zhao, Q., Liu, S., Du, F., Zuo, X., Guo, Y., Xu, Y., Li, J., Xue, T., Zhong, K., Xiong, W., Moderate UV Exposure Enhances Learning and Memory by Promoting a Novel Glutamate Biosynthetic Pathway in the Brain. *Cell.*, 173, 7, 1716–1727, 2018.

174. Wu, M., Ma, B., Pan, T., Chen, S., Sun, J., Silver-nanoparticlecoloredcotton fabrics with tunable colors and durable antibacterial and self-healing superhydrophobic properties. *Adv. Funct. Mater.*, 26, 569–576, 2016. https://doi.org/10.1002/adfm.201504197.
175. Li, L., Fan, T., Hu, R., Liu, Y., Lu, M., Surface micro-dissolution process for embedding carbon nanotubes on cotton fabric as a conductive textile. *Cellulose*, 24, 1121–1128, 2017. https://doi.org/10.1007/s10570-016-1160-2.
176. Kowalczyk, D., Fortuniak, W., Mizerska, U., Kaminska, I., Makowski, T., Brzezinski, S., Piorkowska, E., *Cellulose*, vol. 24, 4057–4068, 2017, https://doi.org/10.1007/s10570-017-1389-4.
177. Wang, L.L., Xintong, Z., Bing, L., Panpan, S., Yang, J., Haiyang, X., Yichun, L., Superhydrophobic and ultraviolet-blocking cotton textiles. *ACS Appl. Mater. Interf.*, 3, 1277–1281, 2011.
178. Wang, R., Xin, J.H., Tao, X.M., Daoud, W.A., UV-Blocking Property of Dumbbell-Shaped ZnO Crystallites on Cotton Fabrics. *Inorg. Chem.*, 44, 11, 3926–3930, 2005. https://doi.org/10.1021/ic0503176.
179. Curiskis, J. and Pailthorpe, M., Apparel textiles and sun protection. *Text. Mag.*, 25, 13–17, 1996.
180. Pailthorpe, M.T., Textiles and sun protection: The current situation. *Australasian Text.*, 14, 54–66, 1994.
181. Zhou, Y. and Crews, P.C., Effect of OBAs and repeated launderings on UVR transmission through fabrics. *Text. Chem. Color.*, 30, 19–24, 1998.
182. Ugur, S.S., Sarıışık, M., Aktaş, H.A., Nano-TiO_2 based multilayer film deposition on cotton fabrics for UV-protection. *Fibers Polym.*, 12, 190–196, 2011.
183. Feng, G., Wang, X., Zhang, D., Xiao, X., Qian, K., Fabrication of ZnO coated nanoTiO_2 and evaluation of its efficiency in stabilization of polypropylene fibers. *Appl. Phys A.*, 125, 359, 2019. https://doi.org/10.1007/s00339-019-2640-7.
184. Lee, S.H. and Youn, J.R., Properties of polypropylene/layered-silicate nanocomposites and melt-spun fibers. *J. Appl. Polym. Sci.*, 109, 2, 1221–1231, 2008.
185. Li, X.W., Song, R.G., Jiang, Y., Wang, C., Jiang, D., Surface modification of TiO_2 nanoparticles and its effect on the properties of fluoropolymer/TiO_2 nanocomposite coatings. *Appl. Surf. Sci.*, 276, 761–768, 2013.
186. Alebeid, O.K. and Zhao, T., Anti-ultraviolet treatment by functionalizing cationized cotton with TiO_2 nano-sol and reactive dye. *Text. Res J.*, 85, 449–457, 2015a.
187. Alebeid, O.K. and Zhao, T., Simultaneous dyeing and functional finishing of cotton fabric using reactive dyes doped with TiO_2 nano-sol. *J. Text. I.*, 107, 625–635, 2016.
188. Zhang, Y., Yanfei, W., Min, C., Limin, W., TiO_2/SiO_2 hybrid nanomaterials: Synthesis and variable UV-blocking properties. *J. Sol–Gel Sci. Techn.*, 58, 326–329, 2011.

189. Mowafi, S., Kafafy, H., Arafa, A. *et al.*, Facile and environmental benign *in situ* synthesis of silver nanoparticles for multifunctionalization of wool fibers. *Environ. Sci. Pollut. Res.*, 25, 29054–29069, 2018. https://doi.org/10.1007/s11356-018-2928-8.
190. Hossam, E.E., Rehan, M., Mashaly, H.M., Ahmed, H.B., Large scaled strategy for natural/synthetic fabrics functionalization via immediate assembly of AgNPs. *Dyes Pigm.*, 133, 173–183, 2016. https://doi.org/10.1016/j.dyepig.2016.06.005.
191. Capjack, L., Kerr, N., Davis, S., Fedosejevs, R., Hatch, K.L., Markee, N.L., Protection of humans from ultraviolet radiation through the use of textiles: A review. *Family Consumer Sci. Res. J.*, 23, 198–218, 1994.
192. Gwendolyn Hustvedt, G. and Crews, P.C., The ultraviolet protection factor of naturally-pigmented cotton. *J. Cotton Sci.*, 47, 47–55, 2005.
193. Shahid, M., -ul-Islam, S., Mohammad, F., Recent advancements in natural dye applications: a review. *J. Clean. Prod.*, 53, 310, 2013.
194. Anum, S. and Ali, S., Treatment of Cotton Fiber with Newly Synthesized UV Absorbers: Optimization and Protection Efficiency. *Fibers Polym.*, 19, 11, 2290–2297, 2018. http://doi.org/10.1007/s12221-018-8517-0.
195. Pisitsak, P., Tungsombatvisit, N., Singhanu, K., Utilization of waste protein from Antarctic krill oil production and natural dye to impart durable UV-properties to cotton textiles. *J. Clean. Prod.*, 174, 1215–1223, 2018. https://doi.org/10.1016/j.jclepro.2017.11.010.
196. Guo, L., Yang, Z.-Y., Tang, R.-C., Yuan, H.-B., Preliminary Studies on the Application of Grape Seed Extract in the Dyeing and Functional Modification of Cotton Fabric. *Biomolecules*, 10, 2, 220, 2020. http://doi.org/10.3390/biom10020220.
197. Shabbir, M., Rather, L.J., Mohammad, F., Economically viable UV-protective and antioxidant finishing of wool fabric dyed with *Tagetes erecta* flower extract: Valorization of Marigold. *Ind. Crop. Prod.*, 119, 277–282, 2018.
198. Paul, R. and Genesca, E., The use of enzymatic techniques in the finishing of technical textiles, in: *Advances in Dyeing and Finishing of Technical Textiles*, M.L. Gulrajni (Ed.), pp. 177–198, Woodhead publishing Ltd., Cambridge, 2013.
199. Chao, Y., Ho, T., Cheng, Z., Kao, L., Tsai, P., A study on combing natural dyes and environmentally-friendly mordant to improve ultraviolet protection of textiles. *Fibres Polym.*, 18, 8, 1523–1530, 2017. https://doi.org/10.1007/s12221-017-6964-7.
200. Grifoni, D., Bacci, L., Lonardo, D.S., Pinelli, P., Scardigli, A., Camilli, F., Sabatini, F., Zipoli, G., Romani, A., UV protective properties of cotton and flax fabrics dyed with multifunctional plant extracts. *Dyes Pigm.*, 105, 89–96, 2014. https://doi.org/10/1016/j.dyepig.2014.01.027.
201. Yusuf, M., Shabbir, M., Mohammad, F., Natural colorants: Historical, processing and sustainable prospects. *Nat. Prod. Bioprospect.*, 7, 123–145, 2017. https://doi.org/ http://doi.org/10.1007/s13659-017-0119-9.

202. Shahidi, S. and Moazzenchi, B., Comparison between Mordant Treatment and Plasma Sputtering on Natural Dyeing and UV Protection Properties of Wool Fabric. *Fibers Polym.*, 20, 1658–1665, 2019. https://doi.org/10.1007/s12221-019-2040-9.
203. Rehan, M., Abdel-Wahed N, A.M., Farouk, A., El-Zawahry, M.M., Extraction of Valuable Compounds from Orange Peel Waste for Advanced Functionalization of Cellulosic Surfaces. *ACS Sustainable Chem. Eng.*, 6, 5, 5911–5928, 2018. https://doi.org/10.1021/acssuschemeng.7b04302.
204. Rodrigues, C.S.D., Madeira, L.M., Boaventura, R.A.R., Synthetic textile dyeing wastewater treatment by integration of advanced oxidation and biological processes e performance analysis with costs reduction. *J. Environ. Chem. Eng.*, 2, 2, 10271039, 2014. https://doi.org/10.1016/j.jece.2014.03.019.
205. Muzaffar, S., Bhatti, I.A., Zuber, M., Bhattia, H.N., Shahid, M., Study of the UV protective and antibacterial properties of aqueous polyurethane dispersions extended with low molecular weight chitosan. *Int. J. Biol. Macromol.*, 94, 51–60, 2017. https://doi.org/10.1016/j.ijbiomac.2016.09.106.
206. Jang, Y.W., Min, B.G., Yoo, K.H., Enhancement in Compressive Strength and UV Ageing-resistance of Poly(p-phenylene benzobisoxazole) Nanocomposite Fiber Containing Modified Polyhedral Oligomeric Silsesquioxane. *Fibers Polym.*, 18, 3, 575–581, 2017. http://doi.org/10.1007/s12221-017-1080-2.
207. Vinisha Rani, K., Hari Prakash, N., Solomon, I. *et al.*, Surface modifications of natural Kanchipuram silk (pattu) fibers using glow discharge air Plasma. *Fibers Polym.*, 17, 52–58, 2016. https://doi.org/10.1007/s12221-016-5513-0.
208. Zhou, Y., Yang, Z.-Y., Tang, R.-C., Bioactive and UV protective silk materials containing baicalin—The multifunctional plant extract from *Scutellaria baicalensis* Georgi. *Mater. Sci. Eng. C*, 67, 336–344, 2016. https://doi.org/10.1016/j.msec.2016.05.063.
209. Kosowska, K., Szatkowski, Piotr, Influence of ZnO, SiO_2 and TiO_2 on the aging process of PLA fibers produced by electrospinning method. *J. Therm. Anal. Calorim.*, 140, 1769–1778, 2020. https://doi.org/10.1007/s10973-019-08890-6.
210. Gardette, M., Thérias, S., Gardette, J.-L., Murariu, M., Dubois, P., Photooxidation of polylactide/calcium sulphate composites. *Polym. Degrad Stab.*, 96, 616–23, 2011. https://doi.org/10.1016/j.polymdegradstab.2010.12.023.
211. Janorkar, A.V., Metters, A.T., Hirt, D.E., Degradation of poly(L-lactide) films under ultraviolet-induced photografting and sterilization conditions. *J. Appl. Polym. Sci.*, 106, 1042–7, 2007. https://doi.org/10.1002/app.24692.
212. Sullalti, S., Totaro, G., Askanian, H., Celli, A., Marchese, P., Verney, V. *et al.*, Photodegradation of TiO_2 composites based on polyesters. *J. Photochem. Photobiol. A Chem.*, 321, 275–83, 2016. https://doi.org/10.1016/j.jphotochem.2015.11.007.
213. Liang, Y., Xie, W., Pakdel, E., Zhang, M., Sun, L., Wang, X., Homogeneous melanin/silica core-shell particles incorporated in poly(methyl methacrylate) for enhanced UV protection, thermal stability, and mechanical

properties. *Mater. Chem. Phys.*, 230, 319–325, 2019. https://doi.org/10.1016/j.matchemphys.2019.03.081.

214. Chen, J., Li, X., Zhu, Y. *et al.*, Storable silicon/shape memory polyurethane hybrid sols prepared by a facile synthesis process and their application to aramid fibers. *J. Sol–Gel Sci. Technol.*, 74, 670–676, 2015. https://doi.org/10.1007/s10971-015-3647-y.
215. Chen, K., Xu, C., Zhou, J., Zhao, R., Gao, Q., Wang, C., Multifunctional fabric coatings with slow-releasing fragrance and UV resistant properties from ethyl cellulose/silica hybrid microcapsules. *Carbohydr. Polym.*, 232, 115821, 2020.
216. Paret, N., Trachsel, A., Berthier, D.L., Herrmann, A., Developing multi stimuli responsive core/shell microcapsules to control the release of volatile compounds. *Macromol. Mater. Eng.*, 304, 1800599, 2019. https://doi.org/10.1002/mame.201800599.
217. Bouazizi, N., Ajala, F., Bettaibi, A., Khelil, M., Benghnia, A., Bargougui, R., Louhichi, S., Labiadh, L., Ben Slama, R., Chaouachi, B., Khirouni, K., Houas, A., Azzouza, A., Metal-organo-zinc oxide materials: Investigation on the structural, optical and electrical properties. *J. Alloy Compd.*, 656, 25, 146–153, 2016.
218. Morshed, M.N., Shen, X., Deb, H., Al Azad, S., Zhang, X., Li, R., Sonochemical fabrication of nanocryatalline titanium dioxide (TiO_2) in cotton fiber for durable ultraviolet resistance. *J. Nat. Fibers*, 17, 1, 41–54, 2020. https://doi.org/10.1080/15440478.2018.1465506.
219. Liu, S., Zhang, L., Zhou, J., Wu, R., Structure and Properties of Cellulose/Fe2O3 Nanocomposite Fibers Spun via an Effective Pathway. *J. Phys. Chem. C*, 112, 12, 4538–4544, 2008. https://doi.org/10.1021/jp711431h.
220. Zeng, J., Liu, S., Cai, J., Zhang, L., TiO_2 Immobilized in Cellulose Matrix for Photocatalytic Degradation of Phenol under Weak UV Light Irradiation. *J. Phys. Chem. C.*, 114, 7806, 2010. https://doi.org/10.1021/jp1005617.
221. Li, C., Xie, Y., Liu, Q., Zheng, Y., Zhang, X., Dong, W., The formation and UV-blocking property of flower-like ZnO nanorod on electrospun natural cotton cellulose nanofibers. *Fibers Polym*, 15, 2, 281–285, 2014. https://doi.org/10.1007/s12221-014-0281-1.
222. Zhang, Y., Xu, J., Guo, B., Photodegradation behavior of poly(butylene succinate-co-butylene adipate)/ZnO nanocomposites. *Colloid Surf. A*, 489, 173–181, 2016.
223. Li, C., Xie, Y., Liu, Q. *et al.*, The formation and UV-blocking property of flower-like ZnO nanorod on electrospun natural cotton cellulose nanofibers. *Fibers Polym.*, 15, 281–285, 2014. https://doi.org/10.1007/s12221-014-0281-1.
224. Li, Y., Hou, Y., Zou, Y., Microwave assisted fabrication of Nano-ZnO assembled cotton fibers with excellent UV blocking property and water-wash durability. *Fibers Polym.*, 13, 185–190, 2012. https://doi.org/10.1007/s12221-012-0185-x.

225. Sricharussin, W., Threepopnatkul, P., Neamjan, N., Effect of various shapes of zinc oxide nanoparticles on cotton fabric for UV-blocking and anti-bacterial properties. *Fibers Polym.*, 12, 1037–1041, 2011. https://doi.org/10.1007/s12221-011-1037-9.
226. Xu, L.H., Zheng, G.G., Miao, J.H., Xian, F.L., Dependence of structural and optical properties of sol-gel derived ZnO thin films on sol concentration. *Appl. Surf. Sci.*, 258, 19, 7760–7765, 2012. http://doi.org/10.1016/j.apsusc.2012.04.137.
227. Montazer, M. and Amiri, M.M., ZnO nano reactor on textiles and polymers: *ex situ* and *in situ* synthesis, application, and characterization. *J. Phys. Chem. B.*, 118, 6, 1453–1470, 2014. http://doi.org/10.1021/jp408532r.
228. Montazer, M., Amiri, M.M., Malek, R.M.A., *In situ* synthesis and characterization of nano ZnO on wool. influence of nano photo reactor on wool properties. *Photochem. Photobiol.*, 89, 5, 1057–1063, 2013. http://doi.org/10.1111/php.12090.
229. Uğur, S.S., Sarıışık, M., Aktaş, A.H., Çiğdem, M.U., Erden, E., Modifying of Cotton Fabric Surface with Nano-ZnO Multilayer Films by Layer-by-Layer Deposition Method. *Nanoscale Res. Lett.*, 5, 1204, 2010. https://doi.org/10.1007/s11671-010-9627-9.
230. Gorjanc, M., Jazbec, K., Šala, M. *et al.*, Creating cellulose fibres with excellent UV protective properties using moist CF_4 plasma and ZnO nanoparticles. *Cellulose*, 21, 3007–3021, 2014. https://doi.org/10.1007/s10570-014-0284-5.
231. Perelshtein, I., Ruderman, E., Perkas, N., Tzanov, T., Beddow, J., Joyce, E., Mason, T.J., Blanes, M., Mollá, K., Patlolla, A., Frenkel, A.I., Gedanken, C.A., hitosan and chitosan–ZnO-based complex nanoparticles: formation, characterization, and antibacterial activity. *J. Mater. Chem. B*, 4, 1, 1968–1976, 2013. https://doi.org/10.1039/C3TB00555K.
232. Zhou, Y., Yang, Z.-Y., Tang, R.-C., Facile and green preparation of bioactive and UV protective silk materials using the extract from red radish (*Raphanus sativus* L.) through adsorption technique. *Arabian J. Chem.*, 13, 3276–3285, 2020. https://doi.org/10.1016/j.arabjc.2018.11.003.
233. Hu, L., Zhang, H., Gao, A., Hou, A., Functional modification of cellulose fabrics with phthalocyanine derivatives and the UV light-induced antibacterial performance. *Carbohydr. Polym.*, 201, 382–386, 2018. https://doi.org/10.1016/j.carbpol.2018.08.087.
234. Qi, H., Pan, J., Qing, F.-L., Yan, K., Sun, G., Anti-wrinkle and UV protective performance of cotton fabrics finished with 5-(carbonyloxy succinic)-benzene-1, 2, 4-tricarboxylic acid. *Carbohydr. Polym.*, 154, 313–319, 2016. https://doi.org/10.1016/j.carbpol.2016.05.108.
235. Baaka, N., Ben Ticha, M., Haddar, W. *et al.*, Upgrading of UV Protection Properties of Several Textile Fabrics by Their Dyeing with Grape Pomace Colorants. *Fibers Polym.*, 19, 307–312, 2018. https://doi.org/10.1007/s12221-018-7327-0.

236. Heliopoulos, N.S., Kouzilos, G.N., Giarmenitis, A.I. *et al.*, Viscose Fabric Functionalized with Copper and Copper Alginate Treatment Toward Antibacterial and UV Blocking Properties. *Fibers Polym.*, 21, 1238–1250, 2020. https://doi.org/10.1007/s12221-020-9578-4.
237. Li, C., Ma, W., Ren, X., Synthesis and Application of Benzotriazole UV Absorbers to Improve the UV Resistance of Polyester Fabrics. *Fibers Polym.*, 20, 2289–2296, 2019. https://doi.org/10.1007/s12221-019-8722-5.
238. Feng, G., Wang, X., Zhang, D. *et al.*, Fabrication of ZnO coated nano-TiO_2 and evaluation of its efficiency in stabilization of polypropylene fibers. *Appl. Phys. A*, 125, 359, 2019. https://doi.org/10.1007/s00339-019-2640-7.
239. Khan, M.Z., Baheti, V., Ashraf, M. *et al.*, Development of UV Protective, Superhydrophobic and Antibacterial Textiles Using ZnO and TiO_2 Nanoparticles. *Fibers Polym.*, 19, 1647–1654, 2018. https://doi.org/10.1007/s12221-018-7935-3.
240. Sahar, A. and Ali, S., Treatment of Cotton Fiber with Newly Synthesized UV Absorbers: Optimization and Protection Efficiency. *Fibers Polym.*, 19, 2290–2297, 2018. https://doi.org/10.1007/s12221-018-8517-0.
241. Rehan, M., Abdel-Wahed, N.A.M., Farouk, A., El-Zawahry, M.M., Extraction of Valuable Compounds from Orange Peel Waste for Advanced Functionalization of Cellulosic Surfaces. *ACS Sustainable Chem. Eng.*, 6, 5, 5911–5928, 2018. https://doi.org/10.1021/acssuschemeng.7b04302.
242. Wang, Y., Tian, M., Qu, L. *et al.*, Enhanced thermal, UV blocking and dye absorptive properties of chitosan/poly(vinyl alcohol)/graphene oxide fibers. *Fibers Polym.*, 16, 2011–2020, 2015. https://doi.org/10.1007/s12221-015-5279-9.
243. Jang, Y.W., Min, B.G., Yoon, K.H., Enhancement in compressive strength and UV ageing-resistance of poly(p-phenylene benzobisoxazole) nanocomposite fiber containing modified polyhedral oligomeric silsesquioxane. *Fibers Polym.*, 18, 575–581, 2017. https://doi.org/10.1007/s12221-017-1080-2.
244. Qu, L., Tian, M., Hu, X., Wang, Y., Zhu, S., Guo, X., Han, G., Zhang, X., Sun, K., Tang, X., Functionalization of cotton fabric at low graphene nanoplate content for ultrastrong ultraviolet blocking. *Carbon*, 80, 565, 2014. http://dx.doi.org/10.1016/j.carbon.2014.08.097.
245. Wang, Q.W., Yoon, K.H., Min, B.G., Chemical and physical modification of poly(p-phenylene benzobisoxazole) polymers for improving properties of the PBO fibers. I. Ultraviolet-ageing resistance of PBO fibers with naphthalene moiety in polymer chain. *Fibers Polym.*, 16, 1–7, 2015. https://doi.org/10.1007/s12221-015-0001-5.
246. Chae, H.G. and Kumar, S., Rigid-rod polymeric fibers. *J. App. Polym. Sci.*, 100, 1, 791–802, 2006. https://doi.org/10.1002/app.22680.

8

Synthetic and Natural UV Protective Agents for Textile Finishing

Iftay Khairul Alam[1], Nazia Nourin Moury[2] and Mohammad Tajul Islam[2]*

[1]European University of Bangladesh, Dhaka, Bangladesh
[2]Ahsanullah University of Science and Technology, Dhaka, Bangladesh

Abstract
Skin cancer is a rising incident worldwide due to excessive exposure to sunlight. Though the ultraviolet rays constitute a very small portion of the solar spectrum, the radiation can cause several effects like sunburn, premature skin aging, allergies, and skin cancer. Effective methods to block the ultraviolet rays from human skin are urgently sought as the intensity of ultraviolet rays are increasing every year. Several means give protection to human skin against ultraviolet radiation. This chapter will discuss the ultraviolet rays and the importance of protection from these rays. The methods of blocking the ultraviolet rays, the measurement system of ultraviolet protective factor (UPF), and the clothing factors affecting UPF will also be discussed. This chapter will also give insight into the available types of ultraviolet absorbers and commercial ultraviolet protective clothing. The nanoparticle coatings for ultraviolet protective finish and the durability of ultraviolet finishes will also be discussed briefly.

***Keywords*:** Ultraviolet protective agent, ultraviolet finishes, ultraviolet protective factor (UPF), ultraviolet absorber

8.1 Introduction

In the solar spectrum, a minimum fraction is constituted by the Ultraviolet rays [1]. Though the amount of that Ultraviolet ray is very low, it has a greater effect on all types of living objects. We may experience simple

**Corresponding author*: tajul.dtt@aust.edu

Luqman Jameel Rather, Mohd Shabbir and Aminoddin Haji (eds.) Sustainable Practices in the Textile Industry, (207–236) © 2021 Scrivener Publishing LLC

suntan to dangerous skin cancers if our skin remains unprotected and get direct exposure to sunlight containing Ultraviolet ray. Getting exposed to a limited amount of sunrays is not harmful to us as it helps the human body in developing the bones and in synthesis of vitamin D; but if the exposure to sunlight and Ultraviolet ray is extreme, it may damage our skin permanently [2], from sunburn to different skin-related diseases (aging, wrinkles, rough skin, blotches, sagging, malignant skin cancers, and DNA damages). Recently, these syndromes caused by ultraviolet radiation are breaking out rapidly [3]. High levels of the discharge of man-made fluorocarbons, especially chlorofluorocarbons (CFCs) are the main reason for Ozone diminution in the earth's atmosphere; and thus, the protection level which protects the earth environment against the incoming solar Ultraviolet Radiation (UVR) has become weaker. It has been revealed that if the thickness of the ozone layer decreased by 1%, it will increase the rate of skin cancer by 2–3%. Zuleyha, Kaynak, and Ali [4] stated that since the beginning of civilizations textiles have been practiced for the protection against solar radiation. Nowadays, many people are suffering from skin cancer and other skin diseases due to the long and frequent contact with UV radiation from sunlight [4].

People who are regularly working outside can't limit the exposure of skin to sunlight, for that reason, blocking UV-light by using textile are counted as the best substitute. Saravanan [5] suggested that though multiple measures can be followed to get protection from ultraviolet radiation; for example: evading extreme exposure to sun rays, using sunglass, putting sunscreen creams, but the higher protection can be received by using textile garments. Clothing made by special fabric can protect skin from solar radiation by reflecting, absorbing, and scattering solar wavelengths. Dying, adding pigments and ultraviolet absorber finishing are some special auxiliaries that are used in those textiles to establish the UV blocking property of cloth which absorb the radiation of ultraviolet and assists in blocking the UV ray's transmission which goes through the cloth to the skin [5].

8.2 Ultraviolet Radiation (UVR)

Electromagnetic radiation comprising the wavelength which is ranging from 100 to 400 nm is addressed as Ultraviolet Radiation (UVR). Moreover, this ray is longer than X-rays in wavelength but shorter than visible light. It would have been almost impossible for the living creatures on earth to sustain if the atmosphere had not filtered the maximum amount of UV rays. In comparison to visible and infrared radiation, UV imparts much higher

Table 8.1 Range and effects of ultraviolet radiation (UVR) [8].

Spectral band (nm)	Wavelength	Effects
UV-C	190–280	Skin burns, skin cancer
UV-B	280–320	Skin wrinkling, sunburn, skin and eye damage
UV-A	320–400	DNA damage, blurry eyes, the human immune system, premature aging of the skin

energy per photon though it has lower intensity [6]. According to Gulrajani [7], ultraviolet radiation can be categorized into three types (Table 8.1):

i. Ultraviolet A (UV-A),
ii. Ultraviolet B (UV-B) and
iii. Ultraviolet C (UV-C).

Among the three types of UV radiation, UV-C is the most dangerous and harmful. Ozone absorbs most of the UV-C photons and helps in prohibiting penetrating on the surface of the earth. Thus the intensity of UV-C on the earth's surface poses no threat to human beings or any living plants. The maximum amount of UV-B is consumed by the stratospheric zone, and a small amount of light reaches on earth. But UV-A has a greater impact on plants and animals as they can penetrate the ozone and stratospheric layer and thus reach the earth's surface [7].

8.3 Importance of Ultraviolet Protective Finish

Protective clothing is known as the clothing, which is particularly invented and treated to impart protection of the human being from hazards which is caused by severe changes in the surrounding environment. UV shielding agent is another name of Ultraviolet (UV) protection finishes and in case of applying to textile materials, these are the most significant clusters of chemical finishing agents. The purpose of these agents is to keep individuals safe from UV radiation's destructive effects [9]. The UV radiation's energy is considerably superior to other visible light and has the potential to instigate different types of chemical reactions that are harmful to human health. Covering materials, sportswear special functional textiles, wearable sensors, high-altitude clothing, and other technical textiles which have higher added value are made with finishes containing UV protection [10]. Though moderate exposure to the sun has advantageous health effects excess exposure to the sun, as well as UV radiation, may cause serious destructive health effects because UVA and UVB both rays to persuade different cellular responses. These are visible as skin aging, skin cancer, pigmentation, sunburn, and damages of DNA [11]. The effectiveness of the immune system has been reduced by UV radiation. The followings are the diseases caused by excessive sun exposure containing UV radiation:

Skin: Up to three million non-melanoma skin cancers takes place per year and more than 130,000 melanomas [12]. Lifestyle and behavioral changes in sunlight are the reasons for increasing skin cancers. Early repetitive sunlight and sunburn seem to be setting the phase for higher rates for melanoma well ahead in life. UV protection filter is provided by the ozone layer but ozone depletion can intensify the crisis even more. Other prolonged skin diseases due to UV radiation impart damages to the cells of the skin, tissue, and blood vessels are recognized as the aging of skin [7].

Eye: The increased effects of UV radiation over the eye area are photokeratitis, corneal and iris irritation, photo conjunctive swelling, and the eyelids skin's internal area. Durable impacts of UV radiation occurred due to the extreme exposure of the eye to the sun, include pterygium expansion (white cloudy growth close to the cornea) and conjunctive squamous cell cancer. Reportedly 16,000,000 people are sightless globally, according to the WHO, as a result of cataracts and around 20% may have been caused by UV exposure [12].

Immune System: Human immune system is sensitive to biological- and chemical substances, such as ultraviolet radiation, which decreases the immune efficiency by varying the action and allowance of immune reaction caused by cells. Many studies reported that natural rates of UV radiation can reduce human immune reactions and increase the chances of infection and reduce the rate of success of human vaccines [13].

Vulnerable Groups: Child needs special protection from UV as they are most delicate to ultraviolet radiation. Fair-skinned people are more sufferers than dark-skinned people in this case. Most of the skin cancers which are non-melanoma take place in people who are fair-skinned and take sunbath deliberately [14].

By avoiding the sunlight and by using protective attire and accessories, protection from UV and sun can be taken. Dropping the exposure to the sun, applying sunscreens to the skin as well as protective wears are the popular ways of getting a shield against the destructive consequences of UV radiation. In recent times, awareness of protective clothing is growing [15].

UV protection through clothing can be done by carrying hats, shoes, umbrellas, special cover for baby carrier, and the fabric materials to manufacture these products. Recently, dermatologists are also informing that the best protective technique to avoid excessive exposure to the sun is to wear clothes for covering body part [16].

8.3.1 Ultraviolet Protection With Textiles

Textiles can be essentially a blockade against UV radiation. Textiles can offer high protection from ultraviolet rays and the potentiality relies on the type, weave, thickness, and construction of materials and chemical finishes of the fabric. Clothing with a UV protective finish can keep the skin safe from solar radiation because the fabric material can absorb, reflect as well as disperse solar wavelengths penetrating through different materials of textile. Higher UV ray reflecting and absorbing properties should be present in UV protective textiles to stops UV rays that causes harm to human skin and body. Transmittance is the key property that determines the quality of clothes that provide UV protection. The ratio between the total amounts of incidental UV rays in a well-defined wavelength range and the number of transmitted UV rays touching the skin is known as the transmittance of UV rays through textile material [17].

When UV radiation hits a textile surface and forces the UV radiation to be wrecked down into many components, different types of effects are seen there. At the borders of the textile surface, a portion of the radiation is reflected. Again, another portion is engrossed when it cracks the sample and later portion transformed into a different energy form. The last portion is known as the "transmission" which enters through the fabric and makes contact with the skin [8].

The reflections, absorbance, and transmissions of relative amounts of radiation depend on different factors such as the smoothness of fiber surface, types of fiber, the fabric's cover factor (the fractions of the surface area of the fabric which are covered by yarns) as well as the presence/absence of fiber delustrants, dyes and UV stabilizers such as UV absorbers, quenchers, and HALs. Protective textile clothing should acquire a high capacity to reflect or absorb the UV rays so that it makes the transmission weaker and thus keep the human skin safe [4, 16].

The degree of solar protection which is provided by fabrics has gained more importance recently. WHO along with several photoprotection associations suggested that the defense given by fabrics as the first line of protection against solar UV and those organizations have also suggested increasing the usage of photoprotective creams, sunglasses, shoes, and hats [5].

Textile companies get new directions from these UV issues to develop new ways of R&D in dyes, detergents, fiber along with finished products to make sure the best of UV protection. That is why UV protected finishing is very important to protect our health in the current polluted environment.

8.4 Methods of Blocking Ultraviolet Rays

Sunlight is a good source of vitamin D and it helps the human body in developing the bones and in absorbing the vitamins, but excessive sunlight containing ultraviolet rays may bring simple suntan to dangerous skin cancers if human skin remains unprotected and get direct exposure to sunlight [18]. People outdoors can't limit the exposure of skins to sunlight, that's why blocking ultraviolet rays is counted as the most preferable alternative to save ourselves from the damage. The following are some methods that can be used to block ultraviolet rays.

Between 10 am to 4 pm, UV rays become more intense. Outdoor planning should be done by checking the peak sunlight hours. It is safe to do outside work before 10 am to after 4 pm and this practice will be helpful to prevent illnesses caused by sun heat and ultraviolet rays. While doing outside activities for a long time, it will be wise to avoid direct exposure to sunlight and ultraviolet rays [19]. Using pavilion and tree shad will be helpful to block the ultraviolet rays. Besides these, an umbrella will also give protection against harmful sunlight. Ultraviolet A and Ultraviolet B both can damage our skin. Using broad-spectrum sunscreen can block UVA and UVB radiation but one should choose the right sunscreen and should apply it properly [20]. Through clothing, more protection can be provided against ultraviolet rays. People should use long pants, long shirts with long sleeves to block harmful ultraviolet rays as much as possible; light-colored, light-weighted clothes and loose-fitting dress up may help us in preventing sunlight [19]. Though we can cover up with by using any type of clothes but not every clothes will give us protection against ultraviolet rays. If light can be seen through a fabric, ultraviolet rays can get through too [5]. The production of lightweight, comfortable, and protection provider against UV rays type fabric is running by many companies. These types of clothes are inclining to be more tightly woven. Many fabrics contain special coatings that absorb UV rays and give protection whether they are dry or wet. A label listing the UV protection factor (UPF) value (the level of protection the garment provides from the sun's UV rays, on a scale from 15 to 50+) is contained by these UV rays' protective fabrics (Table 8.2) [3]. If the fabric contains high UPF, the protection level against ultraviolet rays will also be

Table 8.2 Common UPF ratings [21].

Protection category	UPF rating	Average UV-A and UV-B blocking
Excellent	UPF 50 or UPF 50+	98.0% Blocking
	UPF 40	97.5% Blocking
Very Good	UPF 30	96.7% Blocking
Good	UPF 20	95.0% Blocking

higher. Sunscreens block only UVB rays but fabrics containing Ultraviolet Protection Factor (UPF) blocks both UVA and UVB radiation.

Ultraviolet rays can do damage to our lips by burning them. Lip balm or lipsticks containing SPF should be used to block the harmful sunlight. Using a hat can give protection to our face, neck, ears, and head from the ultraviolet rays. Better protection will come with better coverage. A sunhat with a wide brim should be used to block ultraviolet rays [22]. Ultraviolet rays can damage the eyes too. Using a pair of sunglasses that has UV absorption features, will block UV rays and will give protection to our eyes [23].

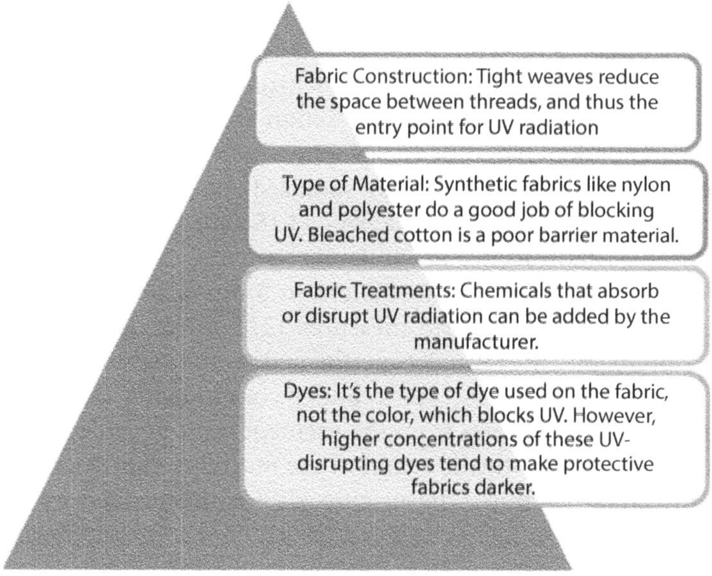

Figure 8.1 How clothing blocks UV radiation [20].

People use tanning beds and tanning lamps to get tanned skin, but these give out both UVA and UVB. So, the usage of tanning beds and sun lamps should be avoided to get rid of the adverse effects of ultraviolet rays [24]. It will be wisest to use all the ultraviolet blocking methods simultaneously to block UV rays. But among all the methods of blocking Ultraviolet rays, clothing is the best method as it gives the highest protection. Figure 8.1 shows how clothing blocks UV radiation.

8.5 Ultraviolet Protection Factor Measurement System

8.5.1 *In Vitro*

The key factor deciding UV safety for textiles is the direct and diffuse UV transmission through a fabric. Basic broadband UV dosimetry for measurements is only useful if a relative alteration of the UPF is needed. Spectroradiometers and spectrophotometers, however, are appropriate for spectral irradiance assessment. The transmitted and dispersed radiation both are collected through these devices via an integrated sphere located behind the textile sample. While a big dynamic range and great accuracy of spectrometers with a dual-monochromator are present, regular UV source scanning such as deuterium or xenon arc lights are mandatory to supply the reference data [25]. While a significant dynamic range and great accuracy of spectrometers with a dual-monochromator are present, usual UV source scanning such as deuterium or xenon arc lights are needed to supply the reference data [26]. The spectrophotometer must be equipped with a fluorescence filter as proposed by standard Australian, American, and European documents, for example, UG-11 (Schott, Mainz, Germany) for optimization of fluorescence in whitening agents.

Wavelengths range from 290–400 nm in 5 nm or less is the usual way to execute spectrophotometric measures [27]. Spectrophotometric textile measurements are usually performed with collimated radiation beams at right angles of the fabric under conditions of the 'worst case.' At least four textile samples from the garment must be collected for UPF determination—two in the direction of the machine and two in the direction of the cross-machine. The *in vitro* UPF is determined by weighing spectral irradiance (source as well as a transmitted spectrum) against erythema and calculating the UPF as the following Equation 8.1:

$$\text{UPF} = \frac{\int E\lambda S\lambda d\lambda}{\int E\lambda S\lambda T\lambda \, d\, \lambda} \tag{8.1}$$

here, Eλ = relative erythemal spectral effectiveness; Sλ = solar spectral irradiance in W/m^2; Tλ = spectral transmission of the sample; dλ = bandwidth in nm and λ = wavelength in nm; the integrals (∫) are measured over the wavelength range of 290–400 nm.

As already stated, the UPF shall be based on the average effective radiance of radiation for exposed skins estimated for skin covered by test fabric with a calculated mean effective radiation for UV radiation. The spectrophotometry is the accurate and consistent UPF testing process for intra- and inter-laboratory comparison studies, especially in specimens with UPF underneath 50 [28, 29]. The average daily UV exposure in Australia is less than 40 minimum erythema doses (MEDs) is of theoretical concern only for UPFs over 50. Besides that, recent findings from an inter-comparison between ten independent research laboratories and 11 separate measuring instruments on UVR were reported by Ref. [30]. This inter-comparison also contains comprehensive scan results from all these ten labs highlights the performance differences between different instrumentation systems in different wavelength regions in addition to comparing the measured UPF. A detailed review of these variations can indicate where device improvements should be made to enhance the performance of the equipment and the end outcome. The UPF measurement differences in the aspects of protection in standards should be expanded.

8.5.2 *In Vivo*

Similar to SPF, sunscreen tests *in vivo* are very infeasible for UPF identification by using human volunteers along with the sun which acts as a UV supply. Usually, to minimize visible and infrared radiation, xenon arc solar modules with collimated beams are applied in wavelength filters underneath 290 nm. Gies *et al.* [31] published non-based *in-vitro* test protocols. The UPF values were measured *in vitro* in the majority of the studies, however, using *in vivo* input [31, 32]. The MED is measured on the top back of a subject with gradual UV-B doses and is analyzed after 24 h, using skin phototype. The textile is placed on the other side of the back to determine the MED of covered skin [33]. The exponential

UV-B dosage for the measurement of the MED of the skin is multiplied by *in vitro* UPF, which leads to accelerated UVB doses for the MED protected skin test. If the *in vitro* process matches the *in vivo* method, the MED-protected skin proportion to the MED-protected skin proportion leads to the original *in vitro* UPF. However, several studies have shown that *in vivo* UPFs have a substantially lower threshold than UPF values attained during the *in vitro* testing of fabric specimens [34]. In comparison, the UPF values derived from *in vitro* experiments did not distinguish. In both studies, the "off-skin" method, which corresponds to a true weary condition, was also tested *in vivo*. UPF values from *in vivo* 'off-skin' testing were shown to differ little from *in vitro*-based UPF values [32]. There is certainly a different method such as UV sources and textile materials, different test protocols, as a result of the data inconsistency of these studies.

8.6 Clothing Factors Affecting Ultraviolet Protection Factor

Several factors associated with clothing including fabric structure, physio-chemical nature of fiber used in clothing, the color of clothes can affect the UPF. Table 8.3 shows the effect of various clothing factors on UPF.

Table 8.3 Clothing factors impact on UPF of textiles [36].

Factors	Effect on UPF
Fiber type	UPF of Nylon, wool, silk fibers is usually higher compared to cotton, viscose rayon, and linen; Polyester fibers show a high UPF value
Porosity, weight, the thickness of fabric	UPF ↑ = Yarn to yarn gap ↓; Fabric weight ↑; Thickness ↑
Color of fabric	For darker colors, the UPF value is higher compared to light colors
UV absorbers	UV absorbers improve UPF value
Stretch	UPF value is decreased when the fabric is stretched
Wetness	When the cotton fabric is wet then UPF value is decreased
Washing	For cotton fabric, UPF increases

8.6.1 Fabric Structure

Woven or knitted fabrics used for protection from UV rays should have a higher cover factor value than traditional fabrics. Researchers from their analysis observed that fabric's cover factor has a positive influence on UPF value. For woven fabric ends/inch (EPI), picks/inch (PPI) and for knit fabric courses/inch, wales/inch are primary determinant components for the cover factor. Usually, woven fabrics show a higher cover factor value than knit fabrics because of the way the yarns are interlaced. In the case of knitted fabric, pores between yarns are usually larger. If the pores between yarns are smaller, then UV radiation cannot transmit through the fabric, so radiation is blocked [35].

$$UPF = \frac{100}{Porosity\%} = \frac{100}{100 - Cover\ Factor} \tag{8.2}$$

The cover factor needs to be high for getting higher UPF. During the time of wearing, the cover factor and UPF value of fabric are reduced because of stretching. UPF is considerably reduced for nylon/lycra garment measured in the stretch state compared to a relaxed state. Washing or laundering greatly impact the cover factor of the fabric. For a 100% cotton long sleeve T-shirt (UPF value 15), it was observed that after one-time washing, the UPF value increased to 35, mainly due to the compression of shrinkage that means for increasing cover factor. Porosity is defined as the number of pores per unit of fabric area which is inversely proportional to the cover factor (Equation (8.2)). As a result, the UPF value is increased when porosity is decreased [37]. When the porosity is 1% then the maximum theoretical UPF value is 100. If porosity% is increased, then the UPF value is decreased. Table 8.4 shows the maximum UPF values that can be obtained for different porosity of fabric. If fabric density and thickness for

Table 8.4 Porosity and maximum theoretical UPF [35].

Porosity (%)	Maximum theoretical UPF
10	10
5	20
2	50
1	100

identical construction are increased, then the UPF value is also increased. UPF value depends on fabric porosity but it is also influenced by the fiber nature. The relative order of importance for ultraviolet protection is as follows: Cover % > Nature of fiber > Fabric thickness. Fabric weight and thickness show a better correlation with UPF compared to porosity. Fabrics making of a higher number of yarns in warp and weft show higher UPF.

8.6.2 Fiber Physio-Chemical Nature

The physical and chemical structure of fiber highly influences the UPF values. Transparency of UV is varied according to fiber's chemical nature. Synthetic fibers like PET have a higher degree of UVR absorption compared to natural fibers such as cotton, silk, and wool. In the region of 280–400 nm, cotton fibers show relatively higher transmission of UV which is found from spectroscopic studies. After bleaching, cotton fabric shows higher permeability to UV radiation. Gray cotton fabrics contain natural pigments, pectin, and waxes which have a UV absorbing property, so they have a higher UPF value compared to bleached fabric. Linen and hemp fibers show UPF values of 20 and 10–15 respectively and despite lignin content, they are not perfect UV protectors. However, Jute fiber has high absorption because it contains lignin which is a natural absorber. In the case of protein fibers, they show mixed results to allow UV radiation. Silk fabrics show medium transparency of UV. Unlike silk, wool fabrics show lower transmission and higher absorption of UV radiation. In the region of 280–400 nm and even above 400 nm, wool fiber highly absorbs UV radiation. In the region of UV-A and UV-B, polyester fibers absorb higher compared to aliphatic polyamide fibers. In the shorter wavelength of the UV region, polyester fibers result in higher absorption. If Titanium dioxide which is a delustering agent is added, then it strongly reduces the permeability of fiber over the UV spectrum. The degree of UV transmission of different materials is given below: Cotton bleached > Cotton gray > Polyamide > Silk > Wool > Polyester [35].

8.6.3 Dyeing

Textile materials ability to give protection from UV is greatly influenced by the dyeing process and variables such as the kind of dye or pigment, absorptive groups of dye, dyeing depth, uniformity, and additives. Bright fibers such as viscose show a higher range of UV radiation transmission compared to dull fibers in a given fabric. Instead of using heavy weight fabrics that are not comfortable for hot conditions, dyeing or printing can be done because these create a protective layer on the fabric. Fabrics of the same type and weave, dyed with a light color like pastel absorb UVR much lesser amount compared to dark color such as black, navy, dark red [5]. Undyed fabrics give lesser protection from UV radiation than dyed ones and when dye concentration is increased then protection level also increased with that [38]. Table 8.5 shows different fabrics dyed with different colored dyes effect on UPF. UPF values increase with the depth of color can be seen in Table 8.6. The K/S values of the dyed fabrics are a measure of color depth which were found to be effective in improving the UPF ratings in dyeing with natural colorants. For improving protection from the sun, vat dyes can be used for cellulosic fibers. If vat dye is used for dyeing

Table 8.5 Different dyes and different substrates impact on UPF [36].

Cotton	UPF	Polyester	UPF
White	12	White	16
Sky blue	18	Light green	19
Black	32	Dark red	29
Navy	37	Black	34

Table 8.6 UPF values of cotton fabric dyed with natural colorants [21].

Colorant	Concentration (%)	UPF	K/S
Madder	2	11.1	0.20
	4	15.8	0.28
	6	16.6	0.38
Cochineal	2	28.5	0.63
	4	34	0.79
	6	36.6	0.99
Indigo	2	43.1	1.78
	4	>50	2.56
	6	>50	3.02

or printing then it shows a protective effect for cellulosic, mainly for cotton fiber [37]. UPF value 50+ can be achieved by using some direct, reactive, and vat dyes. Dyes collected from numerous natural resources result in providing UPF values from 15 to 45 based on the use of mordant. A mixer of disperse-reactive dyes use for P/C blends, can give protection from UV with a UPF value of 50+ [5]. However, getting protection from UV by colored fabric may not be very comfortable in hot weather conditions as colored clothing tends to become hot after absorbing solar radiation.

8.7 Mechanisms of UV Protection

Textile items are primarily consumed for day to day clothing purposes and some of them are used for specific functional and technical purposes as well. UV protection for functional textile items has been introduced to avert the polymers, fibers, and cloths from being photo degraded. There are still some traits of mechanism that could be clarified even though it is impartially well acknowledged about the chemical pathways through which regular polymers are being photo degraded. Nevertheless, it is essential to comprehend the extremely noteworthy impact of additives and finishing substances in the amendment of such pathways, for instance—the effect of pigments, dyes, stabilizers, and extenders is vast [39].

Voluminous tactics have been implemented to defend the polymeric materials from being getting harmed done by UV radiation. The distribution of UV consuming substances in the photodegradable fibers and polymers is amongst the most frequently used techniques in the industry. Inorganic and organic compounds are accessible as UV soaking substances and that can be employed to give protection from UV radiation. Different inorganic materials hinge mostly on blended oxide films or particles, ready to assimilate or disperse light. For this persistence, there are classes of organic atoms as well which skillfully retain UV light as well as can likewise be utilized as added substances [40]. Light radiation can be mostly reflected, absorbed, or transmitted through the fiber or fabric during the strike on its surface (Figure 8.2). These transmitted, absorbed, or reflected radiation rest on numerous parameters, including the nature of the fiber, the surface properties of fiber, fabric cover, and the amount of dyes, fiber delustrants, and UV protective coatings [41].

Table 8.7 refers to the impact of fiber form on UPF of undyed fabrics. Since the radiation can be transmitted without substantial absorption, cotton and silk fibers provide low UV protection. On the other side, wool and polyester have sophisticated UPFs, as the UV radiation is more absorbed

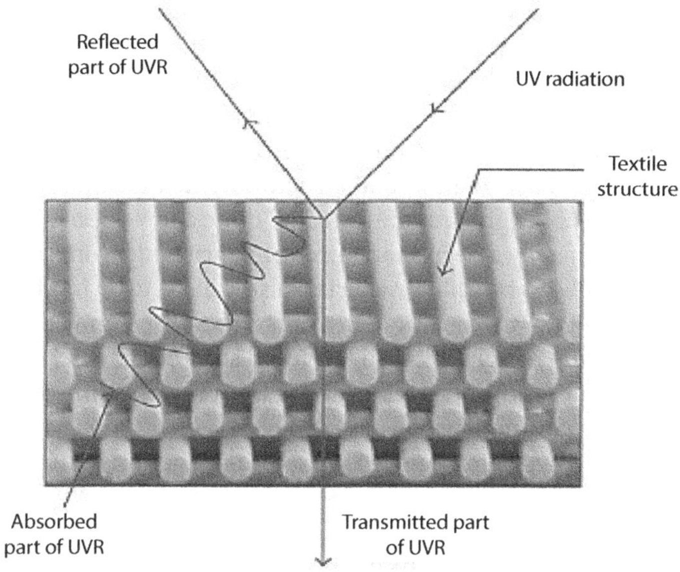

Figure 8.2 Radiation in contact with a textile surface reprinted from [3] under Creative Commons Attribution License.

by these fibers. Nylon falls in between those ends. The existence of the delustrant TiO_2, a substance that absorbs strong UV radiation, is a factor for nylon and polyester absorption [21].

If all the incident radiation is captivated by the fibers, the only source of transmitted rays is the distance between the yarns. By reference, the maximum theoretical UPF is 1 minus the cover factor reciprocal.

$$\text{UPFmax} = 1 / (1 - \text{cover factor}) \quad \ldots \quad \ldots \quad \ldots \quad \ldots \quad \ldots \quad \ldots \tag{8.3}$$

Table 8.7 UV protection factors (UPF) of undyed fabrics [21].

Fiber type	Fabric structure	Approximate UPF
Wool	Tricot	45
Polyester	Tricot	26
Nylon/Spandex	Tricot	12
Silk	Twill	7
Cotton Tricot	Tricot	4

The relation between the maximum UPF along with the cover factor is shown in Figure 8.3. The aim is to provide the wearer with excellent protection from sun UV radiations using the UPF value of 50 a fabric with a coverage factor of 0.98 made of fibers that absorb all non-reflective UV radiation [42]. Naturally, tight micro-fiber fabrics offer better UV protection than the normal size fiber and structurally identical fabrics. The UV radiation and visible light are captivated by quite a few dyes. UPF values of 50+ can be attained by a deep shaded cotton fabric only from the existence of the dye. And UV absorbing materials also can be used in fibers to provide the desired UPF value with light shades, as fashion and comfort regulate the application of light-colored fabrics in summer apparel. By evolving a sum of substances appropriate for use as UV protective finishes, dyestuff and auxiliary producers have reacted to a great extent [43].

To avoid UV rays entering the skin and damage the human body, the UVR-reflecting and absorbing effects of protective clothing should be as high as possible. Its transmission is the distinctive feature to create a determination of clothing protection's quality. UV transmittance through textile material is the ratio of the total incident radiation in a specified range to the

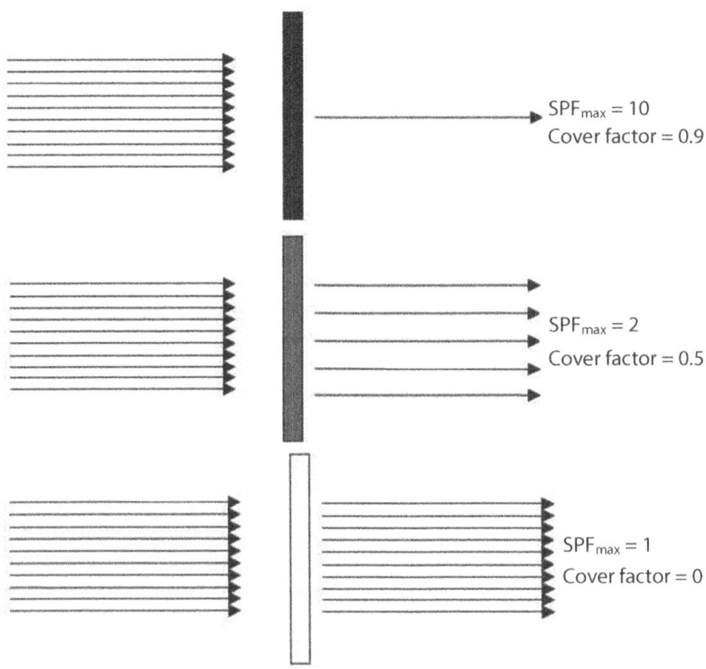

Figure 8.3 Interaction of radiation with fabrics of varying cover factors reprinted from [41] with permission from Elsevier.

amount of UV transmission rays to the skin. Light-weighted both woven as well as knitted fabrics of swimwear, beachwear, sportswear, t-shirts, and shirts are the most common candidates for UV protection. UV-protective material can also be used in products made from synthetic fabrics such as canopies, tents, awnings, and blinds [41].

8.8 Types of Ultraviolet Absorbers

8.8.1 Organic

Organic UV absorbers are found at a comparatively low price and transparent, so they can be used in many-colored textiles, anyhow, the efficiency of UV absorption is decreased over time, as maximum organic absorbers are eventually destroyed by absorbing the UV radiation. Additionally, free radicals are generated during the photodecomposition of organic UV absorbers which are the reasons for degrading other organic molecules. Because of the small size, UV absorbers are getting rid out of textiles and when contaminated with food and drinks, it causes health problems. On the contrary, inorganic UV absorbers e.g. zinc oxide (ZnO), titanium dioxide (TiO_2), and cerium oxide (CeO_2) show excellent fastness to light. The fundamental stability of inorganic UV absorbers gives a UV protective effect over a longer time compared to organic UV absorbers [45].

8.8.2 Inorganic

Inorganic UV absorbers show some other superiority over organic UV absorbers such as ZnO which is safer for topical use because it has anti-irritant and skin-healing properties. Because of the molecular system's typical absorption bands, organic UV absorbers show UV absorption peaks only at a definite wavelength range while zinc oxide has large UV absorption spectra. Amongst many inorganic UV absorbers, ZnO has a wide spectrum absorption range. ZnO has poor chemical stability. In both low and high pH conditions, it can dissolve. TiO_2 show excellent stability to chemical, but its UV absorption range is contracted compare to ZnO, so it usually depends on the scattering of light effects with light absorption effects for blocking UV light. In Figure 8.4, a typical spectrum of transmission for TiO_2 nanoparticle suspension, ZnO nanoparticle suspension, and phenyl acrylate-based organic UV absorber solution is given [45].

Rayosan C (UV absorber) is a heterocyclic compound and it creates a covalent bond when treated with cellulose macromolecule [44]. The

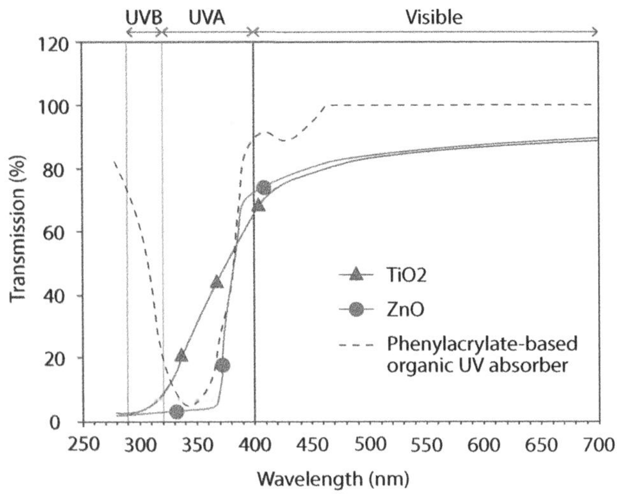

Figure 8.4 Typical spectrum of transmission for a titanium dioxide nanoparticle suspension, zinc oxide nanoparticle suspension, and phenyl acrylate-based organic UV absorber solution (Note: Concentration of UV blocking agents in the system, have a direct influence on the relative transmission values between the curves) reprinted from [45] with permission from Emerald Publishing Limited.

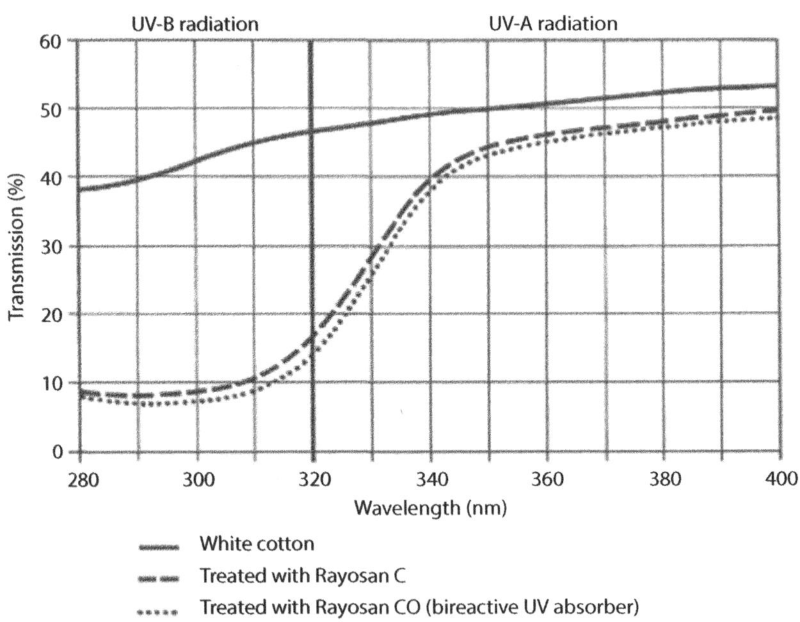

Figure 8.5 An ultraviolet-absorbent finish effects on absorption at different wavelengths reprinted from [46] with permission from Elsevier.

transmission percentage for the white cotton fabric without any treatment with UV absorber, increased steadily as the UV wavelength increased. The UV absorber treated fabrics had a substantially lower rate of transmissions compared to the untreated fabric. Figure 8.5 shows how to block the harmful UVB rays and allow the passing of beneficial UVA rays through the fabric by using a special finish [46]. The UV transmission percentage through Rayosan CO treated fabric is comparatively lower than Rayosan C treated fabric because Rayosan CO is a bi-reactive UV absorber.

8.9 Commercial Ultraviolet Protective Clothing

Rockywoods has woven and stretchy Sun Protective Fabrics to protect human skin. It has carried out a variety of Sun Protective woven fabrics (perfect for shirts, light jackets, pants, sleeves stroller covers, and more) or stretchy fabrics including different weights, finishes, colors, and prints [47].

SunGrubbies Company's mission is to keep skin safe by blocking the sun with stylish and comfortable sun protective hats. They also have UPF clothing and accessories of the highest quality and reliability [48].

SanSoleil UV 50 sun protection tops are beautiful, stylish, and comfortable to wear which are designed for hot and humid days. These tops are provided wicking, moisture control, and ultimate protection from the sun [49].

Their UPF 50+ sun protection garments feature is to encounter harmful UVA and UVB rays. These types of clothing are suitable for optimal wear in a variety of outdoor activities, thick enough to deflect UV rays, and can be used in both humid and dry climates [50].

98% of the sun's harmful UVA and UVB rays are blocked by their UPF 50+ sun protective apparel and providing peace-of-mind in the sun while someone enjoys the outdoors [51].

8.10 Nanoparticle Coatings for Ultraviolet Protective Textiles

Nanotechnology is a part of science and technology which manipulates the atoms, molecules, functional materials, and systems on the scale of nanometer range by using the specific placement of individual atoms of 0.1–100 nm size to create structures of excellent properties. Scientists and researchers are attracted by the rare and different properties of nanomaterials, so the adoption of nanotechnology has increased speedily in the textile industry because textile is one of the best areas for expanding nanotechnology. To produce finished fabrics with diverse functional properties, nanoparticles are applied to textile materials such as TiO_2 nanoparticle is used for blocking of UV radiation and self-cleaning properties, as well as, zinc oxide nanoparticles are imparting for UV-blocking and antibacterial properties.

The advantages of zinc oxide and titanium are that they are not toxic and have chemical stability when exposed to high temperatures and the capability of photocatalytic oxidation. Moreover, nanoparticles provide a large surface area to volume ratio for which the efficiency of photocatalytic oxidation is significantly increased than bulk materials [52]. Irradiation of titanium dioxide by light with higher energy than its band gaps develops electron-hole pairs which activate reactions (redox) at the surface of the TiO_2. As a consequence, electrons present in TiO_2 bounce from the valence band to the conduction band, moreover, on the photocatalyst's surface, the electron (e^-) and electric hole (h^+) pairs are generated. Figure 8.6 shows that the formed negative electrons and oxygen will merge into $O_2^{-\cdot}$ as well as the positive electric holes and water will produce hydroxyl radicals [53].

Cotton yarns treated with ZnO nanoparticles were able to withstand the knitting process. Application of TiO2 nanoparticles on the bleached and reactive dyed cotton fabrics by the sol–gel and linking agent methods were discovered in the original state after numerous domestic washing cycles. Knitted fabrics treated with ZnO nanoparticles offered moderate to high UPF values and TiO_2-coated samples showed 50+ UPF values [54]. Bio-nanocomposite which consists of a polymer matrix (natural) and filler (organic or inorganic) where at least one dimension has to remain on the nanometer scale, is a rising ground in the borderline of nanotechnology, materials science, and life science. To reinforce the multifunctional properties like UV protection and antibacterial activity, cotton fabric was treated with seaweed capped ZnO nanoparticles by pad dry cure method [55]. Cerium oxide nanoparticles have drawn enormous consideration because of their different utilizations as catalysts, superconductor buffer layer, oxygen storage materials, fuel cell electrolyte, and UV absorbents.

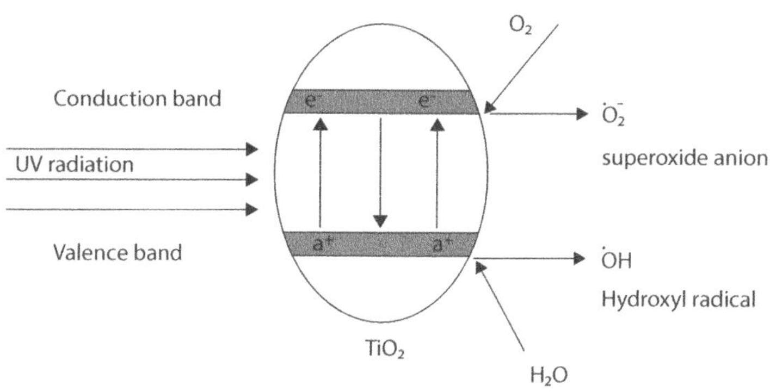

Figure 8.6 TiO_2 nanoparticles' photocatalytic mechanism under the radiation of UV reprinted from [53] with permission from Taylor & Francis.

CeO_2 acquires a relatively limited bandgap compared to ZnO and TiO_2, this property makes CeO_2 a better UV absorbent for the coating of textile fiber. CeO_2 nanoparticles can be a good candidate as per their advantageous properties for functionalizing silk fiber in antibacterial and UV protective applications [56].

When conventional textile finishing materials are applied to fabrics for getting various properties like water repellency and stain repellency, they usually provide temporary effects and lose their functions over time. Nanoparticles have a large surface area and high surface energy which assure affinity towards fabrics and increase the durability of the required textile functions for the treated fabrics. The properties of a material are substantially changed after decreasing the particle size to the nanoscale. Because of the small size and high surface energy, nanoparticles are attached with Van der Waals forces to the fabric surface for which acceptable washing fastness is possible to get. By forming covalent bonds between nanoparticles and fabric surface, washing fastness can be improved and in such cases, it is possible to manage the excellent functional properties after around 55 home laundering [52].

8.11 Durability of Ultraviolet Protective Finish

A study was done where two cotton fabrics that showed limited UV radiation protection measured through spectrophotometric analysis. As the fabric is shrunk after washing with detergent and water so UPF value is improved. During washing, if a UV absorbing agent is added then it substantially reduces the transmission of UV, and UPF value is increased. The UV protection factor determinations for both fabrics, styles '400M' and '437W' washed with only water are shown in Table 8.8.

For the T-shirt made of cotton (437W), the mean value for UPF was 4.70 and after 5 times of washing and drying, the UPF was increased by 50%. A related result in the case of increasing UPF value was also viewed with the mercerized print cloth of style 400M. In Table 8.9, UV protection factor results for the fabrics washed with detergent are shown.

The mean UPF value was 5.00 for the 5 before washed specimens of the cotton T-shirt (437W). As D: 1 to D: 5 specimens had gone through different numbers of washings, so the mean UPF value was not measured for the after washed specimens. Each specimen was delivered as its control for the before wash and after wash results comparison. After the first washing, the UPF value increased to 8% and after five washing, the value increased

Table 8.8 Measurements of UPF for a T-shirt made of cotton (437W) and mercerized print cloth (400M) washed with only water (W) reprinted from [2] with permission from Elsevier.

Specimen ID	No. of washing cycles	Before wash	After wash	Difference between after wash and before wash
T-shirt (Made of Cotton)				
W: 1 specimen	5	4.90	7.10	45.20%
W: 2 specimen	5	4.90	7.20	56.40%
Mean		4.70	7.10	50.60%
Printed cloth (Mercerized)				
W: 1 specimen	5	3.20	4.40	37.70%
W: 2 specimen	5	3.10	4.00	29.90%
Mean		3.10	4.20	33.80%

to 17%, moreover, the related effect for increasing UPF also viewed with the mercerized print cloth of style 400M. The mean value was 3.1 for the before washed specimens. The UPF value increased 36% after the one-time washing and 42.9% after five-time washings. The UV protection factor results for fabrics laundered with detergent and UV absorber are shown in Table 8.10.

For the specimens of the cotton T-shirt (437W), before washing the mean UPF value was 4.8. After one washing, the UPF value was increased to 132% and after five-time washings, the UPF raised by more than 400%. In the case of the mercerized print cloth of style 400M washed with the detergent-UV absorber mixture after one washing the UPF value increased to 127% and after each of the washing treatment, there was viewed a steady increase in UPF value and hence after five times washings, the UPF value increased by 296%. The mean UPF values were 4.94 for the cotton T-shirt fabric and 3.13 for the print cloth before laundering and transmission of UVA (320–400 nm) was higher than UVB (280–320 nm) through these fabrics. After five times washings the cotton T-shirt fabric UPF value increased by 51% when washed with only water and 17% when washed with the only detergent, moreover, washing with detergent-UV absorbing group increased the UPF by 407%. For the print cloth fabric, the same changes in UPF values were viewed [2].

Table 8.9 Measurements of UPF for a T-shirt made of cotton (437W) and mercerized print cloth (400M) washed with only detergent (D) reprinted from [2] with permission from Elsevier.

Specimen ID	No. of washing cycles	Before wash	After wash	Difference between after wash and before wash
T-shirt (Made of Cotton)				
D: 1 specimen	1	5.00	5.30	7.90%
D: 2 specimen	2	4.90	6.00	22.10%
D: 3 specimen	3	5.00	5.80	17.30%
D: 4 specimen	4	4.80	5.90	21.10%
D: 5 specimen	5	5.10	6.00	17.20%
Mean		5.00		
Printed cloth (Mercerized)				
D: 1 specimen	1	3.20	4.40	36.00%
D: 2 specimen	2	3.30	5.00	49.10%
D: 3 specimen	3	3.10	4.00	30.20%
D: 4 specimen	4	3.00	4.00	31.10%
D: 5 specimen	5	3.10	4.40	42.90%
Mean		3.10		

Table 8.10 Measurements of UPF for a T-shirt made of cotton (437W) and mercerized print cloth (400M) washed with detergent and UV absorber (DA) reprinted from [2] with permission from Elsevier.

Specimen ID	No. of washing cycles	Before wash	After wash	Difference between after wash and before wash
T-shirt (Made of Cotton)				
DA: 1 specimen	1	4.70	10.90	132.30%
DA: 2 specimen	2	4.90	17.70	262.00%

(Continued)

Table 8.10 Measurements of UPF for a T-shirt made of cotton (437W) and mercerized print cloth (400M) washed with detergent and UV absorber (DA) reprinted from [2] with permission from Elsevier. (*Continued*)

Specimen ID	No. of washing cycles	Before wash	After wash	Difference between after wash and before wash
DA: 3 specimen	3	4.90	17.70	258.40%
DA: 4 specimen	4	4.90	23.40	380.10%
DA: 5 specimen	5	4.50	23.00	407.70%
Mean		4.80		
Printed cloth (Mercerized)				
DA: 1 specimen	1	3.10	7.00	127.6%
DA: 2 specimen	2	2.90	7.60	160.5%
DA: 3 specimen	3	3.10	10.90	250.3%
DA: 4 specimen	4	3.20	12.30	288.3%
DA: 5 specimen	5	3.00	11.90	296.3%
Mean		3.10		

8.12 Conclusion

The adverse effects of UV radiation are warned by dermatologists, meteorologists, biologists, and others furthermore, textile scientists play a crucial role in providing UV protection through textile material which is a very suitable and effective blockade against UV rays [43]. Because of the thinning of the ozone layer and increasing the number of skin cancer patients, it is needed to protect humans from UV radiation. To fight against skin cancer, textile with a higher UPF value is one of the essential elements [36].

A fabric's UPF value depends upon various factors such as fiber type, fabric construction, nature of chemical processing e.g. dyeing and finishing, presence of additives e.g. UV absorbers or optical brighteners. UV absorption value is decreased after bleaching because of the removal of some natural UV absorbers from cellulosic and lignocellulosic fibers [37]. One of the simplest and cheapest ways for attaining good UV protection is to focus on fabric construction without any additional finishing processes [43]. Fabrics made of plain weave or sateen weave structure in the undyed state provide no protective abilities and after dyeing with natural colorants, the UPF value is considerably increased moreover, the degree of protection depends on the amount of the colorant in the fabric [57]. Because of the porosity of fabric, knit fabrics having open structures show lower UVR protection compared to correspondent woven fabrics [58]. During wear and use, different factors such as stretch, wetness, degradation due to laundering can change a textile's UV protective properties [59].

Textiles without any treatment have limited capabilities of UV light blocking, so treatment with UV absorbing agents is necessary. When the size of particles is reduced to nanoscale then it is possible to obtain a higher range of transparency, as nanoparticles increased the UV shielding effect and have a strong affinity for fiber surfaces. Innovation in the field of nanotechnology is enhanced so, more choices for nanoparticles will be available in the market [45]. UV protective clothing plays an important role in case of preventing skin cancer, photo dermatoses, and premature skin aging. It is necessary to increase awareness for changing people's behavior towards the sun and the use of UV protective clothing. According to the customers' acceptance and demand, there will be created a market for labeled UV protective clothing [18].

References

1. Akgun, M., Becerir, B., Alpay, H.R., *Ultraviolet Protection of Textiles: A Review*, International Scientific Conference, Gabrova, 2010.
2. Wang, S.Q., Kopf, A.W., Marx, J., Bogdan, A., Polsky, D., Bart, R.S., Reduction of ultraviolet transmission through cotton T-shirt fabrics with low ultraviolet protection by various laundering methods and dyeing: Clinical implications. *J. Am. Acad. Dermatol.*, 44, 5, 767–74, 2001.
3. Singh, M.K. and Singh, A., Ultraviolet Protection by Fabric Engineering. *J. Text.*, 2013, 1–6, 2013.
4. Zuleyha, D., Kaynak, H.K., Ali, K., UV protection of naturally colored cotton woven and knitted fabrics in comparison to white and dyed fabrics. *The Fiber Society Spring International Conference*, 2010.
5. Saravanan, D., UV protection textile materials. *Autex Res. J.*, 7, 1, 53–62, 2007.
6. Vázquez, M. and Hanslmeier, A., *Ultraviolet Radiation in the Solar System*, Springer, Dordrecht, Netherlands, 2006.
7. Gulrajani, M., *Advances in the Dyeing and Finishing of Technical Textiles*, Woodhead Publishing Ltd, Cambridge, UK, 2013.
8. Scott, R., *Textiles for Protection*, Woodhead Publishing Ltd, Cambridge, UK, 2005.

9. Gambichler, T., Ultraviolet protection of clothing, in: *Functional Textiles for Improved Performance, Protection and Health,* N. Pan, G. Sun (Eds.), p. 45–63, Woodhead Publishing Limited, Cambridge, UK, 2011.
10. Chakraborty, J.N., Sharma, V., Gautam, P., Enhancing UV Protection of cotton through application of novel UV absorbers. *J. Text. Apparel Technol. Manage.*, 9, 1, 1–17, 2014.
11. Wong, W., Lam, J., Kan, C., Postle, R., Ultraviolet protection of weft-knitted fabrics. *Text. Prog.*, 48, 1–54, 2016.
12. Nasreen, A., Umair, M., Shaker, K., Hamdani, S., Nawab, Y., Development and Characterization of Three-Dimensional Woven Fabric for Ultra Violet Protection. *Int. J. Clothing Sci. Technol.*, 30, 536–547, 2018.
13. Hossain, M. and Rahman, M., A Review of Nano Particle Usage on Textile Material against Ultra Violet Radiation. *J. Text. Sci. Technol.*, 01, 03, 93–100, 2015.
14. Etzel, R. and Balk, S., *Pediatric Environmental Health*, American Academy of Pediatrics, Elk Grove Village, IL, 2012.
15. Reiffenrath, M., Hoerr, M., Gries, T., Jockenhoevel, S., Smart Protective Clothing for Law Enforcement Personnel. *Mater. Sci. Text. Clothing Technol.*, 9, 64, 2015.
16. Coojack, L., Davis, S., Kerr, N., Textiles and UV radiation. *Can. Text. J.*, 111, 3, 14–15, 1994.
17. Gies, H.P., Roy, C.R., Mclennan, A., Diffey, B.L., Pailthorpe, M., Driscoll, C., *et al.*, UV Protection by Clothing: An Intercomparison of Measurements and Methods. *Health Phys.*, 73, 456–464, 1997.
18. Gambichler, T., *Functional textiles for improved performance protection and health*, G. Sun and N. Pan (Eds.), p. 45–57, Woodhead publishing, Cambridge England, 2011.
19. Singh, M.K., Sun protective clothing. *Asian Text. J.*, 14, 1, 91–97, 2005.
20. Hilfiger, R., Improving sun protection factors of fabrics by applying ultraviolet absorbers. *Text. Res. J.*, 66, 2, 61–70, 1996.
21. Zimniewska, M. and Batog, J., *Ultraviolet blocking properties of natural fibres, Handbook of natural fibres,* p. 141–164, Woodhead Publishing, Cambridge, UK, 2012.
22. D'Orazio, J., Jarrett, S., Ortiz, A.A., Scott, T., UV Radiation and the Skin. *Int. J. Mol. Sci.*, 14, 6, 12222–48, 2013.
23. Giese, A., *Living With Our Sun's Ultraviolet Rays*, Springer US, Boston, MA, 1976.
24. Dore and Chignol, Tanning salons and skin cancer. *Photochem. PhotoBiol. Sci.*, 11, 30–37, 2012 2012.
25. Capjack, L., Kerr, N., Fedosejevs, R., Hatch, K.L., Markee, N.L., Protection of Humans from Ultraviolet Radiation Through the Use of Textiles: A Review. *Family Consum. Sci. Res. J.*, 23, 198–218, 1994.
26. Gies, H.P., Roy, C.R., Elliott, G., Zongli, W., Ultraviolet Radiation Protection Factors for Clothing. *Health Phys.*, 67, 131–139, 1994.

27. Laperre, J. and Gambichler, T., Sun Protection Offered by Fabrics: on the Relation Between Effective Doses Based on Different Action Spectra. *Photodermatol. Photoimmunol. Photomed.*, 19, 11–16, 2003.
28. Hoffmann, K., Kesners, P., Bader, A., Avermaete, A., Altmeyer, P., Gambichler, T., Repeatability of *In vitro* Measurements of the Ultraviolet Protection Factor (UPF) By Spectrophotometry with Automatic Sampling. *Skin Res. Technol.*, 7, 223–226, 2001b.
29. Laperre, J., Gambichler, T., Böhringer, B., Driscoll, C., Varieras, S., Gassan, U., Determination of the Ultraviolet Protection Factor of Textile Materials: Measurement Reliability. *Photodermatol. Photoimmunol. Photomed.*, 17, 223–229, 2001.
30. Gies, P., Roy, C., Mclennan, A., Pailthorpe, M., Hilfiker, R., Osterwalder, U., Monard, B., Moseley, H., Sliney, D., Wengraitis, S., Wong, J., Human, S., Bilimis, Z., Holmes, G., Ultraviolet Protection Factors for Clothing: An Intercomparison of Measurement Systems. *Photochem. Photobiol.*, 77, 58–67, 2003.
31. Gies, H.P., Roy, C.R., Holmes, G., Ultraviolet Radiation Protection by Clothing: Comparison of *In vivo* and *In vitro* Measurements. *Radiat. Prot. Dosim.*, 91, 247–250, 2000.
32. Menzies, S.W., Lukins, P.B., Greenoak, G.E., Walker, P.J., Pailthorpe, M., Martin, J.M., A Comparative Study of Fabric Protection against Ultraviolet-Induced Erythema Determined by Spectrophotometric and Human Skin Measurements. *Photodermatol. Photoimmunol. Photomed.*, 8, 157–163, 1991.
33. Gambichler, T., Avermaete, A., Bader, A., Altmeyer, P., Hoffmann, K., Ultraviolet Protection by Summer Textiles. Ultraviolet Transmission Measurements Verified by Determination of the Minimal Erythema Dose with Solar-Simulated Radiation. *Br. J. Dermatol.*, 144, 484–489, 2001a.
34. Greenoak, G.E. and Pailthorpe, M., Skin Protection by Clothing from the Damaging Effects of Sunlight. *Australasian Textiles*, 16, 61, 1996.
35. Das, B.R., UV Radiation Protective Clothing. *Open Text. J.*, 3, 14–21, 2010.
36. Sayed, U., Tiwari, U.R., Dabhi, P., UV Protection Finishes on Textile Fabrics. *Int. J. Adv. Sci. Eng.*, 1, 3, 56–63, 2015.
37. Bajaj, P., Kothari, V.K., Ghosh, S.B., Some Innovations in UV protective clothing. *Indian J. Fibre Text. Res.*, 25, 315–329, 2000.
38. Grifoni, D., Bacci, L., Zipoli, G., Albanese, L., Sabatini, F., The role of natural dyes in the UV protection of fabrics made of vegetable fibres. *Dyes Pigm.*, 91, 279–285, 2011.
39. Andrady, A.L., Hamid, S.H., Hu, X., Torikai, A., Effects of increased solar ultraviolet radiation on materials. *J. Photochem. Photobiol. B.*, 46, 96–103, 1998.
40. Zayat, M., Garcia-Parejo, P., Levy, D., Preventing UV-light damage on light sensitive materials using a highly protective UV-absorbing coating. *Chem. Soc. Rev.*, 36, 1270–1281, 2007.

41. Schindler, W.D. and Hauser, P.J., *Chemical Finishing of Textiles*, p. 157–162, Woodhead Publishing Ltd, Cambridge, UK, 2004.
42. Reinert, G., Fuso, F., Hilfiker, R., Schmidt, E., UV-Protecting Properties of Textile Fabrics and Their Improvement. *Text. Chem. Color.*, 29, 12, 36–43, 1997.
43. Dubrovski, P.D. and Brezocnik, M., Prediction of the ultraviolet protection of cotton woven fabrics dyed with reactive dyestuffs. *Fibres Text. East. Eur.*, 17, 1, 55–59, 2009.
44. Merdan, N., Koçak, D., Şahinbaşkan, B.Y., Yüksek, M., Effects of UV absorbers on cotton fabrics. *Adv. Environ. Biol.*, 6, 7, 2151–2157, 2012.
45. Tsuzuki, T. and Wang, X., Nanoparticle coatings for UV protective textiles. *Res. J. Text. Apparel*, 14, 2, 9–20, 2010.
46. Slater, K., *Environmental impact of textiles: Production, processes and protection*, p. 165, Woodhead Publishing Limited, Cambridge, UK, 2003.
47. Rockywoods, https://www.rockywoods.com/FABRICS/Activewear-Fabrics/Sun-Protective-Fabrics, 2020.
48. Sungrubbies, https://www.sungrubbies.com/pages/sun-hat-specials, 2020.
49. Sansoleil, https://www.sansoleil.com/, 2020.
50. Ibkul, https://ibkul.com/blogs/news/the-future-of-sun-protection, 2020.
51. Uvskinz, https://www.uvskinz.com/pages/about-us, 2020.
52. Kathirvelu, S., D'Souza, L., Dhurai, B., UV protection finishing of textiles using ZnO nanoparticles. *Indian J. Fibre Text. Res.*, 34, 267–273, 2009.
53. Alebeid, O.K. and Zhao, T., Review on: developing UV protection for cotton fabric. *J. Text. Inst.*, 108, 12, 2027–2039, 2017.
54. Paul, R., Bautista, L., Varga, M.D.I., Botet, J.M., Casals, E., Puntes, V., Marsal, F., Nano-cotton Fabrics with High Ultraviolet Protection. *Text. Res. J.*, 80, 5, 454–462, 2010.
55. Pandimurugan, R. and Thambidurai, S., UV protection and antibacterial properties of seaweed capped ZnO nanoparticles coated cotton fabrics. *Int. J. Biol. Macromol.*, 105, 788–795, 2017.
56. Lu, Z., Mao, C., Meng, M., Liu, S., Tian, Y., Yu, L., Sun, B., Li, C.M., Fabrication of CeO_2 nanoparticle-modified silk for UV protection and antibacterial applications. *J. Colloid Interface Sci.*, 435, 8–14, 2014.
57. Sarkar, A.K., An evaluation of UV protection imparted by cotton fabrics dyed with natural colorants. *BMC Dermatology*, 4, 15, 1–8, 2004.
58. Sarkar, A.K., On the relationship between fabric processing and ultraviolet radiation transmission. *Photodermatol. Photoimmunol. Photomed.*, 23, 191–96, 2007.
59. Hoffmann, K., Laperre, J., Avermaete, A., Altmeyer, P., Gambichler, T., Defined UV protection by apparel textiles. *Arch. Dermatol.*, 137, 1089–1094, 2001.

9

Sustainable Orientation of Textile Companies

Gherghel Sabina

University of Oradea, Faculty of Energy Engineering and Industrial Management, Department of Textiles, Leather and Industrial Management, Oradea, Romania

Abstract

The issue of sustainability is a contemporary issue, closely linked to the textile industry, which is one of the largest industrial forces in the world. Through its activity, the textile industry has a negative impact on the environment, throughout the life of the product, starting with the activities of cultivation of textile plants, dyeing and cleaning, the actual production process, the distribution process, but also after the product arrives to the final consumer, through the cleaning and maintenance processes, until the product reaches the landfill at the end of its life. In the context of globalization and economic industrialization, sustainable consumption is becoming an important issue for existence. Thus, the Living Planet Report shows that in the current consumption conditions, in 2030 the inhabitants of the Earth will need another planet in order to exist. At the moment, consumption exceeds the potential of the planet by 50%. The textile industry is governed, in recent years, by the principles of "fast fashion", according to which consumers are encouraged to buy as much as possible, at the lowest possible prices, often with substantial discounts, products designed in a very short time, of dubious quality, designed to meet the needs of the consumer for a short period of time. Also, from a company's perspective, sustainable development does not provide clues as to what the company needs to do or what it needs to transform to highlight the specific characteristics of sustainable development. The concept of sustainability covers exactly this lack from the perspective of economic agents, because their transformation into sustainable economic agents is in fact part of the solution to the problem of sustainable development and within this much needed industry, that of the textile industry.

Email: sgherghel@uoradea.ro

Luqman Jameel Rather, Mohd Shabbir and Aminoddin Haji (eds.) Sustainable Practices in the Textile Industry, (237–252) © 2021 Scrivener Publishing LLC

Keywords: Sustainable development, textile industry, globalization, economy, industrialization

9.1 Introduction

Sustainability or sustainable development are important concepts for brands and manufacturers in the textile industry. In this context, producers are concerned with the development of manufacturing processes that protect the environment and staff. The worldwide "green textiles" market is a market with changing behavior, is constantly expanding and offers huge opportunities to increase production in the textile sector around the world and especially in Europe. The idea of sustainability comes from the concept of sustainable development which became a common language at the First World Earth Summit, Rio, 1992. There is no unanimously accepted definition for "sustainability or sustainable development". Below are some definitions of this concept:

a) "A process of change in which resource exploitation, investment direction, technological development and institutional change are all in harmony and enhance both current and future potential to meet human needs and aspirations" [1];
b) "Sustainable development is a dynamic process that enables people to realize their potential and improve their quality of life through ways that simultaneously protect and improve life support systems" [2];
c) "Essentially, sustainable development is based on five key principles: quality of life; fairness and equity; participation and partnership; caring for the environment and respecting environmental constraints—recognizing "that there are" environmental limits ";" faith in the future on the precautionary principle" [3];
d) "The environment must be protected, the essential functions of the ecosystem and the well-being of future generations must be maintained; environmental and economic policy must be integrated; the aim of the policy must be to improve the overall quality of life, not just the growth of the income; poverty has to be eradicated and resources distributed equally, and all sections of society must be involved in decision-making" [4];

e) "We can not only add sustainable development to our current to-do list, but we must learn to integrate its concepts into everything we do" [5];
f) "A sustainable future is one in which there is a healthy environment, economic prosperity and social justice and all are pursued simultaneously to ensure the well-being and quality of life of present and future generations. Education is crucial to achieving that future" [6];
g) "The first and probably the most difficult problem is that of time: Is a society sustainable if it lasts for a decade, a human life, or a thousand years?" [7].

Through its activity, the textile industry has a negative impact on the environment, throughout the life of the product, starting from the activities of cultivation of textile plants, dyeing and cleaning, the actual production process, distribution, but also after the product reaches the final consumer, through the cleaning and maintenance processes, until the product reaches the landfill at the end of its life [8]. In the context of globalization and economic industrialization, sustainable consumption is becoming an important issue for existence. Thus, the Living Planet Report shows that in the current consumption conditions, in 2030 the inhabitants of the Earth will need another planet in order to exist. At the moment, consumption exceeds the planet's potential by 50% [9]. Sustainable systems include, at the most demanding level, those activities that must be carried out in order to progress towards sustainable development. Sustainability, due to the fact that it requires continuity over time, is difficult to achieve and even more difficult to prove. If external strategic thinking involves negotiations, market penetration, financing, adaptation to legislation, the internal strategic part means product development, organizational structuring, evaluations.

9.2 Textile Industry—Environmental, Social and Economic Issues

The global value of the textile and clothing industry is of several trillion USD, the number of people working in the industry being over 300 million. Annually, the industry consumes more than 70 billion m^3 of water, over 30 billion liters of oil, using over 40 million tons of chemicals in all stages of the value chains and thus becoming one of the most polluting industries. About 25% of hazardous substances are discharged into water, destroying aquatic ecosystems [10, 11]. 14 million employees earn less than $3 a day, more than

180 million employees working in precarious conditions, most of the production is done in low- and middle-income countries that are still developing their environmental regulatory systems, and those for labor protection [12]. Compared to 15 years ago, the use of a clothing product decreased by 36%. A large part of clothing products, especially those bought at a discount, are thrown away without being worn [13]. While for a person a clothing product does not have a high value, for the environment the same product is a destructive element, with a negative effect on the value chain.

In 2017, only 2% of the manufacturing companies in the global textile and clothing industry were involved in a process of certification of sustainability or ethics, while 85% of the fashion brands have a poor performance in the field of sustainability. At the same time, sales volume is extremely high (more than 100 billion garments are sold annually), demonstrating a low level of awareness of these environmental and social issues by consumers [14, 15]. The impact of the linear economic model on the environment is devastating. The following table describes situations, presents figures and proposes various solutions to the problems related to the negative social and ecological impact of the textile and clothing industry:

Problems	Situations and numbers	Proposed solutions for companies
Chemical	8,000 chemicals are used to turn raw materials into finished products, 20% of polluted water comes from textile dyeing and finishing processes; Use of 4% pesticides and 10% insecticides in the fiber growth process; Hazardous chemicals in finished products;	Mapping the likely risks of using chemicals for processing raw materials and finding alternative ways to avoid their use; Mapping probable risks in the supply chain, knowing traceability, controlling the use and management of chemicals; Use of materials, dyes and fabrics that meet third party certification standards, such as the Global Organic Textiles Standard (GOTS), the 100 OEKO-TEX® Standard, Bluesign® or the EU Ecolabel; Use of raw materials that do not require the use of pesticides and insecticides to obtain them; Collaboration in programs, such as Zero Spilled Hazardous Chemicals (ZDHC); Replacing hazardous chemicals with safer alternatives; The use of technologies and processes that do not use many chemicals; Creating efficient liquid waste treatment systems.

(*Continued*)

Problems	Situations and numbers	Proposed solutions for companies
Waste	Every year, for every person on the planet, the fashion industry creates about 13 kg of clothing waste; The average consumer now buys 400% more clothes than 20 years ago; Less than 1% of clothing and 20% of textiles are recycled; Before being discarded, the average lifespan of a garment is 3.3 years.	Product design, manufacturing and marketing activities should be based on the principles of the circular economy, identifying steps that can lead to waste reduction, recycling and reuse of textiles; Reducing errors that cause waste during the production process by improving the forecasting process and by selecting suppliers that offer specifications; Promoting a consumption of clothing and textiles based on recycling—clothing collection points, textile recycling directions, new fashion trends and recycling programs.
Modern slavery	58% of those exploited come from China, India, Pakistan, Bangladesh and Uzbekistan; The US Department of Labor Research has identified 19 countries in which people working in the manufacture of clothing or jewelry are exploited; In 2015, in Uzbekistan, about 1 million adults and children were forced to work in the fields to harvest cotton; Through the work scheme, called the "Sumangali Scheme", many women from Tamil Nadu, southern India are brought to work in spinning and weaving companies.	Mapping the risk or potential of forced labor or modern slavery in their own supply chains and investigating measures to prevent it; Increase the awareness of employees, partners, suppliers or customers about the risk of forced labor and conduct training programs to ensure the prevention, detection and effective remediation of this problem; Supporting initiatives to strengthen local legislation by creating a stimulating framework for the eradication of modern slavery; Use of raw materials whose origin is certified or proven.

(*Continued*)

Problems	Situations and numbers	Proposed solutions for companies
Child labor	In the world, 152 million children are forced to work; In 51 countries, there are children working in the supply chains of cotton, clothing and jewelry.	Mapping in their own supply chains the existence of child labor and potential risks; Collaboration with the providers of direct or online training programs, which are based on quality policies regarding childlabor; Responsible adult employment practices, adequate wages related to purchasing needs; Working children to have access to programs and policies that ensure a decent education; Clear training policies and programs to ensure that workers have access to and understand their laws and regulations; Active monitoring of the implementation of their own policies; Development of collaborations or initiatives with local or sectoral work organizations, in order to build the culture of children's well-being where they carry out their activity.
Waterpolution	20% of the volume of polluted water comes from the dyeing and finishing of textile materials; The volume of plastic microfibers released into terrestrial ecosystems is equivalent to a volume of approximately 4 million to 7 million plastic bags per day; A single garment made of synthetic fabric, depending on its composition, in each washing process, can produce over 1,900 microfibers containing plastic.	In the design phase of the products is essential to consider their impact on the environment; Mapping the probable risks in the supply chain, knowing the traceability, controlling the use and management of wastewater; Selecting the supply chain with the lowest footprint and risk of waterpollution; Selection of raw materials and products containing the least microfibers; Choice of sustainable and certified materials; Selection of the best technologies in the painting and finishing processes, those that use lessharmful chemicals; The water recycling process is recommended to be done in closed loop systems; Encourage customers to reduce the impact of their washing processes on the environment.

9.3 Circular Economy

The existing economic model is based on the linear economy, with a negative economic, social and environmental impact. For this reason, the textile industry needs to move to a different economic model—a circular approach starting with product design and at all stages of the value chain, ensuring appropriate quality from raw materials to finished products, long-term use of products and reintroduction of used products back in the value chain, which reduces the problems in the textile and garment industry mentioned above [16]. The circular economy is based on three principles [17]:

a. Disposal of waste and pollution;
b. Maintaining resources in use (sustainable design, reusable and recyclable products);
c. Regeneration of natural systems (avoiding the use of non-renewable resources, increasing the importance of renewable resources).

The main priorities of the circular economy are:

- Traceability of products during manufacture and use;
- Transparent management of supply chains;
- Efficient use of resources, with emphasis on raw materials/renewable materials;
- Reducing chemicals and pollution at all levels—from raw materials to production processes, throughout the life of products and at the end of it;
- Design and education—from retailers, producers to consumers;
- Supporting innovative technologies;
- Supporting sustainable raw materials;
- Improving and stimulating the global recycling system.

Other challenges:

- Reducing the release of plastic microfibers;
- Use of the best technologies available for production;
- Identification of new sustainable raw materials;

A circular approach to the textile and clothing sector could bring substantial benefits, not only of an economic nature, but also of social and environmental benefits.

Economic benefits include:

- ➢ Reducing material costs by using recycled materials (compared to using virgin materials);
- ➢ Additional profit through new services (rent, individualization, guarantees, maintenance);
- ➢ Improved public image;
- ➢ Environment conducive to innovation;
- ➢ General economic growth.

Environmental benefits:

- ➢ Reducing greenhouse gas emissions;
- ➢ Reduction of virgin and non-renewable materials and energy consumption;
- ➢ Reduction of pollution (water, air, land);
- ➢ Improved crop productivity and soil health;
- ➢ Reducing plastic in the oceans;
- ➢ Reduction of hazardous substances in the environment;
- ➢ Reducing the volume of fresh water used.

Benefits for society:

- ➢ More options, superior quality;
- ➢ Positive impact on health (for the work force and consumers);
- ➢ Respect for all human rights.

In the textile sector there are raw materials that have specific characteristics of the circular economy, such as cotton, wool and other natural fibers, which are recycled from unused products or used to make yarns. In some cases, recycled yarns do not require dyeing, the color resulting from the waste mixture. In the past, this type of business offered the market raw materials; these businesses still operate today, offering to the market raw materials obtained by recycling textile products, with specific characteristics, with a lower recycling capacity, but at a lower price.

9.4 Sustainability Circles

Sustainability circles are a useful way to understand and assess sustainability and to manage projects with socially sustainable results. Although this method is mainly used for cities, a similar approach could be used to assess the level of sustainability in different areas of the textile/garment industry/companies. Defining the four areas—economic (assessment of production, consumption and retail processes, general resource management), political (assessment of business ethics, general policy strategy and social cooperation, specific global and national policies), environmental (environmental impact assessment), the ecological sustainability of the company), cultural (consumer assessment, level of contribution in the field of education and culture). The report entitled Pulse of the Fashion Industry [18], promoted by the Global Fashion Agenda in collaboration with the Boston Consulting Group (2018), contains an assessment of the environmental and social performance of the industry.

"Pulse Score" is calculated based on the evaluated performance for industrial organizations (max value 100). In 2017, the value of the score was 32, but in 2018 it increased to 38, showing some progress and an increase in awareness, although there is still much to be done for the textile and clothing industry to become sustainable and responsible in terms of ecological and social view. For the fashion industry to embark on a path to long-term financial, social and environmental prosperity, the 2018 report highlights the urgency of the collective effort to push the boundaries of what is currently available and possible. To have a real effect, the industry needs systematic change through leadership, innovation and collaboration. Companies involved in the fashion industry need to work with suppliers, investors, NGOs, universities and consumers to create an ecosystem that supports transformative innovation and disruptive business models. The 2018 report aims to provide a system of rules to companies that wish and want to introduce or develop responsible business models. For the first time, the report includes a Pulse Curve chart, which allows companies to compare their results with those of other companies, as well as a Roadmap to Scale guide, which presents good industry practices, suggesting practical business solutions.

9.5 Circularity in the Supply Chain

The supply chain in the textile and clothing industry is long and complex. Different suppliers perform different processing steps specific to garment production, which makes it difficult to maintain process transparency. Although the principles of sustainability must be applied throughout the chain, the greatest impact on the environment is at the beginning of it (raw materials and production), and in these stages further attention is needed to improve the industry's performance on sustainability [19].

Resources:

- Adequate use of natural resources, biodegradable fibers;
- Adequate use of recyclable fibers;
- The use of natural dyes from nature-friendly raw materials and processes that require low energy consumption.

Design:

- Different approach to the creation process of the new product to prolong its life (repairs, recycling, retransformation, etc.);
- Long-lasting design;
- Zero waste design, considering the impact on the environment;
- Design innovative processes appropriate to new technologies and products that follow a circular path.

Production:

- Traceability of all stages of production;
- The impact of the stages of production processes (obtaining fibers, spinning, weaving, dyeing and finishing fabrics, textile manufacturing and finishing of garments) on the environment and social impact;
- Using the best techniques;
- Innovation, new technologies focused on circular processes;
- Production based on the principles of the circular economy in terms of water consumption, energy, raw materials and waste production;
- High quality production, for a longer period of time;
- Products that have sustainable characteristics.

Retail:

- Service-oriented business;
- Online services aimed at reducing social and environmental impact;
- Innovations, new technologies to reduce the social and environmental impact and to obtain more information from the supply chain;
- Clear information on consumer education and without focusing on the ecological side.

Consumption:

- Organization of storage space/cabinets;
- Responsible maintenance of products;
- Re-construction, reconditioning of clothing products, purchase of used products;
- Donations;
- Collection of old clothing products and placing them at recycling points;
- Sustainable attributes of textiles.

Re-use, Recycling:

- Redesign, Recycling, Reproduction;
- Recycling (fibers, materials, clothing);
- Waste collection, creation and promotion of new collection systems;
- Encourage trade and rental of clothing products.

9.6 Consumer Behavior of Sustainable Textile Products

Consumption of sustainable products is defined as the use of goods and services that meet basic needs and increase the quality of life, while minimizing the use of natural resources, toxic materials, toxic gas emissions and pollution throughout the life cycle of the product so as to not jeopardizing the needs of future generations [20]. Consumption of sustainable products supports the reduction and balancing of consumption on three levels, namely: social, environmental and economic, having an impact on the purchase, use, care, repair and recycling of clothing [21]. There are major differences between the consumer of sustainable products and the traditional consumer. The consumer of sustainable products takes into account the conditions under which the product was made and placed in stores, following the link between the product itself and the sustainable environment in which it was made. The main targets for the consumption of sustainable products are to meet the needs of consumers, increase their standard of living and, at the same time, reduce waste, pollution, but also encourage consumers to choose goods and services that meet ethical, social and environmental criteria [22].

The study "METACONSUMER—Sustainability in Global Consumer's View" conducted by Ebeltoft Group, an international association of retail consulting experts, conducted in 17 countries, on a sample of 8,500 people, places Romania in 5th place in the ranking, Romanians are among consumers the most concerned with sustainability. The results of the study regarding Romania are surprising, especially since it reveals that 26% of Romanians are willing to pay up to 10% more for an eco product (a percentage higher than the global average of 24%). However, Romanians' concerns for sustainability are not necessarily translated into consumer behavior, Romania occupying the last place in the top of countries that adopt sustainable practices normally. Young people are interested in purchasing sustainable textiles, usually women, with high incomes and a high level of education [23]. Careful analysis of consumers shows that they are concerned with style and quality, seeing sustainability as bringing pleasure and well-being.

9.7 Decision to Purchase Sustainable Textile Products

Once the consumer identifies a need, he begins to gather information from the market, process it and interpret it, in order to identify the right product. Previous research has shown that, although consumers show a certain degree of interest in sustainable fashion, they are often faced with insufficient information about the prices or quality of products. In most cases, the purchase decision is made based on the design of the goods. When the consumer has an attachment to a certain product, as happens among consumers of sustainable fashion, the purchase decision is much more complex. It starts with a rigorous information related to the characteristics of the product, but also related to the manufacturer and trader, an analysis of the market offer and only then the consumer makes a decision, taking into account the information gathered, but also how the product meets their needs and principles. The consumer will finally choose the brand that will most likely best satisfy his desires [24]. Previous research claims that, in most cases, the Romanian consumer chooses the stores from which he wants to purchase the goods, based on time and budget criteria [25]. Consumers often choose locations by analyzing both the products marketed and the atmosphere in the store [26]. Consumers need to perceive the label, understand it and translate it [27]. For some consumers, the fact that a product is organic can mean a lot, for others the place of origin may be more important, therefore the labels of sustainable products should contain as many details as possible about the sustainability characteristics of the product, so the consumer to make the purchasing decision in full knowledge of the facts.

9.8 Policies and Strategies Used in the Sustainable Textile Industry

In the case of the sustainable textile industry, companies need to fiind the right positioning strategy by focusing first on the product obtained under sustainable conditions, on the maximum quality offered to the consumer and only then on a higher price, but justified by the special features of the product. Strategic sustainability is the active involvement of senior management in the development of products, exceeding the general conditions of efficiency, with the ultimate goal of reducing or even eliminating risk and reducing costs [28]. At the moment, the strategic objective of the textile industry is to increase competitiveness, respecting market conditions, with minimum material and social costs, maximum productivity, efficient and sustainable investments, respecting environmental conditions [29].

Companies that want to develop and grow both in terms of image and in terms of business development, must take into account the impact of their business on the environment and society [30]. Sustainable products are the result of an integrated production process, respecting the requirements of sustainability from the design phase, followed throughout the production, distribution and marketing process, until the product reaches the final customer, in consumption, but also after, by identifying efficient ways of recycling [31]. In the case of the sustainable textile industry, we are talking about a short distribution channel, with a small number of intermediaries, preferring, in many cases, direct distribution, a process through which goods are delivered directly from producer to final consumer [8], without use intermediaries. In order for the distribution system to be sustainable, the conditions of transport, handling, packaging and storage must be respected, so that the processes are optimally carried out and at the same time do not affect social, environmental and economic policies [32]. As for consumers of sustainable products, according to a study by Bly *et al.*, in 2015, they are seen as informed, curious people who feel the social pressure to consume and put sustainability at the forefront of the consumption process [31]. These issues can also be the starting point for formulating future directions of action for policies and strategies used in the sustainable textile industry.

9.9 Conclusions

Manufacturers in the textile industry are increasingly interested in the subject of sustainability [34]. At present, there are companies that make major efforts to adopt sustainable principles. Policies and strategies are meant to contribute to informing, educating and empowering consumers in general, regardless of the products they want to buy. As Lockie pointed out in one of his articles, sustainability can become part of collective common sense [34]. Supply chains in the textile and clothing industry are very complex and fragmented. In order to change the negative impact on the environment and society, globally, there must be a valid governmental and legislative framework that has to be respected by suppliers, producers and traders.

References

1. The World Commission on Environment and Development, https://eur-lex.europa.eu/legalcontent/EN/TXT/?uri=CELEX:52001DC0264.
2. Forum for the Future, https://www.forumforthefuture.org/sustainability-and-system-change.
3. Making London Work by Forum for the Future's Sustainable Wealth London project, 2001, at http://www.ecotexerasmus.eu/wp-content/uploads/Module_1_en_IO4.pdf, 2019.
4. The Politics of the Real World: Meeting the New Century (Real World Coalition), 1996,athttp://www.ecotexerasmus.eu/wp-content/uploads/Module_1_en_IO4.pdf, 2019.
5. *TheDorsetEducationforSustainabilityNetwork*,https://www.sustainabledorset.org/ at http://www.ecotexerasmus.eu/wp-content/uploads/Module_1_en_IO4.pdf, 2019.
6. *Learning for a SustainableFuture—Teacher Centre*, http://www.unesco.org/education/tlsf/ at http://www.ecotexerasmus.eu/wp-content/uploads/Module_1_en_IO4.pdf, 2019.
7. Worster, D., The Shaky Ground of Sustainable Development, in: *The Wealth of Nature*, D. Worster (Ed.), pp. 142–55, Oxford University Press, New York, 1993.
8. Opris, M., Brătucu, G., Palade, A., Distribution policies and strategies for sustainable textile products, in: *Bulletin of the Transilvania University of Brasov. Economic Sciences. Series V*, vol. 8, p. 65, 2015.
9. Korpysa, J., Buyer behaviour in the context of sustainable consumption policy pursued in Poland. *Amfiteatru Econ.*, 15, 7, 703, 2013.
10. https://en.unesco.org/about-us/introducing-unesco
11. Ellen MacArthur Foundation, *A new textiles economy: Redesigning fashion's future*, Isle of Wight, United Kingdom, http://www.ellenmacarthurfoundation.org/publications, 2017.

12. http://www.europarl.europa.eu/RegData/etudes/BRIE/2019/633143/EPRS_BRI(2019)633143_EN.pdf
13. https://www.globalfashionagenda.com/initiatives/pulse/# 15 Ellen MacArthur Foundation, A new textiles economy: Redesigning fashion's future, 2017, http://www.ellenmacarthurfoundation.org/publications)
14. https://www.apparelentrepreneurship.com/your-guide-to-sustainability/
15. https://www.globalfashionagenda.com/initiatives/pulse/
16. https://www.researchgate.net/publication/326546054_Circular_Economy_-
17. Ellen MacArthur Foundation, *A new textiles economy: Redesigning fashion's future*, Isle of Wight, United Kingdom, http://www.ellenmacarthurfoundation.org/publications generabile. Challenges_for_the_Textile_and_Clothing_Industry, 2017.
18. https://www.globalfashionagenda.com/initiatives/pulse/
19. *ECOTEX Material suport Abilități Inovatoare pentru Economia Circulară în Domeniul Textil*, p. 41, at http://www.ecotexerasmus.eu/wp-content/uploads/Module_1_en_IO4.pdf, 2019.
20. Ministrul Norvegian al mediului. Raportul Conferinței Ministeriale de la Oslo. Sustainable Production and Consumption; 1994 www.iisd.ca/consume/oslo004.html Accesat 4 Octombrie 2015
21. Gwozdz, W., Netter, S., Bjartmarz, T., Reisch, L.A., Report Survey. Results on Fashion Consumption and Sustainability among Young Swedes, in: *Mistra Future Fashion. Proiect 7: Sustainable Consumption and Consumer Behaviour*, p. 14, 2012.
22. Barth, M., Fisher, D., Michelsen, G., Nemnich, C., Rode, H., Tackling the Knowledge-Action Gap in Sustainable Consumption: Insights from a Participatory School Programme. *J. Educ. Sustainable Dev.*, 6, 301, 2012.
23. Sima, V., Green Behaviour of the Romanian Consumers. *Econ. Insights—Trends Challenges*, 3, 66, 85, 2014.
24. Gabriel, Y. and Lang, T., *The unmanageable consumer*, Ediția a doua, London Sage, 2006.
25. Bălan, C., Consumer Buying Behavior under the Impact of New Retail Formats in Romania. *J. Econ. Manage.*, 9, 31, 2012.
26. Op.cit.30, p.29.
27. Grunert, K.G., Sustainability in the Food Sector: A Consumer Behaviour Perspective. *Int. J. Food Syst. Dyn.*, 2, 3, 208, 2011.
28. Raderbauer, M., *Strategic Sustainability—Strategic implementation of Sustainable Business practice in Viennese Accomodation*, University of Exeter, United Kingdom, 2011.
29. Visileanu, E., Direcții strategice de dezvoltare a sectorului textile-pielărie. *Buletinul AGIR*, 4, 138, 2008.
30. Epuran, G., Dovleac, L., Ivasciuc, I.S., Sustenabilitatea și marketingul creșterii organice: o abordare exploratorie privind valorificarea principiilor dezvoltării durabile în turism. *Amfiteatru Econ.*, 17, 40, 594, 2015.

31. Bly, S., Gwozdz, W., Reisch, L., Exit From the High Street: An Exploratory Study of Sustainable Fashion Consumption Pioneers. *Int. J. Consum. Stud.*, 39, 2, 125–135, 2015, http://dx.doi.org/10.1111/ijcs.12159.
32. Davies, I., The values and motivations behind sustainable fashion consumption. *J. Consum. Behav.*, 6, Octombrie 2015.
33. Connell, K. and Kozar, J., Environmentally Sustainable Clothing Consumption: Knowledge, Attitudes and Behavior. *Roadmap to Sustainable Textiles and Clothing*, USA, 2014.
34. Lockie, S., Beyond resilience and systems theory: Reclaiming justice in sustainability discourse. *Environ. Sociol.*, 2, 2, 115, 2016.

Part 3

SUSTAINABLE WASTEWATER REMEDIATION

10

Sustainable Application of Ionic Flocculation Method for Textile Effluent Treatment

Hamadia Sultana[1], Muhammad Usman[1*], Abdul Ghaffar[2], Tanveer Hussain Bokhari[1], Asim Mansha[1] and Amnah Yusaf[1]

[1]*Department of Chemistry, Government College University, Faisalabad, Pakistan*
[2]*Department of Biochemistry, Government College University, Faisalabad, Pakistan*

Abstract
Adsorptive Micellar Flocculation (AMF) is one of the promising methods employed for the treatment of textile effluents through applications of surfactants. Critical micelle concentration (CMC) of surfactants plays a vital role in the selection of most suitable media for dye removal. It is necessary to optimize experimental conditions such as temperature, pH, contact time, stirring speed, electrolyte concentration, dosage of surfactant and flocculant for the smooth running of said tool. Removal efficiency has been calculated, at optimized conditions, to ensure the desired results. In this chapter, effluent treatment through solubilization followed by ionic flocculation is discussed briefly. Ionic flocculation has opened a new avenue for industrialists and academicians for sustainable treatment of textile effluent.

Keywords: Dyes, surfactant, flocculation, textile effluents, solubilization

10.1 Introduction

Industrialization, a necessary evil, has imposed many detrimental effects with the deterioration of environmental equilibrium as well as the health of the global community. Environmental pollution has, thus, become an obstacle in industrial growth, and its control is a task of utmost importance [1]. To cope up with its adverse consequences is a matter of great concern.

**Corresponding author*: musman@gcuf.edu.pk

Luqman Jameel Rather, Mohd Shabbir and Aminoddin Haji (eds.) Sustainable Practices in the Textile Industry, (255–272) © 2021 Scrivener Publishing LLC

Treatment of wastewater is an important challenge in this regard. Natural resources of water are becoming scarce and are continuously polluted by non-treated or poorly treated industrial, agriculture, and domestic effluents [2]. Pollution control has become an alarming matter. About 10,000 synthetic dyes are available, and their annual production has reached up to 8×10^5 tons [3, 4]. Textile wastewater is loaded with variety of dyes *viz.* acidic, basic, direct, mordant, reactive, vat and disperse dyes. Said dyes have specific functional groups i.e., azo, nitro, anthraquinone, carbonyl, etc. [5]. These dyes are also applied in paints, leather, paper, printing, cosmetics and food industries [6].

The textile industry is spoiling 17–20% of the freshwater worldwide, which is shedding a lot of pollutants. The wastewater generated by the textile industry is intensely colored, and its acidity, biological oxygen demand (BOD) and chemical oxygen demand (COD) are higher than average values [7]. It contains chlorides, heavy metals, sulfates and hazardous organic compounds such as dyes, phenol [8] etc. The per year dye loss during the dyeing process is about 200,000 tons. Most of the dyes are highly stable in wastewaters and cannot be removed by conventional treatment. Azo dyes, the largest class of colorants, have great structural diversity. About 60–70% of organic dyes in the world are azo, but their degradation produces many carcinogenic and mutagenic compounds. The byproducts of dye degradation pose serious threats to the ecosystem and global health [9, 10]. Many dyes, particularly azo dyes, can cause cancer in human's bladder and aberration of chromosomes [11, 12]. Disperse Blue 291 dye can develop error in genetic code by addition or deletion of base pairs in DNA, whereas Reactive Blue 19 dye can persist in water bodies for a very long time due to its greater half-life [13]. Disperse Blue 291 also has an adverse effect on hepatic cells of humans [14]. Dyes are a great threat to aquatic life as they hinder the penetration of sunlight and reduce the dissolved oxygen in water bodies [15]. The terrible impact of dyes is not only on humans and aquatic species. It can cause the death of the microorganisms by changing the properties of soil. Aromatic amines, produced by the reductive degradation of azo dyes, are the root of many skin and respiratory disorders.

10.2 Conventional Methods for Degradation of Textile Effluents

Different methods have been employed for the removal or degradation of dyes from the wastewater [16]. These are categorized into three main classes.

10.2.1 Biological Methods

Biological methods involve the use of living organisms for the removal of pollutants. Many microorganisms such as fungi, bacteria, and yeast can degrade the contaminants. For example, a white-rot fungus, *P. chrysosporium*, can decolorize textile wastewater up to 99%. Similarly, certain bacteria cultures can degrade azo dyes. It was reported that yeast e.g., *K. marxianus* can remove up to 98% of Remazol Black B dye. There are several reasons which restraints the use of these methods. First, the removal of dyes by these microorganisms takes much time. The process requires days or even weeks for completion. Second, the living organism can only work at specific conditions i.e., ambient temperature, pH, salinity, etc. [17].

10.2.2 Chemical Methods

Chemical methods used to remove dyes are categorized into two classes, advanced oxidation processes (AOP) and chemical oxidation [18]. Advanced oxidation processes proceed by the formation of hydroxyl radical, which acts as a powerful oxidizing agent e.g., Fenton process. Production of iron sludge in this process limits its usage. Chemical oxidation involves the use of several oxidizing agents such as H_2O_2 and O_3. High cost, production of toxic byproducts and less stability of ozone are the drawbacks of using oxidizing agents [19].

10.2.3 Physical Methods

Physical methods employed for the removal of dyes include adsorption [20], filtration [21], coagulation-flocculation [22], ion-exchange [23] and irradiation [24]. A large variety of adsorbents, of organic and inorganic nature, have been used for this purpose, such as activated carbon, wood chips, peat, wheat residue, charcoal, etc. These adsorbents can remove a large amount of dyes, but their usage has become limited due to many reasons such as sludge production, dumping, and high cost of some effective adsorbents [25]. Filtration techniques include microfiltration, nanofiltration, and ultrafiltration. High cost, clogging of membrane and short membrane life are some draw backs of filtration process. Ion exchange involves several resins that can remove dyes and other organic pollutants, but it is unable to tolerate high pressure created by substantial volumes of polluted water. Irradiation methods are highly expensive as compared to the different techniques used for the treatment of wastewater. Coagulation–flocculation (C/F) technique has been widely used in textile effluent treatment because it can effectively remove dyes and reduce COD. Still, the major drawback of this method is the presence of iron or aluminum salts in the sludge produced after coagulation. These salts cause severe harm to aquatic organisms. This technique is also unable to remove many soluble pollutants from wastewater [11, 19].

Comparative analysis revealed that some physical methods are better in the sense of non-toxicity and low-cost. It's a need of time to design a better strategy for minimizing the drawbacks which have been observed in other methods ever since. The proposed technique must be inexpensive, involving accessible bio-degradable material, show maximum removal efficiency in a short period, produce no toxic side product, and can be used on a large scale. Ionic flocculation, also called adsorptive micellar flocculation (AMF), is one of the excellent methods used for dye removal. The details of materials, and the mechanism involved in said technique are described below.

10.3 Surfactants

Surfactants can be classified on the basis of their source and charge, as displayed in Figure 10.1 [26]. Adsorptive micellar flocculation is an important surfactant-based separation technique which involves the use of anionic surfactants and polyvalent metal salts to capture pollutants. For a better understanding of AMF, it's necessary to be aware of the role, and mode of action of the surfactants. Surfactants are distinctive organic compounds that are equally soluble in aqueous as well as in the majority of the organic mediums. The surfactant molecules have hydrophilic head groups and hydrophobic tails. This structural feature is responsible for their

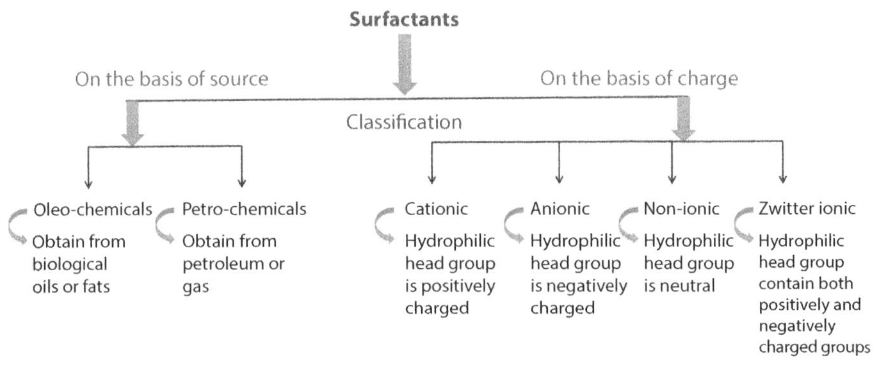

Figure 10.1 Flowsheet diagram for the classification of surfactants.

affinity for both polar and non-polar solvents [27]. Surfactant molecules arrange themselves, at air–solution interface, in such a way that their hydrophilic parts remain in touch with water. The hydrophobic groups are, however, oriented towards air. But when interface is overloaded due to high surfactant concentration and no more molecules can be accommodated in this layer, the surfactant molecules orient themselves in small aggregates such that hydrophilic heads are at the exterior.

Hydrophobic tails form the interior of these nanostructures called micelles. The specific concentration at which surfactant molecules form micelles, critical micelle concentration (CMC), is different for different surfactants depending on their nature and structure [28, 29].

The peculiar behavior of surfactants is the key to their wide-ranging applications. Surfactants are used for the removal of many hazardous pollutants from wastewater *viz.* benzene [30], dyes [31], carbonaceous matter, dioxins [32], polyphenolic compounds, antioxidants [33], pesticides [34], nanoparticles [35], organic matter, heavy metals [36], soil contaminants [37], aniline [38] etc. Other critical applications of surfactants in various industries are given in Figure 10.2. In the textile industry, surfactants have been widely used as washing, wetting, dispersing, and solubilizing agents. They also assist in the dyeing process [39]. The concentration of surfactants is an important parameter to be considered for their usage in different applications. The value of critical micelle concentration (CMC) affects almost all properties of surfactants. The organic nature, electrostatic, and hydrophobic attractions are responsible for interaction of dyes with

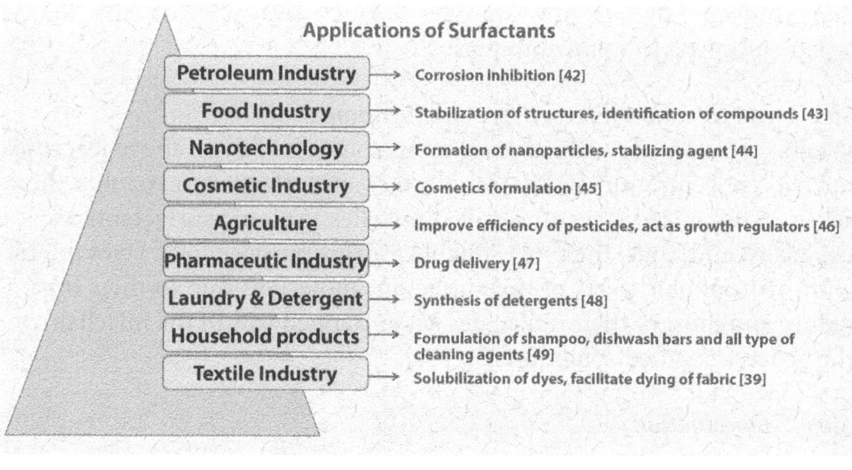

Figure 10.2 Applications of surfactants in different industries.

surfactants. The solubilization of dyes by micelles plays a significant role in their removal from aqueous medium [40]. The solubility of dyes in an aqueous micellar medium can be calculated in terms of binding constant and partition coefficient [41]. Herein, the discussion is related to the use of surfactants as an adsorbent for the withdrawal of dyes from wastewater.

10.4 Adsorptive Micellar Flocculation (AMF)

Coagulation and flocculation (C/F) have been widely used for the treatment of wastewater [50]. Coagulation causes the destabilization of the stable suspended colloidal particles by neutralizing their charges, at the same time flocculation involves agglomeration of these destabilized particles and results in the formation of larger flocs that settle down quickly and can be removed by sedimentation or using a suitable filtration medium [51, 52]. Salts of aluminum and iron are, usually, employed as coagulants [50]. Sometimes polymers are also used to aid flocculation [53]. Micellar flocculation is quite similar to this C/F technique. It involves the entrapment of substrate molecules (dyes) in surfactant micelles that stick together to form flocs on the addition of metal salt. It was employed for the removal of many organic pollutants such as dyes [54, 55], phenol [56], benzene [57], pesticides [58], insecticides [59], etc.

10.5 Mechanism

The removal of dyes from wastewater through adsorptive micellar flocculation completes in following steps;

Step 1: Micellization followed by solubilization

When surfactants are added in aqueous solution of dyes, above its critical micelle concentration (CMC), molecules of surfactants assemble themselves in organized structures called micelles. Anionic surfactants are less expensive and, thus, the most suitable for this purpose [60]. Dyes interact with hydrophobic parts of the surfactant molecules due to their organic nature and most of their molecules get entrapped within the micellar core, the process is called solubilization [61].

Step 2: Flocculation

The polyvalent salts cause flocculation of anionic micelles and dye molecules are entrapped within closed structures of flocculates. The cations,

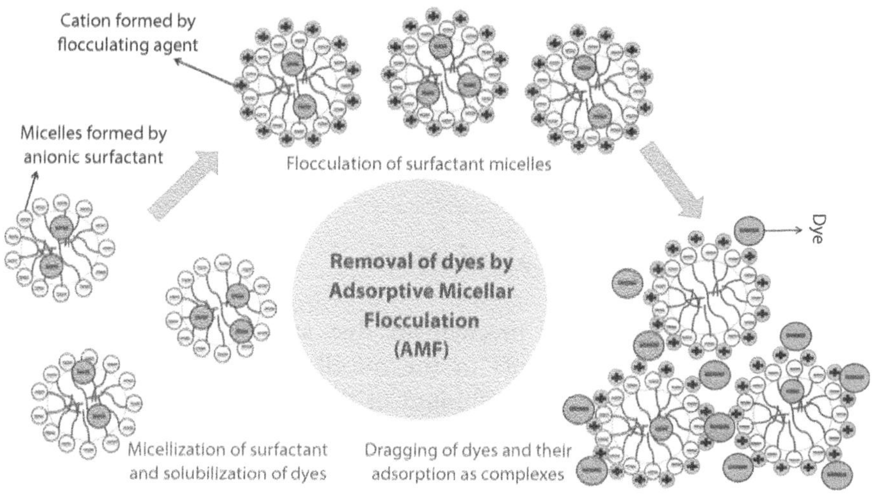

Figure 10.3 Pictorial representation of removal of dyes by adsorptive micellar flocculation.

formed by the ionization of these salts, have high charge density. These cations are adsorbed at the surface of micelles and reduce the inter-micellar repulsion [15]. This phenomenon enables the micelles to come closer and join together to form aggregates called flocs [56, 62].

Step 3: Adsorption

The dyes present within the core of micelle are dragged out during flocculation and involves complexation with these metal cations, as shown in Figure 10.3 [56]. It's due to the electrostatic interactions between the cations and dye molecules. The micellar flocs then act as adsorption surfaces for the attachment of these organic pollutants. The adsorption of dyes takes place in the stern layer and diffuse layer in the form of complexes [63]. This process not only removes dyes but also takes off the maximum amount of surfactants from the solution.

10.6 Choice of Flocculant

A large variety of divalent and trivalent metal salts are available. Flocculation of anionic surfactants has been performed by adding Pb^{2+} [64], Ba^{2+} [65], Ca^{2+} [66], Fe^{3+} and Al^{3+} cations [67, 68]. Salts of aluminum are widely used as a flocculating agent in adsorptive micellar flocculation. The presence of aluminum salt can reduce the pH of the system and work less efficiently in cold water [69]. Therefore, attention is turned toward using other metal salts. Recently, calcium salt has been employed as a flocculant for the removal of pollutants from aqueous medium [15, 55]. The electrolytes are having cations of higher charge density and low price may be the best choice as flocculants.

10.7 Analysis and Calculations

According to the suggested mechanism, dye, surfactant, and metal salt are three primary constituents in the micellar flocculation technique. Determination of the required amount of these reagents is essential for checking the efficiency of the process.

10.7.1 Analysis of Reagents

The concentration of dye can be determined using UV/Visible Spectrophotometer before and after the experiment. The concentration of surfactant can be obtained by performing a double-phase titration. This volumetric analysis involves the standard solution of hyamine, and a mixed indicator consists of dimidum bromide and blue acid 1 [58]. Double-phase titration can also be performed in water–chloroform system with methylene blue as an indicator [60]. Different methods have been employed in determining the amount of salts. The amount of aluminum and iron in the residual medium can be obtained from atomic absorption spectrometry [60], while calcium can be easily calculated by performing EDTA titration [15]. Fourier transform infrared spectroscopy (FT-IR), and X-Ray diffraction (XRD) techniques can assist in the characterization of flocs [57].

10.7.2 Calculated Parameters

Thermodynamic and various other parameters can be calculated to investigate the efficiency, mechanism, and spontaneity of a reaction. Following mathematical expressions can be used for said purpose;

Percentage efficiency (E%)

$$E\% = \frac{C_i - C_f}{C_i} \times 100 \tag{10.1}$$

C_i is the initial, while C_f is the final concentration of the dye. Percentage efficiency is the criteria for checking the efficiency of the process [55].

Separation parameter (α)

$$\alpha = \frac{[Dye]_{sorbed}}{[Dye]_{residual}} \tag{10.2}$$

$[Dye]_{sorbed}$ is the concentration of dye in milligrams per gram of flocculated surfactant. $[Dye]_{residual}$ represents dye concentration in residual solution after separation of flocs. The separation parameters are calculated for different concentrations of surfactants and flocculants [58].

Partition coefficient (K_{FW})

$$K_{FW} = \frac{[dye]_{ads}/[surfactant]_{floc}}{[dye]_w} \tag{10.3}$$

K_{FW} represents the flocculate–water partition coefficient. $[dye]_{ads}$ and $[dye]_w$ are the concentrations of dye adsorbed on flocs and in residual water, respectively. $[surfactant]_{floc}$ is the concentration of surfactant separated in the form of flocs. This term gives the amount of dye removed per unit mass of surfactant [70].

Equilibrium constant (K_o)

$$K_o = {C_{solid}}/{C_{liquid}} \tag{10.4}$$

C_{solid} and C_{liquid} are the solid and liquid phase concentrations (mg/dm^3) at equilibrium [71].

Change in free energy (ΔG), *enthalpy* (ΔH) *and entropy* (ΔS)

$$\Delta G = -RT \ln K_o \tag{10.5}$$

$$\ln K_o = -\Delta G/RT \tag{10.6}$$

$$\Delta G = \Delta H - T\Delta S \tag{10.7}$$

$$\ln K_o = \Delta S/R - \Delta H/RT \tag{10.8}$$

Slope and intercept of the graph plotted between $\ln K_0$ and 1/T gave the values of change in enthalpy and entropy (Equation (10.8)). Change in Gibbs free energy is calculated by putting the value of ΔH and ΔS in Equation (10.7) [71].

10.8 Optimization of Conditions for Better Removal of Dye Using AMF

Different factors affect the mode of action of AMF, such as temperature, pH, surfactant dosage, flocculant/surfactant ratio, electrolyte, contact time, and stirring speed. The selection of optimum conditions can guarantee smooth processing and better results.

10.8.1 Effect of Temperature

The extraction of dye by AMF depends on the formation of micellar flocs and adsorption of dye molecules on their surface. Temperature is one of the crucial factors that significantly affect the micellar flocculation [66]. Higher temperatures can reduce rate of flocculation because the solubility of the surfactant increases with temperature. Hence, the removal efficiency will be low at the higher temperatures. It has been reported that the temperature between 30-45°C is suitable for achieving high removal efficiency [55].

10.8.2 Effect of pH

Adjusting the pH of the system is a necessary step in many chemical reactions. Micellar flocculation can be affected by pH as it alters the stability of the surfactant. Low pH impairs the formation of the floc due to protonation of surfactants at acidic conditions, and it can be observed by the appearance of an oil-in-water emulsion. The pH of the dye solution, after addition of some surfactant (e.g., surfactants obtained from base soap), is in the range of 9–10, which makes it clear that there is no need to adjust the pH of the medium. The increase of pH can increase the extraction of dye but up to a specific limit. Interaction between flocculant and hydroxyl ions inhibits the formation of flocs at higher pH. Best pH range reported for this experiment was from 9 to 12 [54, 55].

10.8.3 Surfactant Dosage

In micellar flocculation, surfactants are the necessary ingredients which increase the solubility of the insoluble organic species by encapsulating them within their micellar structure and subsequently remove them by adsorbing onto their surfaces. A large number of micelles are formed at a higher concentration of surfactant. The removal efficiency increases with surfactant concentration till certain value and then becomes constant. The presence of excess of surfactant can cause re-solubilization of the adsorbed pollutants [57].

Dye molecules can be accommodated at various sites within micelle. These sites are (1) micelle-water interface (2) between hydrophilic head groups (3)in palisade layer (between the hydrophobic tails) and (4) in core of micelle. Increase in surfactant concentration changes the structure of micelle, which disturbs the solubilization process that, in turn, affects the removal efficiency [56]. The concentration of surfactants is, therefore, kept within a specific range for maximum removal of dye.

10.8.4 Flocculant/Surfactant Ratio

The extent of flocs formation and size depends on the amount of polyvalent salt added for flocculation of micelles. The concentration of surfactant is the scale for deciding the amount of salt required for this purpose. As cations (formed by ionization of salt) are responsible for flocculation of micelles, the influence of the amount of flocculant has been studied in terms of flocculant/surfactant ratio. Flocculation and size of flocs both increase with an increase in the flocculant/surfactant ratio. This ratio was varied up to its maximum value i.e. 1. It has been reported that the high flocculant/surfactant ratio can give a large number of well-defined bigger sized flocs [56, 66].

10.8.5 Addition of Electrolyte

The flocculation of micelles in the presence of polyvalent salt can be explained by following general reaction.

$$M^{x+} + x(RCOO^{-}Na^{+}) \longrightarrow M(RCOO^{-})_x + xNa^{+}$$

According to this reaction, polyvalent metal cation (M^{x+}) reacts with surfactant molecules ($RCOO^{-}Na^{+}$) and replace Na^{+}. It reveals that the addition of electrolytes such as NaCl in the reaction mixture gives monovalent

Na^+ ions. The presence of Na^+ ions reduces the dissociation of the surfactant molecules. It also disturbs the interaction of flocculating agents (polyvalent cations) with surfactant molecules that are responsible for the flocculation of micelles. Hence, the presence of electrolyte can reduce the removal efficiency during micellar flocculation [55, 72].

10.8.6 Contact Time and Stirring Speed

Contact time significantly affects the capacity of micellar flocs to hold the pollutants. Due to the affinity of dyes for water molecules, removal efficiency decreases with the passage of contact time until the system attains equilibrium. Stirring is done for thorough dispersion of reagents in the medium. During the addition of reagents, slow stirring followed by rapid stirring for 4–5 min is suitable for achieving expected results. Further stirring can disturb the size of flocs and adsorption of dyes on their surfaces [54, 55].

10.9 Potential Advantages of AMF

This technique has been performed for the removal of a variety of organic pollutants. Following features of AMF ensure its successful implementation for treatment of textile effluent [57].

i. Operational simplicity
ii. High efficiency
iii. Low cost
iv. Harmless and accessible reactants
v. Use of biodegradable surfactants *viz.* carboxylate surfactants
vi. Non-toxic byproduct
vii. Take less time to accomplish
viii. Can be performed at room temperature
ix. Recyclability of chemicals

10.10 Application to Wastewaters

Organic substances have an extensive presence in our environment. Industries are releasing a large variety of pollutants. Organic materials are necessary to ensure the better life. Their presence in the environment, on the other hand, is offering severe threats to the ecosystem. The removal of their excess amount is, therefore, as necessary as their production.

Removal of different hazardous substances of organic nature, by utilizing ionic flocculation, is already reported in the literature. Highly toxic pesticides e.g., pyrethrin [59], herbicide e.g., 2,4-dichlorophenoxyacetic acid (2,4-D) [73], antibiotic tetracycline [74, 75] and many other aromatic pollutants e.g., phenylamine, catechol [76], polychlorinated biphenyls (PCBs) [60], phenol [56], derivatives of benzene [57], etc. have been efficiently eliminated from aqueous medium using AMF.

10.11 Conclusion

Removal of dyes by adsorptive micellar flocculation (AMF) is a surfactant mediated phyico-chemical adsorption. The surfactants have great significance in this technique because their molecules have excellent potential for removal of dyes. Conventional methods employed for this purpose suffer from severe limitations. Anionic surfactants and polyvalent salts are the only chemicals required to perform this method. Micellization, solubilization of dyes, flocculation of micelles and adsorption of dyes on micellar flocs (as metal complexes) are the processes taking place in the dye-containing aqueous medium during micellar flocculation. Electrostatic and hydrophobic interactions play a significant role in said processes. Any parameter that can affect micellization and formation of flocs can consequently affect the removal efficiency e.g., surfactant concentration, salt concentration, temperature, pH, stirring speed, etc. The best conditions for performing micellar flocculation are standard room temperature and alkaline medium. This technique has some unique advantages that are important for its industrial-scale applications such as low cost, high efficiency, and sustainability.

10.12 Future Prospective

From the critical analysis of this technique, we can predict that surfactants could be the future of water treatment technology. The use of bio-surfactants will satisfy the environmental regulations and make this technique more reliable and human friendly. Only a few experiments have been, so far, reported with nontoxic salts. It is, therefore, need of the hour to search nontoxic or less toxic flocculants. To enhance the holding capacity of surfactant flocs, researchers have to search the substances that can work as a flocculant aid. We can make this technique practically more viable, for the treatment of a large volume of industrial effluent by selecting inexpensive, reliable and more efficient filtration medium. The search of biodegradable surfactants is equally important.

References

1. Yusaf, A., Adeel, S., Usman, M., Mansha, A., Ahmad, M., Chapter 11, Removal of Heavy Metal Ions from Wastewater Using Micellar-Enhanced Ultrafiltration Technique (MEUF): A Brief Review. *Textiles and clothing*, pp. 289–315, 2019.
2. Adeel, S., Rehman, F., Rafi, S., Kamran, M., Zuber, M., Bhatti, I., Akhtar, N., *Current innovation in synthetic dyes and dyeing*, Stadium Press LLC Press, USA, 2018.
3. Gupta, V., Khamparia, S., Tyagi, I., Jaspal, D., Malviya, A., Decolorization of mixture of dyes: A critical review. *Global J. Environ. Sci. Manage.*, 1, 71, 2015.
4. Yusaf, A., Adeel, S., Usman, M., Akram, N., Bokhari, T.H., Removal of dyes from wastewater by micellar enhanced ultrafiltration. *Bioremediation and biotechnology*, vol. 2, pp. 189–206, Springer, 2020.
5. Saeed, M., Adeel, S., Muneer, M., ul Haq, A., Photo catalysis: An effective tool for treatment of dyes contaminated wastewater. *Bioremediation and biotechnology*, vol. 2, pp. 175–187, Springer, 2020.
6. Saeed, M., Adeel, S., Attaulhaq, Kiran, S., Khan, S., Siddique, M., Shahzad, M.A., Degradation of methyl orange catalysed by Pt/Al2O3 in aqueous medium. *Oxid. Commun.*, 40, 13, 2017.
7. Saeed, M., Adeel, S., Abdur-raoof, H., Usman, M., Mansha, A., Ahmad, A., Amjed, M., ZnO catalyzed degradation of methyl orange in aqueous medium. *Chiang Mai J. Sci.*, 4, 1646, 2017.
8. Verma, Y., Acute toxicity assessment of textile dyes and textile and dye industrial effluents using *Daphnia magna* bioassay. *Toxicol. Ind. Health*, 24, 491, 2008.
9. Bae, J.-S. and Freeman, H.S., Aquatic toxicity evaluation of new direct dyes to the. *Daphnia magna. Dyes Pigm.*, 73, 81, 2007.
10. Chequer, F.D., de Oliveira, G.A.R., Ferraz, E.A., Cardoso, J.C., Zanoni, M.B., de Oliveira, D.P., Textile dyes: Dyeing process and environmental impact. *Eco-Friendly Text. Dyeing Finish.*, 6, 151, 2013.
11. Singh, K. and Arora, S., Removal of synthetic textile dyes from wastewaters: A critical review on present treatment technologies. *Crit. Rev. Environ. Sci. Technol.*, 41, 807, 2011.
12. Gottlieb, A., Shaw, C., Smith, A., Wheatley, A., Forsythe, S., The toxicity of textile reactive azo dyes after hydrolysis and decolourisation. *J. Biotechnol.*, 101, 49, 2003.
13. Bhatia, D., Sharma, N.R., Singh, J., Kanwar, R.S., Biological methods for textile dye removal from wastewater: A review. *Crit. Rev. Environ. Sci. Technol.*, 47, 1836, 2017.
14. Tsuboy, M., Angeli, J., Mantovani, M., Knasmüller, S., Umbuzeiro, G., Ribeiro, L., Genotoxic, mutagenic and cytotoxic effects of the commercial dye CI Disperse Blue 291 in the human hepatic cell line HepG2. *Toxicol. in Vitro*, 21, 1650, 2007.

15. Farooqi, Z.H., Sultana, H., Begum, R., Usman, M., Ajmal, M., Nisar, J., Irfan, A., Azam, M., Catalytic degradation of malachite green using a crosslinked colloidal polymeric system loaded with silver nanoparticles. *Int. J. Environ. Anal. Chem.*, 1–17, 2020.
16. Kiran, S., Adeel, S., Nosheen, S., Hassan, A., Usman, M., Rafique, M.A., Recent trends in textile effluent treatments: A review. *Adv. Mater. Wastewater Treat.*, 29, 29, 2017.
17. Robinson, T., McMullan, G., Marchant, R., Nigam, P., Remediation of dyes in textile effluent: A critical review on current treatment technologies with a proposed alternative. *Bioresour. Technol.*, 77, 247, 2001.
18. Saeed, M., Adeel, S., Ilyas, M., Shahzad, M.A., Usman, M., Haq, E-u., Hamayun, M., Oxidative degradation of Methyl Orange catalyzed by lab prepared nickel hydroxide in aqueous medium. *Desalin. Water Treat.*, 57, 12804, 2016.
19. Holkar, C.R., Jadhav, A.J., Pinjari, D.V., Mahamuni, N.M., Pandit, A.B., A critical review on textile wastewater treatments: Possible approaches. *J. Environ. Manage.*, 182, 351, 2016.
20. Seow, T.W. and Lim, C.K., Removal of dye by adsorption: A review. *Int. J. Appl. Eng. Res.*, 11, 2675, 2016.
21. Zheng, Y., Yao, G., Cheng, Q., Yu, S., Liu, M., Gao, C., Positively charged thin-film composite hollow fiber nanofiltration membrane for the removal of cationic dyes through submerged filtration. *Desalination*, 328, 42, 2013.
22. Wu, J.-S., Liu, C.-H., Chu, K.H., Suen, S.-Y., Removal of cationic dye methyl violet 2B from water by cation exchange membranes. *J. Membr. Sci.*, 309, 239, 2008.
23. Bhatti, I.A., Adeel, S., Abbas, M., Effect of radiation on textile dyeing. *Textile dyeing*, vol. 1, pp. 1–16, New York, 2011.
24. Verma, A.K., Dash, R.R., Bhunia, P., A review on chemical coagulation/flocculation technologies for removal of colour from textile wastewaters. *J. Environ. Manage.*, 93, 154, 2012.
25. Yusuf, M., *Handbook of textile effluent remediation*, CRC Press, New York, 2018.
26. Cowan-Ellsberry, C., Belanger, S., Dorn, P., Dyer, S., McAvoy, D., Sanderson, H., Versteeg, D., Ferrer, D., Stanton, K., Environmental safety of the use of major surfactant classes in North America. *Crit. Rev. Environ. Sci. Technol.*, 44, 1893, 2014.
27. Iram, M., Sultana, H., Usman, M., Ahmad, B., Akram, N., Siddiq, M., Rashid, S., Interaction of sulphone based reactive dyes with cationic surfactant: A spectroscopic and conductometric study. *Z. Phys. Chem.*, 2021.
28. Krstonošić, V., Dokić, L., Milanović, J., Micellar properties of OSA starch and interaction with xanthan gum in aqueous solution. *Food Hydrocolloids*, 25, 361, 2011.
29. Pal, A. and Chaudhary, S., Thermodynamic and aggregation behavior of aqueous tetradecyltrimethylammonium bromide in the presence of the

hydrophobic ionic liquid 3-methyl-1-pentylimidazolium hexafluorophosphate. *J. Mol. Liq.*, 207, 67, 2015.

30. Weschayanwiwat, P., Kunanupap, O., Scamehorn, J.F., Benzene removal from wastewater using aqueous surfactant two-phase extraction with cationic and anionic surfactant mixtures. *Chemosphere*, 72, 1043, 2008.
31. Dalali, N., Khoramnezhad, M., Habibizadeh, M., Faraji, M., Magnetic removal of acidic dyes from wastewaters using surfactant-coated magnetite nanoparticles: Optimization of process by Taguchi method. *Proc. Int. Conf. Environ. Agric. Eng. IPCBEE*, 15, 89–94, 2011.
32. Liu, H.-Q., Liu, F., Wei, G.-X., Zhang, R., Zhu, Y.-W., Effects of surfactants on the removal of carbonaceous matter and dioxins from weathered incineration fly ash. *Aerosol Air Qual. Res.*, 17, 2338, 2017.
33. Hosseinzadeh, R., Khorsandi, K., Hemmaty, S., Study of the effect of surfactants on extraction and determination of polyphenolic compounds and antioxidant capacity of fruits extracts. *PloS one*, 8, e57353, 2013.
34. Koner, S., Pal, A., Adak, A., Use of surface modified silica gel factory waste for removal of 2, 4-D pesticide from agricultural wastewater: A case study. *Int. J. Environ. Res.*, 6, 995, 2012.
35. Khan, R., Inam, M.A., Khan, S., Jiménez, A.N., Park, D.R., Yeom, I.T., The Influence of Ionic and Nonionic Surfactants on the Colloidal Stability and Removal of CuO Nanoparticles from Water by Chemical Coagulation. *Int. J. Environ. Res. Public Health*, 16, 1260, 2019.
36. Li, F., Li, X., Zhang, J., Peng, L., Liu, C., Removal of organic matter and heavy metals of low concentration from wastewater via micellar-enhanced ultrafiltration: An overview. *IOP Conf. Ser. Earth Envir. Sci.*, 1, 012077, 2017. IOP Publishing.
37. Mao, X., Jiang, R., Xiao, W., Yu, J., Use of surfactants for the remediation of contaminated soils: A review. *J. Hazard. Mater.*, 285, 419, 2015.
38. Fu, H.-Y., Zhang, Z.-B., Chai, T., Huang, G.-H., Yu, S.-J., Liu, Z., Gao, P.-F., Study of the removal of aniline from wastewater via meuf using mixed surfactants. *Water*, 9, 365, 2017.
39. Naeem, K., Shah, S.S., Shah, S.W., Laghari, G.M., Solubilization of cationic hemicyanine dyesin anionic surfactant micelles: A partitioning study. *Monatsh. Chem.*, 131, 761, 2000.
40. Bielska, M., Sobczyńska, A., Prochaska, K., Dye–surfactant interaction in aqueous solutions. *Dyes Pigm.*, 80, 201, 2009.
41. Tunç, S., Duman, O., Kancı, B., Spectrophotometric investigation of the interactions between cationic dye (CI Basic Yellow 2) and anionic surfactant (sodium dioctylsulfosuccinate) in the premicellar and micellar region. *Dyes Pigm.*, 94, 233, 2012.
42. Migahed, M. and Al-Sabagh, A., Beneficial role of surfactants as corrosion inhibitors in petroleum industry: A review article. *Chem. Eng. Commun.*, 196, 1054, 2009.

43. Kralova, I. and Sjöblom, J., Surfactants used in food industry: A review. *J. Dispersion Sci. Technol.*, 30, 1363, 2009.
44. Morsy, S.M., Role of surfactants in nanotechnology and their applications. *Int. J. Curr. Microbiol. App. Sci.*, 3, 237, 2014.
45. Myers, D., Physical properties of surfactants used in cosmetics. *Surfactants in cosmetics*, vol. 68, pp. 49–102, Routledge, 2017.
46. Castro, M.J., Ojeda, C., Cirelli, A.F., Surfactants in agriculture. *Green materials for energy products and depollution*, vol. 3, pp. 287–334, Springer, 2013.
47. Jiao, J., Polyoxyethylated nonionic surfactants and their applications in topical ocular drug delivery. *Adv. Drug Delivery Rev.*, 60, 1663, 2008.
48. Scheibel, J.J., The evolution of anionic surfactant technology to meet the requirements of the laundry detergent industry. *J. Surfactants Deterg.*, 7, 319, 2004.
49. Uphues, G., *Chemistry of amphoteric surfactants. Lipid/Fett*, vol. 100, p. 490, Wiley-VOH Verlag GmbH, Germany, 1998.
50. Prakash, N., Sockan, V., Jayakaran, P., Wastewater treatment by coagulation and flocculation. *Int. J. Eng. Sci. Innov. Technol.*, 3, 479, 2014.
51. Gregory, J., *Coagulation and flocculation with an emphasis on water and wastewater treatment*, vol. 1980, p. xi+ 354, John Bratby Published by Uplands Press, Elsevier, Croydon, UK, 1981.
52. Bratby, J., *Coagulation and flocculation in water and wastewater treatment*, IWA Publishing, London, UK, 2016.
53. Haydar, S. and Aziz, J.A., Coagulation–flocculation studies of tannery wastewater using combination of alum with cationic and anionic polymers. *J. Hazard. Mater.*, 168, 1035, 2009.
54. Melo, R., Neto, E.B., Moura, M., Dantas, T.C., Neto, A.D., Oliveira, H., Removal of direct Yellow 27 dye using animal fat and vegetable oil-based surfactant. *J. Water Process. Eng.*, 7, 196, 2015.
55. Melo, R., Neto, E.B., Nunes, S., Dantas, T.C., Neto, A.D., Removal of Reactive Blue 14 dye using micellar solubilization followed by ionic flocculation of surfactants. *Sep. Purif. Technol.*, 191, 161, 2018.
56. Talens-Alesson, F.I., Anthony, S., Bryce, M., Removal of phenol by adsorptive micellar flocculation: multi-stage separation and integration of wastes for pollution minimisation. *Colloids Surf., A*, 276, 8, 2006.
57. Wang, H., Wang, D., Tian, T., Ren, W., Removal of Organic Compounds Containing a Benzene Ring from Water by Adsorptive Micellar Flocculation. *J. Surfactants Deterg.*, 22, 161, 2019.
58. Porras, M. and Talens, F., Removal of 2, 4-D from aqueous solutions by micellar flocculation. *Sep. Sci. Technol.*, 34, 2679, 1999.
59. Kuipa, P.K. and Kuipa, O., Removal of pyrethrin from aqueous effluents by adsorptive micellar flocculation. *J. Chem.*, 2015, 1–6, 2015.
60. Sultana, H., Bokhari, T.H., Usman, M., Adsorptive micellar flocculation (surfactant-based phase separation technique): Theory and applications. *J. Mol. Liq.*, 323, 115001, 2021.

61. Irshad, S., Sultana, H., Usman, M., Saeed, M., Akram, N., Yusaf, A., Rehman, A., Solubilization of direct dyes in single and mixed surfactant system: A comparative study. *J. Mol. Liq.*, 321, 114201, 2021.
62. Paton-Morales, P. and Talens-Alesson, F., Flocculation of anionic surfactant micelles in the presence of hydrocarbons. *Colloid Polym. Sci.*, 278, 697, 2000.
63. Talens-Alesson, F.I., Anthony, S., Bryce, M., Complexation of organic compounds in the presence of Al3+ during micellar flocculation. *Water Res.*, 38, 1477, 2004.
64. Pereira, R.F., Valente, A.J., Burrows, H.D., The interaction of long chain sodium carboxylates and sodium dodecylsulfate with lead (II) ions in aqueous solutions. *J. Colloid Interface Sci.*, 414, 66, 2014.
65. Sahinkaya, H.U. and Ozkan, A., Investigation of shear flocculation behaviors of colemanite with some anionic surfactants and inorganic salts. *Sep. Purif. Technol.*, 80, 131, 2011.
66. Cavalcante, P., Melo, R., Dantas, T.C., Neto, A.D., Neto, E.B., Moura, M., Removal of phenol from aqueous medium using micellar solubilization followed by ionic flocculation. *J. Environ. Chem. Eng.*, 6, 2778, 2018.
67. Talens, F., Paton, P., Gaya, S., Micelar flocculation of anionic surfactants. *Langmuir*, 14, 5046, 1998.
68. Pereira, R.F., Valente, A.J., Burrows, H.D., Thermodynamic analysis of the interaction between trivalent metal ions and sodium dodecyl sulfate: An electrical conductance study. *J. Mol. Liq.*, 156, 109, 2010.
69. Ndabigengesere, A. and Narasiah, K.S., Use of *Moringa oleifera* seeds as a primary coagulant in wastewater treatment. *Environ. Technol.*, 19, 789, 1998.
70. Porras-Rodriguez, M. and Talens-Alesson, F.I., Removal of 2, 4-dichlorophenoxyacetic acid from water by adsorptive micellar flocculation. *Environ. Sci. Technol.*, 33, 3206, 1999.
71. Mohammad, A., Studies on cloud point extraction and adsorptive micellar flocculation of some dyes, 2014.
72. Anthony, S. and Talens-Alesson, F., Effect of an electrolyte on Adsorptive Micellar Flocculation (I): Increased selectivity in the presence of monovalent–monovalent electrolyte. *Colloids Surf., A*, 301, 1, 2007.
73. Talens-Alesson, F.I., Binding of pesticide 2, 4-D to SDS and AOS micellar flocculates. *Colloids Surf., A*, 180, 199, 2001.
74. Saitoh, T., Shibata, K., Fujimori, K., Ohtani, Y., Rapid removal of tetracycline antibiotics from water by coagulation–flotation of sodium dodecyl sulfate and poly (allylamine hydrochloride) in the presence of Al (III) ions. *Sep. Purif. Technol.*, 187, 76, 2017.
75. Saitoh, T., Shibata, K., Hiraide, M., Rapid removal and photodegradation of tetracycline in water by surfactant-assisted coagulation–sedimentation method. *J. Environ. Chem. Eng.*, 2, 1852, 2014.
76. Almeida, T.D.O. and Talens-Alesson, F.I., Removal of phenylamine and catechol by adsorptive micellar flocculation. *Colloids Surf., A*, 279, 28, 2006.

11

Remediation of Textile Wastewater by Ozonation

Astha Gupta, Suhail Ayoub Khan and Tabrez Alam Khan*

Department of Chemistry, Jamia Millia Islamia, Jamia Nagar, New Delhi, India

Abstract

The textile wastewater laden with synthetic dyes is a key contributor to water pollution. Discharges from the textile sector encompass an enormous quantity of recalcitrant dyes that have a negative impact on biotic life which necessitated their removal to safeguard the ecosystem. Ozonation is a powerful way to uptake these pollutants by employing the high oxidizing potential of ozone. This book chapter illustrates the efficacy of ozonation for the decolorization of dyes. The impact of various parameters on ozonation such as inlet ozone concentration, initial dye concentration, pH, and time was extensively summarized. The effect of ozonation on dye degradation in combination with other methods was also provided. Additionally, the possible decolorization mechanism was furnished.

***Keywords*:** Ozonation, textile wastewater, dissolved organic carbon (DOC), dyes, decolorization

11.1 Introduction

Fondness towards colored clothes by humans has led to the development of the textile industry which is one of the major industries in the world. The charismatic colors of synthetic dyes on textiles witnessed their dramatic usage from the last few decades [1]. The discharge of effluents from these industries into the aquatic sources made them a rampant source of water pollution [2]. The effluents contain a high concentration of various inorganic and organic chemicals like chlorides, nitrates, metals like

**Corresponding author*: takhan@jmi.ac.in; takhan501@yahoo.com

Luqman Jameel Rather, Mohd Shabbir and Aminoddin Haji (eds.) Sustainable Practices in the Textile Industry, (273–284)

manganese, sodium, lead, copper, chromium, iron, and dye residues which are carcinogenic, mutagenic and teratogenic [3] and non-biodegradable due to their complex aromatic structure and artificial fabrication [4]. These dyes even in low concentrations can affect the aquatic life as they deplete the oxygen level [5], they also hinder the way of sunlight into the water which causes a fatal condition for algae as there will be no sunlight for photosynthesis [6]. Despite all the disadvantages of dyes on water, environment, and ecological system textile industry is growing day by day due to the enormous demand, therefore it is important to find an effective way for their treatment. The pH values of these textile water oscillate in between a large range from 2 to 12 due to the presence of a variety of dyes used in it, this variation causes a challenge in their removal due to the finite endurance of the pH range of various wastewater treatment [7]. The coloring or dyeing process operates at a high temperature of about 90 °C which is another obstacle for their removal form textile water as the temperature should be 30 °C for any treatment [8].

There are many physical and chemical methods to treat textile water. The most used ones are electrochemical oxidation, flocculation, phytoremediation, crystallization, adsorption, and electrolysis [9–11]. Although being advantageous in some ways, there are several drawbacks associated with these methods such as high operational cost, technical limitations, production of toxic sludge, and lack of effective treatment. Ozonation remediation is one of the most efficient and promising methods for the treatment of textile water based on the oxidation process. Ozone is the most powerful oxidizing agent due to its high value of oxidation potential (2.07) compare to other powerful oxidizing agents such as chlorine and hydrogen peroxide (1.78) [12]. Ozonation of dye residues occurs via direct reaction with ozone or free radicles. Among the various advantages, one of them is that it can be applied in its gaseous state, and therefore does not increase the volume of water and sludge. It leaves the effluent with no color and low COD and it also increases the level of dissolved oxygen of sewage [13] suitable for discharge into water sources. Additionally, it modifies the molecular structure of pollutants by transforming them into the products which are effortlessly assimilated biologically [14]. This chapter will discuss the ozonation remediation for the treatment of textile water.

11.2 Sources of Wastewater

Wastewater contains a large range of pollutants such as suspended particles, microorganisms, organic macromolecules, heavy metals, dissolved inorganic, and organic molecules. The range of pollutants of untreated water is so vast that an individual characterization of each pollutant is difficult, and it also varies with the source of wastewater. Among the wide source of wastewater, some main sources are domestic wastewater, industrial wastewater, agriculture discharge, and textile wastewater, etc. In the textile industry coloring or dyeing of fibers to get the desired colored fabric and printing on it are two major stages. Adsorption of dye into fabric and spreading of dye into the fabric are two steps in the dyeing process, although the mechanism of dyeing of fabric completely depends on the type of fibers and dyeing mode [15]. Textile wastewater can have a COD value of 800 to 1,600 mg/l [16]. The value of COD (carbon oxygen demand), TSS (total oxygen demand), TOC (total organic carbon), conductivity, and turbidity are major characteristics of textile wastewater [5]. Dyes employed for coloration of fibers are usually applied in aqueous solution. A huge amount of water is required in this step, it has been found that 100 L of water is required for 1 kg of dyed fabric [16] thus, generating an enormous amount of effluents. Among all the textile operation dyeing process of polyester involves the highest BOD values from 480 to 27,000 mg/L.

11.3 Ozonation Remediation for Textile Water

Ozone is the most powerful oxidizing agent its oxidation power is 1.5 times greater than chlorine and has been used for industrial application for wastewater treatment since 1990 [17]. Ozonation treatment for wastewater is a more promising method [18] than any other methods due to the following factors

- It treats the wastewater without the formation of any sludge thus there will be no residue left after the treatment.
- This method reduces both dye and organic compound present in textile water, there will be no need to the application of two different processes to treat dye and organic compound individually, these dyes have a chromophore group which consists of conjugated double bonds presence of hydroxide ion, and ozone in aqueous solution open the ring and breaks it into smaller molecules which lead to the depolarization or reduction of dye [19].
- The whole experimental setup requires a small space therefore it can be easily installed.

- Ozone left after the process can be easily decomposed into oxygen.
- In aqueous solution, ozone reacts with different organic and inorganic compounds by two possible reaction mechanisms either by direct reaction i.e. reaction of molecular ozone with compounds, or with indirect reaction i.e. free radical mechanism [20]. Now the pH of the solution governs the mode of the reaction mechanism, in acidic solution reaction follows a direct mechanism and attack on some specific functional groups as an electrophile while in basic pH reaction follows an indirect mechanism; it decomposes into hydroxide ion and in other radicals and reacts with compounds [21]. Though ozonation in the basic medium is more effective than in acidic medium [22].

11.3.1 Impact of pH on Uptake of Organic Pollutants

Various investigations have revealed that the confiscation of dye-laden wastewater is *pH*-dependent [23]. The dissociation of ozone in the liquid phase is sparked by electron transfer reaction with aromatic moieties [e.g. dissolved organic carbon (DOC)] that are rich in electron content [24]. The reaction furnishes hydroxyl radicle and hydroxyl anion (Equations 11.1–11.3), among which hydroxyl anion is greatly affected by the pH resulted in the dissociation of ozone in water.

$$\mathrm{DOC} + \mathrm{O_3} \longrightarrow \mathrm{DOC^+} + \mathrm{O_3^-} \tag{11.1}$$

$$\mathrm{O_3^-} \longleftrightarrow \mathrm{O_2} + \mathrm{O^-} \tag{11.2}$$

$$\mathrm{O^-} + \mathrm{H_2O} \longleftrightarrow \mathrm{OH} + \mathrm{OH^-} \tag{11.3}$$

The decomposition of ozone is relatively faster at higher *pH* with a short half-life with the generation of an ample amount of hydroxyl radicles leading to higher competition time [25]. At lower pH ozone becomes more stable and generates a smaller number of hydroxyl radicles which leads to shorter competition between molecules with ample electrons in the DOC. Peng *et al.* [21] in their study revealed that at basic pH, the decomposition of ozone is rapid with the release of hydroxyl radicle which limits its dye uptake while at acidic pH, the direct reaction between ozone and organic moieties is very efficient in the discoloration of the dye as the ozone attacks on conjugated double bonds which are often related with coloration property of the dye. Sevimli *et al.* [26] from their investigation demonstrated that in basic pH maximum decolorization, COD, and DOC were obtained. Under acidic medium, these properties were slight less which depicted that the impact of pH on the ozone employment ratio is minor. Soares *et al.* [27] revealed that the decolorization at buffered solutions was 90, 88, and 82%, while at unbuffered solutions the removal of dyes was 92, 93, and 93% at initial pH 5, 7, and 9, respectively. This behavior can be explained that in unbuffered solutions the direct reaction prevails which leads to higher removal efficacy. Venkatesh *et al.* [28] demonstrated the effect of pH on the decolorization of a dye via ozonation. It was found that 93% of acid Red 14 was removed from the unbuffered solution having acidic pH while in the buffered solution the decolorization was 84.5% exhibiting acidic media favors ozonation process.

11.3.2 Impact of Initial Dye Concentration

A significant impact on the economical utilization of ozone for decolorization purposes is dependent on the inlet dye concentration [29]. The time required for decolorization depends on the concentration of dye as well as on the consumption of ozone. Soares *et al.* [27] depicted that when dye concentration increases both the decolorization and TOC removal decreases at steady-state and consumption of ozone increases. On increasing the dye concentration from 50 to 100 mg/L utilization of ozone increases from 21 to 27 go3/m3 and color removal and TOC removal decreases from 93 to 76% and 10 to 4% respectively. Muniyasamy *et al.* [30] their study showed that a decline in the decolorization of DR 81 dye with an increase in initial concentration. This may be ascribed to the seizure effect developed by the intermediates generated during the ozonation process and to the decline in ozone to DR 81 molecules in solution [31]. Lovato *et al.* [32] from their investigation on the degradation of anthraquinone dye using ozonation revealed that decolorization of RB 19 dye decreased with elevation in initial concentration. After 5 min of ozonation 100, 96, and 90% decolorization was achieved at initial concentrations of 111, 235, 466 mg/L, respectively. The decrease might be due to the exhaustion of ozone molecules at higher concentration and the generation of various inorganic anions which compete with DOC [33]. Quan *et al.* [34] demonstrated that with the usage of the $O_3/Ca(OH)_2$ system, the decolorization of Acid Red 18 with every concentration was almost the same which indicated higher ozone utilization efficacy. The reason for this trend might be the elimination of mineralization products by Ca^{2+} in the solution.

11.3.3 Impact of Inlet Ozone Concentration

Consumption of ozone per volume is affected by the flow rate of gas and consequently, color and TOC removal get affected too. On increasing the gas flow rate consumption of ozone per volume increases which leads to the larger removal of color and TOC. Thus, removal efficiency increases on increasing the gas flow rate. The considerable influence of ozone concentration on decolorization and TOC was shown by Quan *et al.* [34] when the concentration of ozone was increased from 30 75 mg/L, a higher removal and decolorization rate was achieved. The escalation of dye decolorization and TOC removal with increment in ozone concentration is due to the more ozone available per molecule of dye solution and the increase in hydroxyl radicles along with mass transfer driving force [35]. Ghuge *et al.* [36] varied the ozone dosage from 1 to 7.5g/m^3 and showed higher decolorization and removal of TOC with an increase in ozone dosage. As the ozone concentration was increased from 1 to 5 g/m^3, decolorization from 38.5 to 100% was achieved after 21 min. Similarly, TOC removal increased from 19 to 86% within 1 to 5 g/m^3 after 60 min. Buyukada [37] in his study of ultrasound-assisted ozonation revealed that acidic pH condition along with lower ozone concentration resulted in higher decolorization characteristics of dye laden water. The ultrasonic irradiation generated more radicles and the dyestuff can easily react with O_3 resulting in which leads to increment in decolorization even at low ozone concentration.

11.3.4 Impact of Ozonation Time

Time is a critical factor that influences the decolorization of dyes during the ozonation process [38]. Rekhate *et al.* [39] demonstrated that decolorization efficacy increase with an escalation in reaction time. Keeping the concentration constant, they revealed that within 10 min decolorization as achieved in the acidic medium than in basic. Venkatesh *et al.* [40] revealed that longer reaction time is effective for the degradation of congo red. 90% of decolorization of congo red was observed at an initial concentration of 1,500 mg/L for 25 min. Wu *et al.* [41] revealed 87.4% removal within 10 min of ozonation. After the time was extended to 30 min, the decolorization percentage reached 96% implying that longer reaction time favors decolorization.

11.4 Impact of Various Techniques in Combination Ozonation Process for Treatment of Textile Wastewater

This combined action generates an adequate amount of hydroxyl radicals which helps in the fast and more efficient removal of the toxic pollutant from textile water [42]. There are two ways to produce hydroxyl radicals, Electrolysis and Photolysis. To generate the hydroxyl radical via electrolysis ozone is used. Ozone decomposes into water and gives hydroxyl radical [43], this hydroxyl radical acts as a powerful oxidizing agent and helps in the mineralization of organic compounds. Decolorization takes place by breaking of chromophore groups of dye by ozone molecules. In the photolysis method hydroxyl radicals generated with the help of a semiconductor, on illumination with the sufficient amount of energy semiconductor generates electron-hole pair now the photogenerated holes in aerated aqueous solution give hydroxyl radical by the oxidation of water molecules [44].

$$H_2O + h^+ = OH^{\cdot} + H^+$$

Whereas dissolved O_2 molecules reduced into superoxide ion-

$$O_2 + e^- = O_2^-$$

Several studies have suggested that photo degradation of dyes on a semiconductor surface is an effective method for the treatment of textile dye [45]. These hydroxyl radicals and superoxide serve as an oxidizing agent and leads to the mineralization of organic compounds [46]. In the photochemical process toxic organic compounds degraded into CO_2 and H_2O.

11.5 Degradation of Dyes via Ozonation

The positions with strong electron density on an aromatic moiety are most vulnerable to the electrophilic attack by ozone [47]. Apart from the electrophilic reaction, ozone leads to the formation of carbonyl and nitroso groups via undergoing cycloaddition reaction addition with unsaturated bonds. The degradation pattern of dye via ozone is shown in Figure 11.1. Ozone attacks the azo position of a dye leading to the decolorization of the

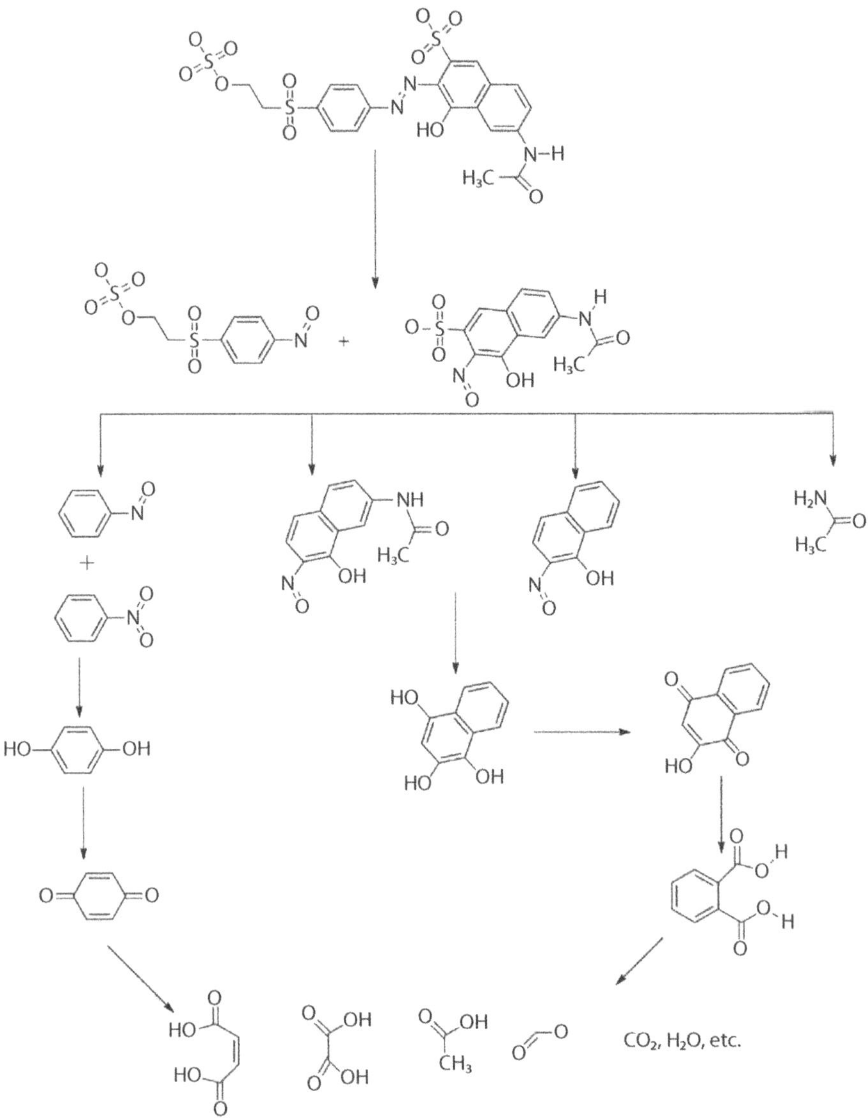

Figure 11.1 Degradation of a dye by ozonation.

solution with the breakage of azo bond and generation of various intermediates. The cleavage of the sulfonic group, nitro group via ozonation leads to the release of sulfur and nitrogen from the dyes. The extensive ozonation of intermediated resulted in complete mineralization leading to the production of carbon dioxide and water.

11.6 Conclusion

Ozonation is a powerful method to decolorize the recalcitrant dyes from textile wastewater due to its higher oxidative potential. The aromatic moieties which are tenacious to biological oxidation and pose a substantial threat to the ecosystem are converted to biologically assimilated form by ozonation. Owing to its higher efficacy to degrade micropollutants from a myriad of water sources, it has become a prime factor of water reclamation programs for potable reuse. The pH of the solution has a considerable impact on ozonation. At acidic pH, the direct reaction occurs resulting in higher removal while at basic pH hydroxyl ions are generated which lessens the removal capacity. The higher ozone dosage and time influence the removal of pollutants in a positive way while the higher concentration of dyes diminishes the decolorization ability.

References

1. Khattab, T.A., Abdelrahman, M.S., Rehan, M., Textile dyeing industry: Environmental impacts and remediation. *Environ. Sci. Pollut. Res.*, 27, 4, 3803–3818, 2020.
2. Liang, Z., Wang, J., Zhang, Y., Han, C., Ma, S., Chen, J., An, T., Removal of volatile organic compounds (VOCs) emitted from a textile dyeing wastewater treatment plant and the attenuation of respiratory health risks using a pilot-scale biofilter. *J. Clean. Prod.*, 253, 120019, 2020.
3. Wazir, M.B., Daud, M., Ali, F., Al-Harthi, M.A., Dendrimer assisted dye-removal: A critical review of adsorption and catalytic degradation for wastewater treatment. *J. Mol. Liq.*, 315, 113775, 2020.
4. Hynes, N.R.J., Kumar, J.S., Kamyab, H., Sujana, J.A.J., Al-Khashman, O.A., Kuslu, Y., Suresh, B., Modern enabling techniques and adsorbents-based dye removal with sustainability concerns in textile industrial sector—A comprehensive review. *J. Clean. Prod.*, 272,122636, 2020.
5. Holkar, C.R., Jadhav, A.J., Pinjari, D.V., Mahamuni, N.M., Pandit, A.B., A critical review on textile wastewater treatments: Possible approaches. *J. Environ. Manage.*, 182, 351–366, 2016.
6. Weldegebrieal, G.K., Synthesis method, antibacterial and photocatalytic activity of ZnO nanoparticles for azo dyes in wastewater treatment: A review. *Inorg. Chem. Commun.*, 120, 108140, 2020.
7. Chauhan, A.K., Kataria, N., Garg, V.K., Green fabrication of ZnO nanoparticles using Eucalyptus spp. leaves extract and their application in wastewater remediation. *Chemosphere*, 247, 125803, 2020.
8. Lin, S.H. and Peng, C.F., Treatment of textile wastewater by electrochemical method. *Water Res.*, 28, 2, 277–282, 1994.

9. Chowdhury, M.F., Khandaker, S., Sarker, F., Islam, A., Rahman, M.T., Awual, M.R., Current treatment technologies and mechanisms for removal of indigo carmine dyes from wastewater: A review. *J. Mol. Liq.* 318, 114061, 2020.
10. Vimala, R.T.V., Escaline, J.L., Murugan, K., Sivaramakrishnan, S., An overview of organic matters in municipal wastewater: Removal via self-assembly flocculating mechanism and the molecular level characterization. *J. Environ. Manage.*, 266, 110572, 2020.
11. Khan, S.A., Siddiqui, M.F., Khan, T.A., Ultrasonic-assisted synthesis of polyacrylamide/bentonite hydrogel nanocomposite for the sequestration of lead and cadmium from aqueous phase: Equilibrium, kinetics and thermodynamic studies. *Ultrason. Sonochem.*, 60, 104761, 2020.
12. Kolosov, P., Peyot, M.L., Yargeau, V., Novel materials for catalytic ozonation of wastewater for disinfection and removal of micropollutants. *Sci. Total Environ.*, 644, 1207–1218, 2018.
13. Khataee, A., Fazli, A., Zakeri, F., Joo, S.W., Synthesis of a high-performance Z-scheme 2D/2D WO3@ CoFe-LDH nanocomposite for the synchronic degradation of the mixture azo dyes by sonocatalytic ozonation process. *J. Ind. Eng. Chem.*, 89, 301–315, 2020.
14. Shokouhi, S.B., Dehghanzadeh, R., Aslani, H., Shahmahdi, N., Activated carbon catalyzed ozonation (ACCO) of Reactive Blue 194 azo dye in aqueous saline solution: Experimental parameters, kinetic and analysis of activated carbon properties. *J. Water Process. Eng.*, 35, 101188, 2020.
15. Kobya, M., Can, O.T., Bayramoglu, M., Treatment of textile wastewaters by electrocoagulation using iron and aluminum electrodes. *J. Hazard. Mater.*, 100, 1–3, 163–178, 2003.
16. Amutha, K. and Sudhapriya, N., Dyeing of textiles with natural dyes extracted from *Terminalia arjuna* and *Thespesia populnea* fruits. *Ind. Crops Prod.*, 148, 112303, 2020.
17. Baig, S. and Liechti, P.A., Ozone treatment for biorefractory COD removal. *Water Sci. Technol*, 43, 2, 197–204, 2001.
18. Chu, W. and Ma, C.W., Quantitative prediction of direct and indirect dye ozonation kinetics. *Water Res.*, 34, 12, 3153–3160, 2000.
19. Sevimli, M.F. and Kinaci, C., Decolorization of textile wastewater by ozonation and Fenton's process. *Water Sci. Technol.*, 45, 12, 279–286, 2002.
20. Wang, C., Yediler, A., Lienert, D., Wang, Z., Kettrup, A., Ozonation of an azo dye CI Remazol Black 5 and toxicological assessment of its oxidation products. *Chemosphere*, 52, 7, 1225–1232, 2003.
21. Peng, R.Y. and Fan, H.J., Ozonalytic kinetic order of dye decolorization in aqueous solution. *Dyes Pigm.*, 67, 2, 153–159, 2005.
22. Robinson, T., McMullan, G., Marchant, R., Nigam, P., Remediation of dyes in textile effluent: A critical review on current treatment technologies with a proposed alternative. *Bioresour. Technol.*, 77, 3, 247–255, 2001.
23. Hien, N.T., Nguyen, L.H., Van, H.T., Nguyen, T.D., Nguyen, T.H.V., Chu, T.H.H., Aziz, K.H.H., Heterogeneous catalyst ozonation of Direct Black 22

from aqueous solution in the presence of metal slags originating from industrial solid wastes. *Sep. Purif. Technol.*, 233, 115961, 2020.
24. Li, B., Li, C., Qu, R., Wu, N., Qi, Y., Sun, C., Wang, Z., Effects of common inorganic anions on the ozonation of polychlorinated diphenyl sulfides on silica gel: Kinetics, mechanisms, and theoretical calculations. *Water Res.*, 186, 116358, 2020.
25. Tizaoui, C. and Grima, N., Kinetics of the ozone oxidation of Reactive Orange 16 azo-dye in aqueous solution. *Chem. Eng. J.*, 173, 2, 463–473, 2011.
26. Sevimli, M.F. and Sarikaya, H.Z., Ozone treatment of textile effluents and dyes: Effect of applied ozone dose, pH and dye concentration. *J. Chem. Technol. Biotechnol.*, 77, 7, 842–850, 2002.
27. Soares, O.S.G., Orfao, J.J., Portela, D., Vieira, A., Pereira, M.F.R., Ozonation of textile effluents and dye solutions under continuous operation: Influence of operating parameters. *J. Hazard. Mater.*, 137, 3, 1664–1673, 2006.
28. Venkatesh, S. and Venkatesh, K., Ozonation for degradation of acid red 14: Effect of buffer solution. *Proc. Natl. Acad. Sci. India Section A: Phys. Sci.*, 90, 2, 209–212, 2020.
29. El Hassani, K., Kalnina, D., Turks, M., Beakou, B.H., Anouar, A., Enhanced degradation of an azo dye by catalytic ozonation over Ni-containing layered double hydroxide nanocatalyst. *Sep. Purif. Technol.*, 210, 764–774, 2019.
30. Muniyasamy, A., Sivaporul, G., Gopinath, A., John, J., Achary, A., Chellam, P.V., Fractional factorial design modelling on degradation of Direct Red 81 dye by advanced oxidation process–ozonation: Reaction kinetics. *Water Sci. Technol.*, 80, 11, 2037–2046, 2019.
31. Tehrani-Bagha, A.R., Mahmoodi, N.M., Menger, F.M., Degradation of a persistent organic dye from colored textile wastewater by ozonation. *Desalination*, 260, 1–3, 34–38, 2010.
32. Lovato, M.E., Fiasconaro, M.L., Martín, C.A., Degradation and toxicity depletion of RB19 anthraquinone dye in water by ozone-based technologies. *Water Sci. Technol.*, 75, 4, 813–822, 2017.
33. Shen, Y., Xu, Q., Wei, R., Ma, J., Wang, Y., Mechanism and dynamic study of reactive red X-3B dye degradation by ultrasonic-assisted ozone oxidation process. *Ultrasonf. Sonochem.*, 38, 681–692, 2017.
34. Quan, X., Luo, D., Wu, J., Li, R., Cheng, W., Ozonation of acid red 18 wastewater using O_3/Ca $(OH)_2$ system in a micro bubble gas-liquid reactor. *J. Environ. Chem. Eng.*, 5, 1, 283–291, 2017.
35. Hu, E., Wu, X., Shang, S., Tao, X.M., Jiang, S.X., Gan, L., Catalytic ozonation of simulated textile dyeing wastewater using mesoporous carbon aerogel supported copper oxide catalyst. *J. Clean. Prod.*, 112, 4710–4718, 2016.
36. Ghuge, S.P. and Saroha, A.K., Ozonation of Reactive Orange 4 dye aqueous solution using mesoporous Cu/SBA-15 catalytic material. *J. Water Process. Eng.*, 23, 217–229, 2018.
37. Buyukada, M., Modeling of decolorization of synthetic reactive dyestuff solutions with response surface methodology by a rapid and efficient process

of ultrasound-assisted ozone oxidation. *Desalination Water Treat.*, 57, 32, 14973–14985, 2016.
38. Mecha, A.C., Onyango, M.S., Ochieng, A., Momba, M.N., Impact of ozonation in removing organic micro-pollutants in primary and secondary municipal wastewater: Effect of process parameters. *Water Sci. Technol.*, 74, 3, 756–765, 2016.
39. Rekhate, C.V. and Shrivastava, J.K., Decolorization of Azo Dye Solution by Ozone Based Advanced Oxidation Processes: Optimization Using Response Surface Methodology and Neural Network. Ozone. *Sci. Eng.*, 1–15, 2020.
40. Venkatesh, S., Quaff, A.R., Pandey, N.D., Venkatesh, K., Decolorization and mineralization of CI direct red 28 azo dye by ozonation. *Desalination Water Treat.*, 57, 9, 4135–4145, 2016.
41. Wu, J., Gao, H., Yao, S., Chen, L., Gao, Y., Zhang, H., Degradation of crystal violet by catalytic ozonation using Fe/activated carbon catalyst. *Sep. Purif. Technol.*, 147, 179–185, 2015.
42. Brillas, E., Mur, E., Sauleda, R., Sanchez, L., Peral, J., Domènech, X., Casado, J., Aniline mineralization by AOP's: Anodic oxidation, photocatalysis, electro-Fenton and photoelectro-Fenton processes. *Appl. Catal.*, 16, 1, 31–42, 1998.
43. De los Santos Ramos, W., Poznyak, T., Chairez, I., Remediation of lignin and its derivatives from pulp and paper industry wastewater by the combination of chemical precipitation and ozonation. *J. Hazard. Mater.*, 169, 1–3, 428–434, 2009.
44. Hoffmann, M.R., Martin, S.T., Choi, W., Bahnemann, D.W., Environmental applications of semiconductor photocatalysis. *Chemical Reviews*, 95, 1, 69–96, 1995.
45. Nasr, C., Vinodgopal, K., Fisher, L., Hotchandani, S., Chattopadhyay, A.K., Kamat, P.V., Environmental photochemistry on semiconductor surfaces. Visible light induced degradation of a textile diazo dye, naphthol blue black, on TiO2 nanoparticles. *J. Phys. Chem.*, 100, 20, 8436–8442, 1996.
46. De Moraes, S.G., Freire, R.S., Duran, N., Degradation and toxicity reduction of textile effluent by combined photocatalytic and ozonation processes. *Chemosphere*, 40, 4, 369–373, 2000.
47. Rivera-Utrilla, J., Sánchez-Polo, M., Zaror, C.A., Degradation of naphthalenesulfonic acids by oxidation with ozone in aqueous phase. *Phys. Chem. Chem. Phys.*, 4, 7, 1129–1134, 2002.

12

Design of a New Cold Atmospheric Plasma Reactor Based on Dielelectric Barrier Discharge for the Treatment and Recovery of Textile Dyeing Wastewater: Profoks/CAP Reactor

Lokman Hakan Tecer[1]* and Ali Mutlu Gündüz[2]

[1]*Tekirdağ Namik Kemal University, Çorlu Engineering Faculty, Environmental Engineering Department, Çorlu, Republic of Turkey*
[2]*Pargino Elektronik Yazilim Sağlik İmalat San. Tic. A.Ş., Ankara, Republic of Turkey*

Abstract

Protection of water resources, treatment and recovery of wastewater have become the most important environmental policy all over the world. The efforts to develop water recovery and treatment technologies and new approaches are increasingly continued. Although the activation potential of Reactive Oxygen Species (ROS) has been studied, it has not been brought to an economically and technologically widespread level in water recovery and treatment. In this chapter, a newly developed Cold Atmospheric Plasma (CAP) system called Profoks, which can realize organic matter mineralization in wastewater and flue gas environments by producing "ROS" *in situ*, is introduced. The Profoks/CAP System includes specific system designs depending on the contaminant and industrial process characteristics. With the Profoks/CAP reactor, the recovery of the textile wastewater and its reuse in processes by performing water recycling have been achieved. Chemical oxygen demand (COD) could be reduced by 94.76% after treatment. 92.35% efficiency in removing of the suspended solids was obtained. The amount of color and total nitrogen were removed at 99.92% (up to 1.5 Pt-Co) and 96.4% efficiency, respectively. The conductivity was reduced to 51.6 Us/cm. Most importantly,

**Corresponding author*: lhtecer@nku.edu.tr

Luqman Jameel Rather, Mohd Shabbir and Aminoddin Haji (eds.) Sustainable Practices in the Textile Industry, (285–306) © 2021 Scrivener Publishing LLC

ROS production in desired amounts and continuity has been achieved with Profoks CAP.

Keywords: Textile wastewater recovery, CAP, profoks

12.1 Introduction

Physical, chemical and biologically contaminated wastewater created by the waters used for the needs affects negatively the ecological balance in the environmental ecosystems in which they are discharged. For this reason, protecting human health and the environment has become the most important environmental policy by protecting water resources, treating and discharging wastewater and stabilizing sludge all over the world. In line with this policy, countries have enacted and implemented laws and regulations to ensure and maintain the chemical, physical and biological quality integrity of their national waters. However, current legal regulations, wastewater treatment technologies and conventional treatment practices are unsuccessful and inadequate in water recovery in terms of economical, technological and efficiency aspects. Similar problems are observed in the treatment of domestic/urban and industrial wastewater, disposal of the sewage sludge formed and the protection of receiving environments. For this reason, the efforts to develop water recovery and treatment technologies and to introduce new approaches are increasingly continued. Scientific research on the supply, recovery and treatment of water, which is one of the most basic needs of humanity, has focused on the introduction of new technologies and approaches. No technology or a single system can solve all environmental problems, but every successful new technique is a step towards the goal of environmentally friendly and safe industrial activities. The Profoks-Cold Atmospheric Plasma (CAP) system is currently one of these steps and offers economically and technologically sustainable solutions for water recovery, wastewater treatment, treatment sludge stabilization and control of air pollutant emissions.

In this study, a new technology, which has been developed by different specialties for a long time, and which takes its power and effectiveness from the activation processes of oxygen in nature, has been introduced. Reactive Oxygen Species (ROS) activation potential in various fields has been known all over the world for a long time and there are many scientific studies, but it could not reached an economically and technologically widely applicable level in water recovery, wastewater treatment and control of air pollutants. In this study, a Cold Atmospheric Plasma (CAP) system, which is called as "Profoks", which can perform treatment in organic matter mineralization efficiencies up to 99% by producing "reactive oxygen types" with UV frequency adjusted in the presence of a capacitor *in situ*, was developed. The developed Profoks/CAP System includes specific system designs, including CAP-ROS generator and filtration systems (UF, NF), depending on the contaminant characteristics and industrial process characteristics in water recovery and wastewater treatment. On the other hand, in flue gas environments, it includes specific system designs including aqueous phase scrubber.

12.2 Advanced Oxidation Processes (AOP) in Wastewater Treatment

Basically, wastewater treatment is usually carried out in two stages, a pretreatment stage that includes mechanical and physicochemical treatments to reduce the different pollutant components of wastewater, and then an advanced treatment process. The pretreatment process increases the effectiveness of advanced treatment by enabling agglomeration of pollutants to larger sizes for flowing filtration or removal [1, 2]. However, in advanced treatment processes, the degradation of calcitrant components by membrane and bioremediation is complicated. Therefore, in solving this problem, the advanced oxidation process (AOP) has become great importance due to its potential to degrade a wide variety of organic contaminants [2–4].

This process allows reactive hydroxyl radicals to be producted with photon energy and without additional chemicals. The main purpose of AOPs is the production of strong non-selective hydroxyl radicals that break down pollutants, and also UV light, several molecular and ionic species are produced in these processes [5]. In general, UV/O_3 and UV/H_2O_2 processes require large amounts of oxidant consumption, which is not economical in terms of operating costs [6]. However, CAPs are highly advantageous in terms of operating at atmospheric temperature and pressure, using a low cost and chemically stable catalyst, and fully mineralizing contaminants and by-products. Other advantages include not large amounts of sludge, rapid reaction speed, low cost and ability to operate under atmospheric temperature and pressure conditions [1, 7, 8]. Non-thermal plasma (NTP) advanced oxidation processes have been shown to produce several types of reactive oxygen species (ROS) such as O_3, H_2O_2, O_2, O, OH, etc which oxidize water pollutants directly or indirectly. Therefore, NTP/AOPs have proven to be sufficiently effective in water and wastewater treatment in recent years [9].

12.2.1 Cold Atmospheric Plasma Technology (CAP)

Plasma is considered to be the fourth state of matter and is defined as a wholly or partially ionized gas. Irvine Langmuir (1928) was the first to call ionizing gas "plasma". In plasma electrons, positive and negative ions can be defined as neutral atoms and neutral or charged molecules. It is also characterized by different types of radiation, temperatures and electric fields. Plasmas can be seen in everyday life as well as thunderstorms, northern lights, neon lights or plasma screens.

Plasmas can be "thermal/hot" and "non-thermal/cold". While ionization occurs almost entirely in thermal plasma, there is only partial ionization in non-thermal plasma. By adding atmospheric or lower pressure energy to a gas such as air, oxygen, argon or helium, artificially plasma can be produced. Plasma applications are found in different fields of technology and industry, for example, they are used in automotive or metallurgy [10, 11]. In cold atmospheric plasma (CAP), temperature is not thermally stable and differs greatly between electrons and other particles such as atoms, ions and molecules. For this reason, CAP is also called "non-thermal plasma (NTP)". Due to the small masses of electrons whose temperature is typically between 10,000 and 250,000 K (1–20 eV), electrons can be accelerated very easily within an electrical field and by its enfluence. High energy electrons also produce free radicals from parent molecules as a result of multi-stage physical and chemical processes. These free radicals react with target contaminants and form decomposed products. In this processes, direct interactions between electrons and pollutants are usually negligible, as the typical concentration range of pollutants of relevant is several hundred ppmv [12]. Since CAPs use oxygen and water vapor to generate reactive radicals that carry out subsequent chemical reactions, it is not necessary to add any other chemicals as oxidants. Chemical reactions driven by free radicals usually end quickly in less than 10^{-3} s, and these rapid reactions make the system compact. Extensive studies have been carried out in the fields of air pollution control system, sterilization, water treatment and water recovery for environmental applications of CAP [12, 13]. At least 3 different CAPs have been developed for different applications [14–16]:

a) Plasma Jet
b) Corona discharge plasma
c) Dielectric barrier discharge (DBD) plasma

The dielectric barrier discharge (DBD) plasma has at least one substance that will be the dielectric barrier between the electrodes. Glass, quartz, mica and alumina are commonly used materials for dielectric barrier. The physicochemical properties of DBD have been summarized in detail in many studies [17, 18]. There are also a number of studies on DBD plasma applications [19, 20]. Discharge properties in DBD largely depend on gas composition, type of dielectric material, voltage and frequency operating conditions [21]. In atmospheric pressure-like mixtures, the dominant discharge mode is short-lived filaments called micro-charge.

Non-thermal plasma configurations based on DBDs produce a large amount of free radicals. These oxidizing agents oxidize organic pollutants directly or indirectly into dissolved CO_2, H_2O and simpler compounds. Although O_3 has proven to be selective against organic pollutants [22], it is also known that H_2O_2, one of the main precursors of non-selective OH radicals, is also a well-known disinfectant. Non-selective OH radicals have also been shown to oxidize target pollutants, oxidizing them to CO_2, H_2O and inorganic salts. Many studies have shown that the DBD reactor configuration can be effectively used to break down organic pollutants and microorganisms. Researchers have demonstrated that the DBD system produces several types of reactive oxygen species (ROS) that form a chemical mixture such as O_3, H_2O_2, OH, O•, O_2• that can break down pollutants [5, 11, 23, 24].

12.2.2 Formation and Chemical Reactivity of Reactive Oxygen Species (ROS)

Reactive Oxygen Species (ROS) are oxidant species that can be derived using single or two electron oxidation. Radical ROS types are derived by single electron oxidation, and non-radical ROS types by two electron oxidation. The basic situation, triple molecular oxygen, is paramagnetic, with two unpaired electrons occupying separate π^* orbitals with parallel spin. Most of the non-radical organic molecules are diamagnetic with opposing spin electron pairs. Spin restriction accomplished by the oxygen accepting one electron at a time, preferably through redox reactions. Thus, the molecular oxygen accepting the electron can react quickly with other radicals with a single electron transfer. It can also react with other species carrying unpaired electrons, such as transition metals such as Fe, found in the [Fe–S] clusters. The delivery of oxygen by an electron causes the formation of a superoxide radical anion ($O_2^{\cdot -}$). Electron thinning of O_2• leads to other ROS formation, such as hydrogen peroxide (H_2O_2), a closed shell molecule (Figure 12.1). Reduction of hydrogen peroxide, respectively, forms the hydroxyl radical (OH), which is reduced to give water (or hydroxide OH^-).

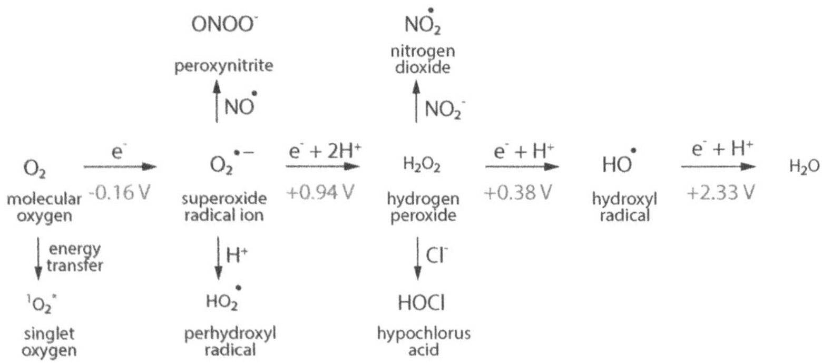

Figure 12.1 ROS formation during energy and electron reactions [25].

However, the basic state oxygen (O_2) can be converted into more reactive oxygen-containing forms. Thus, energy transfer to O_2 leads to the formation of more reactive molecular oxygen form, such as singlet oxygen ($^1O_2$). Singlet oxygen has paired electrons in opposite spins. For this reason, the tightening restriction is eliminated, which increases the oxidation ability of $^1O_2$ compared to the basic state O_2 [25].

The chemical effects that occur in electrical discharge are basically the result of energy injection into a gas stream. This energy injection is realized by electron impact processes under the influence of an electric field. The collision of neutral electrons with energized electrons provides ionization, disintegration of molecules and electronic, rotational and vibrational stimulation of neutral gas. The basic processes in the CAP, which occur due to based on time-dependent streamer spread, can be divided into primary and secondary processes. Figure 12.2 summarizes the typical timeline of key operations in CAP.

In the primary process, ionization, excitation, separation, light scattering and charge transfer take place. Methods of energizing and their parameters such as pulse, DC Pulse, AC, AC pulse or DC, voltage rise time and frequency largely determine the efficiency of the primary process. The efficiency of the secondary process is determined by the chemical reactions of electrons, radicals, ions and excited molecules, which are the primary process products. In addition, some additional radical types and reactive molecules such as O_3, HO_2 and H_2O_2 are formed by radical-neutral recombination in secondary processes. The typical time interval for secondary transactions is very fast. Therefore, gas residence time in the CAP reactor has little or no effect on overall performance. The CAP process is very effective since selective energy transfer to electrons occurs in the primary process. However, the overall responses in the CAP process are not just the first process but a few basic processes. Therefore, the overall

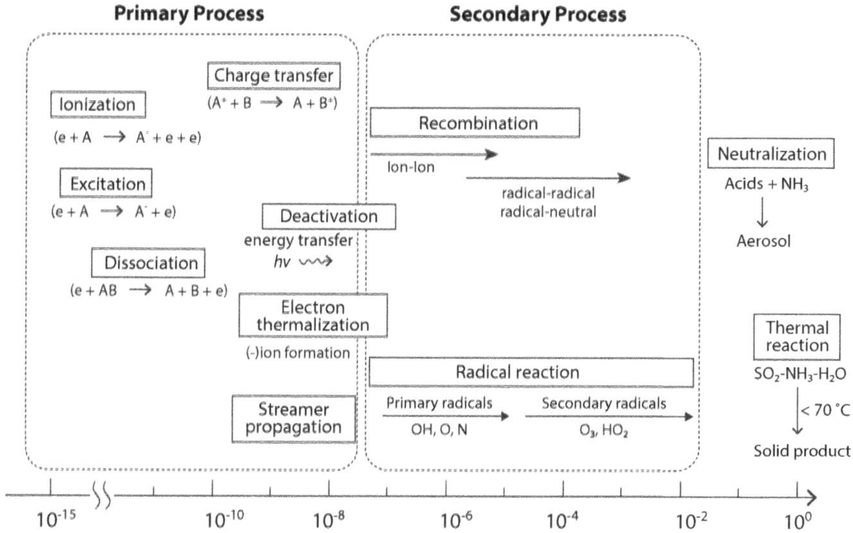

Figure 12.2 Time scale of basic processes taking place in a cold atmospheric plasma [12].

effectiveness of the CAP process will be a product of chemical reactions in the secondary process, in addition to the efficiency of the primary process. The effectiveness of the reactive species involved in the decomposition of pollutants depends largely on the nature of the pollutants. Therefore, determining the reactive species and their reaction paths with different contaminants is important to optimize CAP processes for real applications. For example, halogen compounds are easily decomposed by the addition of dissociated electrons due to their high electron affinity, while chlorinated hydrocarbons are very resistant to OH radical attack [26]. Similarly, olefin compounds and unsaturated hydrocarbons are highly reactive to radical species and ozone and easily decompose in the CAP process [27].

In many studies, numerical calculations have been made for the formation of time-dependent radical species and it has been tried to understand and characterize chemical reactions in CAP reactors. Most simulation studies take into account that electrons only collide with ground level gas molecules. However, unlike simulation studies, the chemical effects caused by electron molecule collisions are largely dependent on the electron energy and the energy state of the molecules [28].

12.2.3 CAP/AOP Application in Textile Wastewater Treatment

The main pollutants specifically found in textile wastewater are suspended solids, high chemical oxygen demand, intense color dyes and other soluble substances. Removal of color from the textile industry and wastewater from the dyestuff manufacturing industry poses a major environmental challenge. The strongest effects of the textile industry on the environment are caused by high amounts of primary water consumption (80–100 m^3/ton of finished textile) and thus high amounts of wastewater discharge (115–175 kg COD/ton of finished textile), a wide range of organic chemicals, low biodegradability, color and salinity [29]. Therefore, reuse of wastewater is an economic and ecological imperative for the entire sector. Textile companies use a variety of chemicals and these chemicals are selected depending on the nature of the raw material and product [30]. Wastewater from these processes differs mainly due to differences in process, fabric and machinery. Typical textile industry wastewater can be characterized with a COD (between 150 and 12,000 mg/L), total suspended solids (between 2,900 and 3,100 mg/L), total Kjeldahl nitrogen between (70 and 80 mg/L), and BOD (range from 80 to 6,000 mg). It has a large amount of biodegradable organic matter content with a BOD/COD ratio of around 0.25.

Textile industry is a sector where water is used extensively. Water is used to clean the raw material and in many washing stages throughout the production. Production in the textile industry is becoming an environmental problem as textile wastewater contains a wide variety of dyes and chemicals. Textile industry dyes are deliberately designed to remain photolytic, chemically and biochemically stable and are therefore generally not conducive to biodegradation [31]. Most pollution in textile wastewater results from dyeing and finishing processes. Textile finishing includes bleaching, dyeing, printing and curing during the processing of textile products. The effect of the textile industry on the environment in terms of both the discharge of pollutants and water and energy consumption has been known for some time. Another important problem of the textile industry wastewater is color. If colored water is not treated properly, these dyes can remain in the environment for a long time [32]. The color wastewater problem has been an integral part of the textile wastewater treatment process as a result of environmental regulations. Dyes originating from textile wastewater can be easily released in receiving environments even at low concentrations. This is not just visual pollution, but dyes in wastewater can have a serious inhibitory effect on water ecosystems [33]. Studies in this area are generally new treatment strategies for synthetic wastewaters, mainly containing azo-dyes, the largest dye class used in the textile industry, and the surfactants responsible for anomalies, the growth of algae (eutrophication) and toxicity in some aquatic organisms focused on its development [34]. Therefore, it has been widely shown that AOPs have the greatest promises for the treatment of textile wastewater [35].

12.3 Profoks/CAP Wastewater Treatment and Water Recovery System

In this study, a Dielectric Barrier Discharge (DBD) based Cold Atmospheric Plasma (CAP) reactor called "Profoks" was designed, developed and commercialized. The developed Profoks/CAP is a remote discharge reactor. With the developed Profoks reactor, treatment and water recovery studies of textile wastewater were carried out.

The methodology of remote discharge reactors for wastewater treatment is not new. This Profoks system is a ROS reactor where O_2 stimulated by the gas phase plasma discharge is produced and then transported to the wastewater to be treated. Electrical discharge reactors have been developed recently, in which bubbles of plasma gas produced in the distance are introduced into wastewater. The main difference of the Profoks system from other reactors is that only pure O_2 is used as the plasma feed gas, and with a new design developed, the desired amount of reactive oxygen species (ROS) is produced (Figure 12.3). With purified O_2 taken from the atmosphere, ROS is produced in addition to UV for wastewater treatment in dielectric barrier discharged plasma. The Profoks reactor has an electrode configuration, oxygen is used as the feed gas and is energized by a positive DC voltage source. Treatment and water recovery in a selected textile wastewater was accomplished by the production of reactive oxygen species that act as the main oxidant during bubble discharge. With the Profoks/CAP, the stimulated reactive oxygen species produced inside the tube with the electric field power are pumped into the wastewater from the quartz tube and then delivered to the solution through a gas diffuser (Figure 12.3). Working gas determines the properties of many plasmas such as voltage, electron density and temperature, plasma homogeneity and density, and reactive species produced for a given input voltage waveform. Due to the abundant availability of air, it is a common gas for plasma reactors used in wastewater treatment. However,

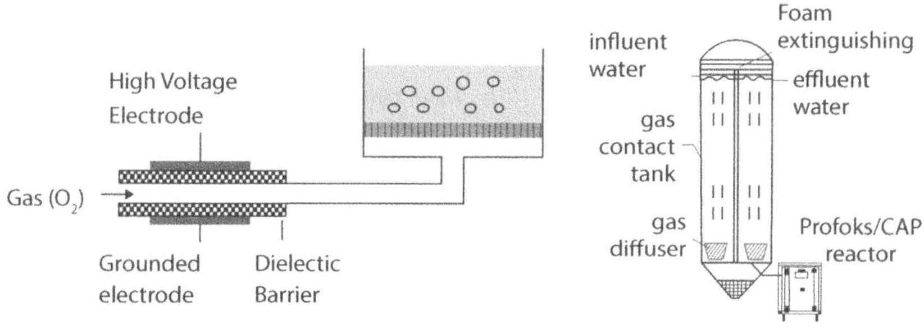

Figure 12.3 Profoks/CAP remote discharge reactor.

it has been reported that pure oxygen gas generally provides more efficient organic decomposition than nitrogen gas [9, 36]. This is partly explained by the use of oxygen, with more effective OH radicals and O_3 formation. Sometimes noble gases such as helium and argon are used in discharge reactors. Generally, argon leads to faster degradation of phenols, but for other compounds it performs lower than oxygen [9, 11, 37, 38].

The Profoks/CAP reactor has been modified with a membrane system that follows it. As is known, membrane systems for water recovery after mineralization of organic pollutants in wastewater give successful results. The developed Profoks CAP/DBD plasma system was designed using ultrafiltration (UF), nanofiltration (NF) and high pressure reverse osmosis (RO) after plasma reactor as shown in Figure 12.4.

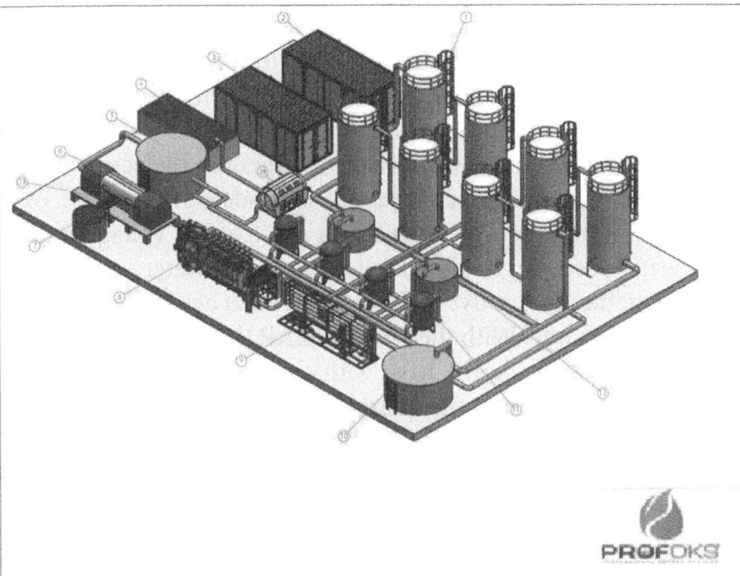

1	Profoks/CAP contact tank	8	Ultrafiltration
2	O_2 generator	9	Reverse Osmosis
3	Profoks/CAP reactor	10	Water tank
4	Wastewater	11	Sand filter
5	Sludge dewatering tank	12	Balancing tank
6	Decantor	13	Conveyor
7	Sludge tank	14	Tanbur sieve

Figure 12.4 Pilot scale Profoks/CAP treatment system.

Membranes are developed for size separation in wastewater treatment plants and are often integrated with chemical and biological treatment in secondary treatment or used as stand-alone systems [8, 39, 40]. Figure 12.5 shows a typical membrane mechanism. Typical membranes have a driving force that acts as a barrier that controls the movement of components through semi-permeable pores of different sizes. Permeability and selective retention are a function of membrane pore size and chemical affinity; this helps ensure a product stream free of target components [41]. Membrane technology has relatively low energy demand and high wastewater treatment efficiency. Therefore, membrane technologies have been greatly improved for industrial applications. Organic and inorganic components removal, disinfection and desalination are some of the applications of membrane technologies [42, 43]. Most of the membranes available on the market and industrially used are pressure and energy powered membranes [44]. Pressure-operated types are microfiltration (MF), ultrafiltration (UF), reverse osmosis (RO) and nanofiltration (NF).

Ultrafiltration (UF) is a separation mechanism that uses the principle of physical sieving. The pore size of a UF membrane ranges from 0.05 to 1 nm and its pressure in the range of 1–500 kDa and 1–7 bar [42]. Therefore, UF with designed molecular weight breaker (MWCO) is impermeable to compounds of molecular weight that exceed MWCO. In such a MWCO application, colloids, viruses and coliform bacteria are eliminated. The use of MF and UF as pretreatments to RO has slowly become an industry standard. Their use in this form allows to reduce membrane clogging, as well as organic chemical extraction and pH adjustment, chemical precipitation for phosphorus, hardness and metals and is applied as a final treatment step for wastewater treatment [39, 42, 43, 45]. Nano filtration is a process based on molecular size pressure to remove dissolved micro-pollutants

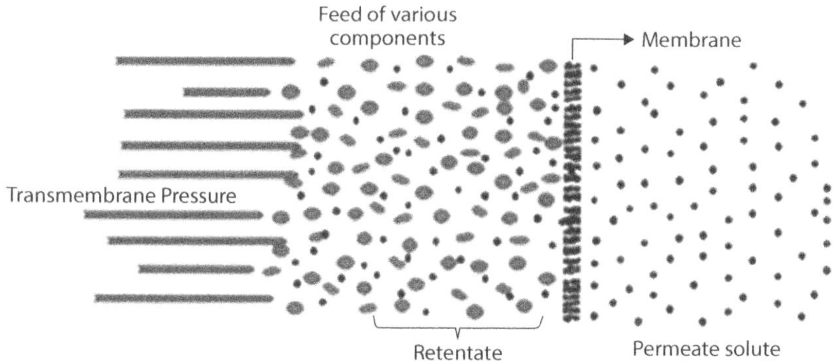

Figure 12.5 Membrane [45].

and multivalent ions of the separation mechanism. NF is a complex process that can be characterized by solvent diffusion, transport and electrostatic repulsion effects on the membrane surface and nanopore [39]. Recent applications are used as a pretreatment that facilitates the operation of NF and reduces operating and maintenance costs in RO [41, 44, 45]. Reverse osmosis is widely used in wastewater treatment, in removing salinity against the traditional thermal multi-stage methods.

12.3.1 Profoks/CAP Wastewater Treatment and Water Recovery System and Textile Wastewater Recovery Studies

The 5-liter wastewater sample taken for the purpose of investigating the treatability and water recovery of wastewater formed in the textile dyeing process with the Profoks/CAP treatment system. Study has been carried out with a pilot scale version of the Profoks treatment system. Relevant analyzes of raw process waters and process waters that have been tretaed with Profoks were evaluated. In the laboratory work, a Profoks reactor, an ultrafiltration unit and a nanofiltration unit were used, producing 100 g of ROS per hour. In the study, first of all, 5 L of wastewater samples were poured into a stainless steel reactor and ROS from the Profoks reactor were given to this leakage water for 90 min through two diffusers. After 90 min, the wastewater was first passed through the ultrafiltration unit and then through the nanofiltration unit. Schematic representation of this pilot scale system is given in Figure 12.4.

12.3.2 Profoks/CAP Wastewater Treatment and Water Recovery System and the Results of Treatability of Textile Wastewater and the Study of Water Recovery

In Table 12.1, the inlet water characterization of the raw wastewater taken at the process and the pollutant concentration values measured at the exit of the treatment system and the treatment efficiencies realized are presented. In Figure 12.6, the photos of raw and purified process water are given. All values measured according to the results given in Table 12.1 provide discharge standards. When the treatment performance is evaluated, the following results are achieved.

- COD is one of the most important pollution parameters in industrial wastewater. COD parameter could be reduced by 94.76% at the end of treatment and decreased below

Table 12.1 Wastewater characterization of textile wastewater before and after treatment with Profoks/CAP wastewater treatment and water recovery system.

Parameter	Unit	Before treatment	After treatment with Profoks	Efficiency, %
pH	–	6.05	6.53	–
Suspended Solid	mg/L	39.2	<3	92.35
Total Nitrogen	mg/L	13.88	<0.5	96.40
Color	Pt-Co	1,974.6	<1.5	99.92
COD	mg/L	954.3	<50	94.76
Turbidity	NTU	198.5	<1	99.50
Ammonium Nitrogen	mg/L	0.575	<0.1	82.61
Total Phosphorus	mg/L	5.55	<0,3	94.59
Conductivity	Us/cm	703	51.6	92.66
Oil and Grease	mg/L	<5	<5	–
TOC	mg/L	301.2	<5	

Figure 12.6 Appearances of raw and treated process water.

50 mg/L. These yields are at the level of yields obtained in advanced wastewater treatment plants. These results mean that the treatment method provides advanced organic mineralization. The COD leaving water concentration in the example of wastewater taken from the facility provides the national discharge standards more than enough.

- As a result of the Profoks/CAP treatment process, 92.35% efficiency was achieved in the Suspended Solid (SS) parameter. Effluent water SS concentration meets the relevant regulation limit value.
- Color is one of the most important problems in textile wastewater. Textile dyes that are soluble in nature and can be found in colloid waters can generally be classified as azo-dyes or pigment dyes. Due to its low biodegradability, physicochemical techniques or advanced oxidation processes are often used in the treatment of these wastewater. After the oxidation process with the Profoks system, a 99.92% treatment was realized in the color parameter. In the wastewater sample, the effluent color amount was reduced to 1.5 units in Pt-Co. This result shows that with the proposed Profoks/CAP method, a high level of leaving water color quality has been reached.
- Nitrogen is one of the difficult to reduce pollutants in textile wastewater. In waters, nitrogen can be of organic and inorganic origin. Especially in nitrogenous water of inorganic origin, it is present in the form of NH_3 or NH_4 (high rate of NH_4 at low pH) depending on the pH [46]. After the treatment, the dissolved NH_4 present in the medium was oxidized and turned into a stable end product NO_3. The resulting NO_3 concentration is even below the drinking water standards of the Turish Standart, World Health Organization (WHO) and the American Environmental Protection Agency (EPA). Other nitrogen formations (organic and inorganic nitrogen and/or nitrogen dyes) were purified by Total Nitrogen (TN) by 96.4% and decreased below 0.5 mg/L. This value is even below the value given for the Kjeldahl nitrogen in the regulation. As is known, all organic and inorganic nitrogens are included in the TN value and are always higher than the Kjeldahl nitrogen value.
- High electrical conductivity values are found in textile wastewater depending on the processes. High conductivity is due to its organic and inorganic chlorinated compounds

(organo chlorides such as chlorine amines, chlorine, chlorates, hypochlorites, vinyl chloride), which exist in different forms, especially in wastewater. After treatment, the effluent conductivity value has been reduced to about 51 µs/cm. The conductivity level observed in the effluent is within the limits that can be observed in natural waters (especially groundwater from karstic aquifers) [47].

According to the results obtained from the laboratory study carried out to investigate the wastewater produced as a result of the textile industry activities with the Profoks/CAP wastewater treatment and water recovery system, based on the measured parameters of the wastewater passed through the Profoks reactor and the ultrafiltration and nanofiltration units operating accordingly. It has been found that it provides a advanced treatment rate. The textile wastewater treated in this system has the qualities that can be reused in the process. With the developed Profoks/CAP wastewater treatment and water recovery system, textile wastewater was not only treated according to the discharge standards, but also "water recovery" was realized. It was also revealed in another study that "recycled water" brought to qualities that can be reused in all processes at the mill did not have a negative effect on product quality.

12.3.3 Profoks/CAP Wastewater Treatment and Water Recovery System Investment and Operating Costs

For textile wastewater, investment and operating costs were calculated by designing a real-size system on the basis of water recovery works with the Profoks/CAP wastewater treatment and water recovery system. In the calculations, the works and services to be performed for the manufacturing, equipment supply, installation, commissioning works of the Profoks/CAP Wastewater Treatment and Water Recycling Process for the plant wastewater were taken as basis. The design of the Profoks/CAP system was made according to the worst scenario that can occur in the reactor tank with a capacity of 1,500 m^3/day. With these assumptions, the parameters based on the design and the target quality parameters regarding the water to be obtained at the exit of the facility as a result of treatment and water recovery with the Profoks/CAP system are given in Table 12.2.

Profoks/CAP wastewater treatment and water recovery system investment costs and operating costs of the system designed according to the pollutant concentrations determined in Table 12.2 are given in Table 12.3. When the effluent flow is taken as maximum 1,500 m^3/day, the electricity

Table 12.2 Profox/CAP wastewater treatment and water recovery plant design inlet water and pollutant parameters in the targeted outlet water.

Parameter	Unit	Design (inlet water)	Target (effluent water)
Suspended Solid (SS)	mg/lt	100	<1
Chemical Oxygen Demand (COD)	mg/lt	1,000	<10
Color	Pt-Co	2,000	<1
Conductivity	uS	1,000	<50

Table 12.3 Profoks/CAP wastewater treatment and water recovery facility investment and operating costs.

Investment cost	
Explanation	**Total price**
Profoks/CAP wastewater treatment and water recovery plant (1,500 m^3/day)	**1,100,000 €**
Investment cost for 1 m^3 of wastewater treatment and water recovery	**733 €**
Plant Components	
• O_2 Gas Contact Tank • Backwash Tanks • Precipitation Tank • Foam Extinguisher • Drum Screen Particle Filtration • Sand Filter • Ultra Filtration 0.02 μm • Nano Filtration 0.001 μm • High Pressure RO	• Liquid Transfer Pump System • Electric, Electronic Board Automation • Container System 40 foot • Air Conditioning and Air Conditioning System • Profoks Pro System 1,000 g (ROS)/h • O_2 Generator • Gas Transfer and Diffuser System • Dewatering System

(Continued)

Table 12.3 Profoks-CAP wastewater treatment and water recovery facility investment and operating costs. (*Continued*)

Investment cost	
Explanation	**Cost**
For 1 m^3 of wastewater treatment and water recovery	
• Electricity cost	0.19 €
• Consumable cost	0.07 €
• Sludge Disposal Cost (Approximately 3 tons of sludge will be formed daily)	0.01 €
• Personal	0.02 €
Total	**0.29 €**

requirement of the plant will be 210 KWh. The area needed for the installation of the system is approximately 40 m × 25 m = 1,000 m^2. Application areas should be at least 8 m high. The temperature of this area should be between +4 and +25 °C. Containers with Profoks/CAP systems should be air conditioned with air conditioning systems. The temperature of the wastewater entering the system should be between the lowest +4 °C and the highest +80 °C

12.4 Conclusion

With the development of a Profoks/CAP reactor prototype, the results of this study, which investigated the recovery of the textile wastewater and its reuse in processes by making water recycling under laboratory conditions, achieved the desired treatment efficiency. Most importantly, ROS production in desired amounts and continuity has been achieved with the developed Profoks/CAP. With the ROS, which can be produced in sufficient quantities and *in situ*, the desired treatment efficiency has been achieved in wastewater, water recycling has been realized and usability targets have been achieved in the processes. Chemical oxygen demand, color and suspended solids, which are the main pollutants in textile wastewater, have been met with very high yields. By reducing the conductivity, salinity removal from the wastewater is provided at the desired level. In addition, textile wastewater has been recycled and its usability in processes has been successfully achieved. The history and future potential of plasma technology, on which intensive research and studies have been carried out in recent years, are remarkable. In this process, it is predicted that plasma can be a solution to environmental problems and traditional treatment approaches cannot compete with plasma technique. Commercially viable plasma solutions already exist to address some important environmental issues. For example, water sterilization and removal of organic pollutants using ROS and/or UV radiation; There are waste disposal, electrostatic filters, particulate removal, NOx and SOx reduction applications in flue gas. In addition, plasma technology has significant potential in solving environmental problems such as air sterilization and disinfection (removal of low concentration volatile organic compounds from high volume flue gases), control of air pollutants such as dioxins and mercury, water disinfection and water softening, surface sterilization, reducing harmful emissions for internal combustion engines, H_2S removal and carbon capture.

References

1. Thiruvenkatachari, R., Vigneswaran, S., Moon, I.S., A review on UV/TiO2 photocatalytic oxidation process (Journal Review). *Korean J. Chem. Eng.*, 25, 1, 64-72, 2008.
2. Sievers, M., Advanced oxidation processes. *Treatise Water Sci.*, Wilderer, P., Vol. 4, 377–408, Amsterdam, 2010.
3. Bethi, B., Sonawane, S.H., Bhanvase, B.A., Gumfekar, S.P., Nanomaterials-based advanced oxidation processes for wastewater treatment: A review. *Chem. Eng. Processing: Process Intensification*, 109, 178–189, 2016.
4. Cuerda-Correa, E.M., Alexandre-Franco, M.F., Fernández-González, C., Advanced Oxidation Processes for the Removal of Antibiotics from Water. An Overview. *Water*, 12, 102, 2020.
5. Liu, N. and Sun, G., Production of reactive oxygen species by photoactive anthraquinone compounds and their applications in wastewater treatment. *Ind. Eng. Chem. Res.*, 50, 9, 5326–5333, 2011.
6. Dezotti, M., Lippel, G., Bassin, J.P., *Advanced Biological Processes for Wastewater Treatment*, Springer, International Publishing AG, Dio de Janeiro, 2018.
7. Ameta, S.C. and Ameta, R. (Eds.), *Advanced oxidation processes for wastewater treatment: Emerging green chemical technology*, Academic Press, Elsevier Inc, London, 2018.
8. Tetteh, E.K., Rathilal, S., Chetty, M., Armah, E.K., Asante-Sackey, D., Treatment of Water and Wastewater for Reuse and Energy Generation-Emerging Technologies. *IntechOpen*, 84474, 1-21, 2019. http://dx.doi.org/10.5772/intechopen.

9. Vanraes, P., Nikiforov, A.Y., Leys, C., Electrical discharge in water treatment technology for micropollutant decomposition. *Plasma Sci. Technol. Progress Phys. States Chem. Reactions*, 429–78, 2016.
10. Von Woedtke, T., Reuter, S., Masur, K., Weltmann, K.D., Plasmas for medicine. *Phys. Rep.*, 530, 4, 291–320, 2013.
11. Mouele, E.S.M., Tijani, J.O., Fatoba, O.O., Petrik, L.F., Degradation of organic pollutants and microorganisms from wastewater using different dielectric barrier discharge configurations—A critical review. *Environ. Sci. Pollut. Res.*, 22, 23, 18345–18362, 2015.
12. Kim, H.H., Nonthermal plasma processing for air-pollution control: A historical review, current issues, and future prospects. *Plasma Process. Polymers*, 1, 2, 91–110, 2004.
13. Lu, X., Naidis, G.V., Laroussi, M., Reuter, S., Graves, D.B., Ostrikov, K., Reactive species in non-equilibrium atmospheric-pressure plasmas: Generation, transport, and biological effects. *Phys. Rep.*, 630, 1–84, 2016.
14. Weltmann, K.D., Brandenburg, R., von Woedtke, T., Ehlbeck, J., Foest, R., Stieber, M., Kindel, E., Antimicrobial treatment of heat sensitive products by miniaturized atmospheric pressure plasma jets (APPJs). *J. Phys. D Appl. Phys.*, 41, 19, 194008, 2008.
15. Grzegorzewski, F., Ehlbeck, J., Schlüter, O., Kroh, L.W., Rohn, S., Treating lamb's lettuce with a cold plasma—Influence of atmospheric pressure Ar plasma immanent species on the phenolic profile of *Valerianella locusta*. *LWT-Food Sci. Technol.*, 44, 10, 2285–2289, 2011.
16. Wu, D., You, H., Du, J., Chen, C., Jin, D., Effects of UV/Ag-TiO2/O3 advanced oxidation on unicellular green alga *Dunaliella salina*: İmplications for removal of invasive species from ballast water. *J. Environ. Sci.*, 23, 3, 513–519, 2011.
17. Kogelschatz, U., Dielectric-barrier discharges: Their history, discharge physics, and industrial applications. *Plasma Chem. Plasma Process.*, 23, 1, 1–46, 2003.
18. Braun, D. and Pietsch, G., Microdischarges in air-fed ozonizers. *J. Phys. D Appl. Phys.*, 24, 4, 564, 1991.
19. Eliasson, B. and Kogelschatz, U., Modeling and applications of silent discharge plasmas. *IEEE Trans. Plasma Sci.*, 19, 2, 309–323, 1991.
20. Eliasson, B., Hirth, M., Kogelschatz, U., Ozone synthesis from oxygen in dielectric barrier discharges. *J. Phys. D Appl. Phys.*, 20, 11, 1421, 1987.
21. Falkenstein, Z. and Coogan, J.J., Microdischarge behaviour in the silent discharge of nitrogen–oxygen and water–air mixtures. *J. Phys. D Appl. Phys.*, 30, 5, 817, 1997.
22. Glaze, W.H., Reaction products of ozone: A review. *Environ. Health Perspect.*, 69, 151–157, 1986.
23. Othmer, K., *Encyclopedia of Chem. Tech*, vol. 18, p. 448, John Wiley & Sons Inc, New Jersey, 1996.

24. Zimmer, R.J., *Ergodic theory and semisimple groups*, vol. 81, Springer Science & Business Media, New York, 2013.
25. Nonell, S. and Flors, C. (Eds.), *Singlet Oxygen: Applications in Biosciences and Nanosciences Volume 1*, Royal Society of Chemistry, London, 2016.
26. Penetrante, B.M., Hsiao, M.C., Bardsley, J.N., Merritt, B.T., Vogtlin, G.E., Wallman, P.H., Bayless, J.R., Electron beam and pulsed corona processing of carbon tetrachloride in atmospheric pressure gas streams. *Phys. Lett. A*, 209, 1–2, 69–77, 1995.
27. Futamura, S., Zhang, A., Yamamoto, T., Mechanisms for formation of inorganic byproducts in plasma chemical processing of hazardous air pollutants. *IEEE Trans. Ind. Appl.*, 35, 4, 760–766, 1999.
28. Medodovic, S. and Locke, B.R., Primary chemical reactions in pulsed electrical discharge channels in water. *J. Phys. D Appl. Phys.*, 42, 4, 049801, 2009.
29. Savin, I.-I. and Butnaru, R., Wastewater characteristics in textile finishing mills. *Environ. Eng. Manage. J. (EEMJ)*, 7, 859–64, 2008.
30. Aslam, M.M., Baig, M.A., Hassan, I., Qazi, I.A., Malik, M., Saeed, H., Textile wastewater characterization and reduction of its COD & BOD by oxidation. *Electron. J. Environ. Agric. Food Chem.*, 3, 804–11, 2004.
31. Yonar, T., *Decolorisation of textile dyeing effluents using advanced oxidation processes*, Uludag University, Environmental Engineering Department, Gorukle, Bursa, Turkey, 2010.
32. Yonar, T., Yonar, G.K., Kestioglu, K., Azbar, N., Decolorisation of textile effluent using homogeneous photochemical oxidation processes. *Coloration Technol.*, 121, 258–264, 2005.
33. Nigam, P., Armour, G., Banat, I.M., Singh, D., Marchant, R., Physical removal of textile dyes and solid state fermentation of dye-adsorbed agricultural residues. *Bioresour. Technol.*, 72, 219–226, 2000.
34. Burgos, A.J., Rodriguez, P.U., Lopez, J.S., *Advanced Oxidation Processes (AOPs) Series*, Water and Environmental Engineering Group, FS-AVA-001, Technology Fact Sheets For Effluent Treatment Plants Of Textile Industry, Inditex, Universidade Da Coruna, 2015.
35. Oller, I., Malato, S., Sánchez-Pérez, J.A., Combination of advanced oxidation processes and biological treatments for wastewater decontamination—A review. *Sci. Total Environ.*, 409, 4141–4166, 2011.
36. Malik, M.A., Water purification by plasmas: Which reactors are most energy efficient? *Plasma Chem. Plasma Process.*, 30, 1, 21–31, 2010.
37. Xiong, Z., Cold atmospheric pressure plasmas (CAPs) for skin wound healing. *Plasma Med.: Concepts Clin. Appl.*, 121, 2018.
38. Li, Y., Chen, R., Zhou, H., Shi, Y., Qin, C., Gao, Y., Zhang, G., Gao, Y., Xiao, L., Jia, S., Observation of Singlet Oxygen with Single-Molecule Photosensitization by Time-Dependent Photon Statistics. *J. Phys. Chem. Lett.*, 9, 5207–5212, 2018.

39. Diallo, M.S., Water treatment by dendrimer-enhanced filtration: Principles and applications, in: *Nanotechnology applications for clean water*, pp. 143–155, William Andrew Publishing, Elsevier Inc., London, 2009.
40. Warsinger, D.M., Chakraborty, S., Tow, E.W., Plumlee, M.H., Bellona, C., Loutatidou, S., Padhye, L.P., A review of polymeric membranes and processes for potable water reuse. *Progress Polymer Sci*, 81, 209–237, 2018.
41. Van der Bruggen, B., Integrated membrane separation processes for recycling of valuable wastewater streams: Nanofiltration, membrane distillation, and membrane crystallizers revisited. *Ind. Eng. Chem. Res.*, 52, 31, 10335–10341, 2013.
42. Bozdek, M., Membrane technologies for the removal of micropollutants in water treatment, in: *Advances in Membrane Technologies for Water Treatment: Materials, Processes and Applications*, pp. 465–517, Elsevier Publishing, London, 2015.
43. Shahmansouri, A. and Bellona, C., Nanofiltration technology in water treatment and reuse: Applications and costs. *Water Sci. Technol.*, 71, 3, 309–319, 2015.
44. Tedesco, M., Hamelers, H.V.M., Biesheuvel, P.M., Nernst–Planck transport theory for (reverse) electrodialysis: II. Effect of water transport through ion-exchange membranes. *J. Membrane Sci.*, 531, 172–182, 2017.
45. Hamouda, S.B., Touati, K., Amor, M.B., Donnan dialysis as membrane process for nitrate removal from drinking water: Membrane structure effect. *Arabian J. Chem.*, 10, S287–S292, 2017.
46. Kim, Y.M., Park, D., Lee, D.S., Park, J.M., Inhibitory effects of toxic compounds on nitrification process for cokes wastewater treatment. *J. Hazard. Mater.*, 152, 3, 915–921, 2008.
47. WHO, *Guidelines for Drinking Water Quality. 4th ed*, World Health Organization (WHO) Press, Geneva, Switzerland, 2011.

13

Nanotechnology and its Application in Wastewater Treatment

Nitu Singh, Manzoor Ahmad Malik* **and Athar Adil Hashmi**

Bioinorganic Chemistry Lab, Department of Chemistry, Jamia Millia Islamia, New Delhi, India

Abstract

There is a continuing effort from scientists all over the world to provide clean and affordable water in order to meet human needs which otherwise is a big challenge. The increased human population has resulted in water shortage and the amplified release of pollutants that contaminate available water sources. Nanotechnology has furnished ground-breaking solutions for the purification of water. Several new techniques and procedures are adopted for fabricating nanomaterials, which ultimately are employed in the wastewater treatment. The unique characteristic properties of nanomaterials like high surface area and others can be proficiently used in removing toxic metal ions, disease causing microbes, inorganic and organic solutes from water. Nanotechnology has led to countless effective ways for wastewater treatment in a more precise way where the special and beneficial properties of novel nanostructured materials may be utilized to provide more functional and sustainable solutions to the prevalent water-related problems. This chapter reviews nanotechnology enabled wastewater-treatment procedures, presenting how they transform our water supply and wastewater treatment. Furthermore, the use of carbon- based nano-adsorbents, metal-based nano-adsorbents, polymeric nano-adsorbents, mixed oxides and zeolites in wastewater treatment are also reviewed.

***Keywords*:** Nanotechnology, wastewater, polymers, zeolites, nano-adsorbents

**Corresponding author*: manzoormalik.kash@gmail.com

Luqman Jameel Rather, Mohd Shabbir and Aminoddin Haji (eds.) Sustainable Practices in the Textile Industry, (307–332) © 2021 Scrivener Publishing LLC

13.1 Introduction

For the existence of life on earth, water is the most essential substance that covers almost two-thirds of the earth surface. Access to clean water has become a global concern today. Providing safe drinking water is the major challenge of this century. Supply of safe potable water has become a global concern today because of uncontrolled human population growth. As per report of United Nations more people have died because of contaminated water than any sort of violence [1]. There are many physical and chemical methods for the purification of water like distillation, filtration, chlorination, reverse osmosis etc. There are several reasons of the water contamination like various toxins, industrial wastes, pathogenic microbes; untreated man made wastes, agricultural wastes [2, 3]. Water contaminants like Arsenic and other heavy metals cadmium, mercury, lead, nickel, chromium and so on are highly toxic and carcinogenic [4–6] and produce adverse effects on our ecosystem [7]. High concentration of ions like fluorides, chromates, sulfates, phosphates show deleterious effects. Pharmaceuticals and personal care products originated from hospital and household water are immune to natural biodegradation [8–10]. Volatile contaminants such as aromatic hydrocarbons get easily evaporate in the atmosphere and show hazardous effects. The traditional approaches cannot efficiently remove dissolved organic compounds at low concentrations. WHO report stated that about 1.7 million people died from contaminated water and because of waterborne diseases, four billion cases of various health problems have been reported annually [11]. Also, more than 0.78 billion people worldwide do not currently have access to healthy water resources, contributing to significant health problems [12].

Water-related issues are also projected to escalate more in the world due to severe climate change and increasing population over the decades [13]. The lack of the availability of fresh water also stems from the exploitation of water sources in many areas of the world for residential, agricultural and irrigation purposes [14]. Rising global demand for food has increased the burden on freshwater resources and also the danger of climate change [15, 16]. Furthermore, polluting surface/ground water bodies have resulted in the decreased availability of freshwater [17–19]. Owing to numerous issues of saltwater infiltration, soil depletion, poor drainage, pollution of ground/surface waters by contaminants, fertilizers, algal blooms, detergents, heavy metals, and so on, aquifers around the world are depleting and becoming polluted [20–23]. Conventional water management systems are not capable of effectively handling the elimination of a wide variety of toxic substances and pathogenic microorganisms. Several technologies have been employed to manage wastewater before exposing it to the atmosphere. High capital investment, huge plant areas, high energy needs, operation and maintenance costs are needed by existing wastewater treatment technologies. Therefore, it is a challenge for research, growth and technology organizations to come up with cost-effective, small-area alternative wastewater treatment systems in order to resolve these problems. Nanotechnology provides the opportunity for alternative wastewater treatment technologies to be developed. The evolving field of nanotechnology presents a potential bid for low-cost water purification, high working performance in pollutant removal and reusable capacity [24]. Nanomaterials have specific properties such as nano scale, large surface area, porosity, high degree of reactivity etc., rendering them applicable for wastewater disposal and water purification [25–27]. Groundbreaking developments in the field of nanoscience and nanotechnology inspired the scientific community over the last few decades to pursue this extremely promising area of study where the special and beneficial assets of novel nanostructured materials may be utilized to provide more functional and sustainable solutions to the prevalent water-related problems.

13.2 Nanotechnology

Nanotechnology studies, examination and implementation of nanoscale materials in which, relative to the same larger equivalents, molecules and atoms conform to their particular characterization [28]. Usually, nanomaterials are classified as materials smaller than 100 nm in one dimension, at least. Materials have innovative size-dependent properties on this scale, distinct from their large counterparts, many of which have been investigated for water and wastewater treatment applications. Most of the applications use required nanomaterial properties such as high adsorption surface area, high photocatalytic reactivity, strong antimicrobial assets for disinfection, and to assess water quality strong sensing characteristics [29].

13.2.1 Adsorption

Adsorption is the tendency of solid objects to attract gas molecules or solutions on their surfaces that they are in direct contact with. The solids are termed as adsorbents and the adsorbed molecules are referred to as adsorbate [30]. Adsorption usually takes place by physical forces, although it can also be due to weak chemical bonds [31]. Adsorption is one of the key methods for isolating contaminants from water where the contaminants are adsorbed to the nanoparticles and then isolated or eliminated from the water after reaching equilibrium. In wastewater treatment adsorption process is the most preferred solution to other approaches owed to the high performance, low cost, ease of service, versatility and efficiency towards a wide variety of water contaminants [32].

Nanoadsorbents have significantly higher adsorption rate for organic compounds relative to conventional adsorbents owing to their high specific surface area [33]. They have tremendous potential for the elimination of inorganic and organic contaminants by novel, more effective and quicker decontamination methods. The extraordinary properties of nanoadsorbents, for example, small scale, high porosity, high reactivity, synergist potential and huge number of active contaminant contact sites, entrust them as suitable adsorbent materials for the treatment of wastewaters [34].

Following nanoadsorbents are presently used in the water treatment:

- carbon-based nanoadsorbents
- metal-based nanoadsorbents
- polymeric nanoadsorbents
- zeolites.

13.2.1.1 Carbon-Based Nanoadsorbents

Carbon nanomaterials are commonly used in water to adsorption different organic and inorganic contaminants. Allotropes of carbon can be categorized into four distinct dimensional classifications. In the zero-dimensional material group, carbon dots and fullerenes are included while carbon nanotubes (CNTs) fall in group of one-dimensional material and Graphene is a two-dimensional planar structure. Carbon nanotubes are cylindrical shaped structure and two kinds of carbon nanotubes are present depending on their preparation. Carbon nanotubes are classified as single-walled carbon nanotubes, consisting of a single rolled-up grapheme sheets whereas multi-walled carbon nanotubes are composed of multiple rolled-up graphene sheets. Activated carbon is the most common carbon material among these nanomaterials because of its high adsorption capability, high thermal strength, and low cost. Activated carbon has been used in the elimination of numerous different organic toxins and water pollutants [35–38]. Furthermore, toluene and benzene adsorbed on activated carbon revealed high adsorption limit with regards to toluene and benzene [39]. It was also observed that activated carbon was efficient for the removal of heavy metal ions, like as cadmium, nickel, copper, lead, mercury and chromium respectively [40–42]. In combination with activated carbon, carbon nanotubes are attracting huge interest as impressive adsorbents due to their high surface area, high adsorption efficiency and tunable surface chemistry [43–47]. As a result of this high surface area, hollow and porous structure, the interaction of contaminant molecules with carbon nanotubes is strong. Acid or alkali treatment or surface alterations or inclusion of functional groups, metals etc. may improve the adsorption potential and selectivity of CNTs [48]. In aqueous medium CNTs form loose bundles and aggregates because of their hydrophobic surface that decreases its active surface area. The interstitial spaces and grooves are the main sites for adsorption [49]. The occurrence of larger pores in bundles and more available sorption sites is the responsible for the adsorption of large organic contaminants [50]. Polar organic molecules may also be adsorbed by CNTs due to various interactions between contaminants—CNT like hydrophobic effect, hydrogen bonding, π–π interactions, electrostatic interactions and covalent bonding [51]. The CNT surface rich in π electrons makes π–π interactions with organic molecules such as polycyclic aromatic hydrocarbons and polar aromatic compounds [52, 53]. Organic compounds with acids, amines, alcoholic functional groups also form hydrogen bond with CNT surface that donates electrons [54].

The main adsorption sites for metal ions are the surface functional groups (like carboxylic acid, hydroxyl, and phenol) of CNTs, primarily by electrostatic attraction and chemical bonding [55]. Consequently, the adsorption potential of CNTs will be improved by surface oxidation. CNTs are reported to be stronger adsorbents for heavy metals ions, like Cu^{2+}, Pb^{2+}, Cd^{2+}, and Zn^{2+} than activated carbon [56, 57]. In spite of fact CNTs have substantial advantages, in the treatment of water by adsorption process, they have some limitations. Because of the high cost of development, the usage of CNTs for wastewater treatment on an industrial scale is not anticipated [58]. Toxicity of CNTs is well reported and several papers are available on their health problems [59]. Another drawback is that CNTs display a phenomenon of coagulation with certain algae and organic toxins, and lose their structure composition [60]. Recently a low-cost absorbent, graphite oxide coated sand granules was effective in the removal of Hg^{2+} and a bulky dye molecule (Rhodamine B) and its overall performance was corresponding to activated carbon [61].

A major aspect assessing the cost-effectiveness of adsorbents is regeneration. By reducing the pH of solution, metal ion adsorption on CNTs can be reversed in a simple manner. In addition, after regeneration CNTs adsorbing capability remains relatively stable. It has been reported that after 10 cycles of regeneration and reuse, Zn^{2+} adsorption potential by CNTs decreased below 25%, while that of activated carbon loses its function by more than 50% after one regeneration [62].

13.2.1.2 Metal-Based Nanoadsorbents

The use of nanoadsorbents based on metal and metal oxide for wastewater treatment has gained a lot of interest in recent years [63]. Metal oxide like iron oxide, magnesium oxide, titanium dioxide, alumina and zinc oxide are low cost, effective adsorbents used in heavy metal removal during water decontamination. These metal oxides show some favorable characteristics, such as high performance, better adsorption kinetics and easy recovery after adsorption [64]. The adsorption potential increases several times as the particle size reduces. For instance, magnetite particles of 300 and 20 nm have similar degrees of adsorption for arsenic. The specific surface area of the nanoparticles increases with decrease in the particle size to less than 20 nm and the adsorbing capability increases three times more for arsenic [65]. This is due to the magnetite "nanoscale effect" in which new adsorption sites in metal ions are generated by the change in surface structure [66].

Some iron oxide nanoparticles, e.g., maghemite, hematite and magnetite, can be superparamagnetic in addition to high adsorption ability. By the introduction of an external magnetic field, they can be conveniently isolated from the reaction media. The literature contains enough reports on the applications of these magnetic nanoadsorbents for the reduction of a number of elements, such as arsenic, cobalt, copper, chromium, lead, and nickel [67–71]. It is found that maghemite nanotubes and nanorods are highly efficient in binding iron, nickel, copper, zinc, cadmium and lead. For the unique binding of mercury ions from water, graphene oxide coated hematite nanoparticles have been used effectively [72, 73].

Nanostructured ZnO, TiO_2 and MgO are also well researched and are described to be highly effective in extracting different toxic elements from water [74–76]. In addition to copper, lead and cadmium, another metal oxide TiO_2 is found to be highly successful for the removal of arsenic. Magnesium oxide nano flowers with high surface area are highly efficient adsorbents for arsenic, lead and cadmium [75]. Zinc oxide and alumina are some low-cost nanoadsorbents commonly used for the adsorption and isolation of cadmium, mercury and chromium ions from water [76, 77].

By modifying the solution pH, metal oxide nanoadsorbents may be regenerated [78]. The adsorption potential of metal oxide nanoadsorbents is not altered after many cycles of regeneration and reuse [79]. However, in some situations due to the creation of aggregates, the adsorption potential is decreased after regeneration [49]. High adsorption potential, low cost, easy recovery and regeneration make these metal based nanoadsorbents more beneficial.

13.2.1.3 Polymeric Nanoadsorbents

Recently, polymeric nanoadsorbents have gained popularity, arisen as new class of water treatment materials [80]. Different organic and organic-inorganic hybrid polymers are designed and used to eliminate various forms of water contaminants by adsorption. The most significant aspect of organic–inorganic hybrid polymer is their high adsorption capacity, higher recyclability and good thermal tolerance over a large pH spectrum [81].

It is reported in a study that modified Fe_3O_4 magnetic nanoparticles with 3-aminopropyltriethoxysilane, acrylic acid and crotonic acid copolymers have been used to extract heavy metal ions like Zn^{2+}, Cd,$^{2+}$, Pb $^{2+}$ and Cu^{2+} at varying pH levels from aqueous solutions [82]. Metal ion absorption potential was also analyzed as a function of contact time and metal ion concentration. This polymer modified MNPs demonstrated proficient metal ion adsorption even under acidic pH.

In a research, micro- and nanomultifunctional polymer doped with bimetal, an adsorbent was developed for the removal of fluoride and arsenic. A bimetal-doped nanoadsorbent was obtained by combining Al and Fe salts during polymerization. In contrast with fluoride, Fe-doped nanoadsorbents displayed good arsenic adsorption, while Al-doped nanoadsorbents exhibited good fluoride adsorption relative to arsenic [83].

A polymer hybrid sorbent, namely ZrPS-001, was formed by impregnating $Zr(HPO_3S)_2$ within a permeable polymeric cation exchanger D-001 for the improved sorting of heavy metal ions, such as Zn^{2+}, Cd^{2+} and Pb^{2+}. The negative loaded groups connected to D-001 polymer matrix result in preconcentration and better permeation of metal ions before sequestration and ZrPS nanoparticles would sequestrate heavy metals selectively through an ion exchange mechanism [84].

While polymeric nanoadsorbents are distinguished materials regarding their shapes, sizes and functional groups that can be modified, it is a challenging job to extract heavy metals from wastewater to make them selective for a given pollutant. These materials have relatively low adsorption potential and need regeneration at a high cost [85].

Dendrimers (highly branched macromolecules) are polymeric nanoadsorbents are capable of eliminating both organics and heavy metals. They are composed of central nucleus, repeating units and terminal functional groups are composed of them [86, 87]. The internal hydrophobic shells can adsorb organic compounds, while the tailored external branches can adsorb heavy metals [88]. Sorption may be due to electrostatic interactions, complexation, hydrophobic effect and hydrogen bonding [89]. To extract copper ions from aqueous solutions, a dendrimer-ultrafiltration device was designed [90]. The use of this method has recovered almost all copper ions. On decreasing the pH to 4, the metal ion laden dendrimers were recovered, thereby allowing the bound copper ions to be retrieved and the dendritic polymer recycled.

13.2.1.4 Zeolites

Zeolites are inorganic microporous aluminosilicate crystalline materials with a highly ordered structure [91, 92]. Zeolites, because of their porous structure can accommodate a wide range of metal ions, are among the most common natural adsorbents. For the adsorption-based wastewater treatment, both synthetic as well as naturally occurring zeolites have been commonly used. Zeolites that occur naturally are essentially three-dimensional structures in which small cations such as Na, K, Ca etc. are loosely attached and help these cations to be easily exchanged with other pollutant cations found in water [93]. For the degradation of quinoline and pyridine, a biologically modified zeolite was developed. Quinoline and pyridine biodegradation creates ammonium ions in water, that are adsorbed by biologically modified zeolite. Therefore, in these zeolite biodegradations of quinoline and pyridine and adsorption of ammonium ions occur simultaneously [94]. Naturally occurring zeolite such as chabazite and clinoptilolite showed better adsorption potential for lead and cadmium, with more than 99% removal efficiency [95].

Zeolites embedded with silver and lead were used for eliminating hazardous anions and pathogens from water by formation of insoluble complexes [96, 97]. Porous structure of zeolite provides it greater adsorption ability and its photocatalytic reduction capacity helps in reducing the valence of metal ions, thereby decreasing their toxicity.

13.2.2 Membrane-Based Techniques

One of the main strategies to remove undesired components in water is membrane filtration. Membranes acted as a physical barrier for contaminants reliant on their size and shape. Whenever a membrane is employed for the filtration of water, it prevents the pollutants from going in, sending out cleaner water. Membrane technology has evolved quite quickly in recent times, from the desalination process to antimicrobial filtration. To restrict the passage of microorganisms, micropollutants and other organic materials, highly efficient membranes have been created. They have a high degree of automation, need less land and chemical usage as the main component of water treatment [98]. High energy consumption is a huge obstacle to the extensive implementation of pressure driven membrane processes. Fouling of membrane leads to energy consumption, makes the operation very difficult and decreases the lifespan of membranes. The effectiveness of membrane systems is primarily dictated by the membrane content. The integration of functional nanomaterials into membranes provides a good opportunity to enhance membrane permeability, resistance to fouling, mechanical and thermal stability.

Membranes can be classified into many forms depending on their porosity, structure and mode of application, such as:

- Nanofiber membranes
- Nanocomposite membranes
- Thin film nanocomposite membranes
- Nanofiltration membranes
- Aquaporin-based membranes.

13.2.2.1 Nanofiber Membranes

Electrospinning is simple, economical and powerful nanofiber processing technique using different materials such as polymers or ceramics [27, 99]. These nanofibers are in the nanometer range and have high specific surface porosity and form complicated pore structures of nanofiber mats. Nanofibers could be modified to specific applications such as membrane thickness, composition, structure and spatial orientation [99, 100]. Electrospinning method is a completely proven air treatment technique that is not so popular in the treatment of wastewater [101]. Without any noticeable fouling, nanofiber membranes extract micron-sized particles from water, so they can be used before ultrafiltration or reverse osmosis in pretreatment [101].

NanoCeram® (Argonide Corporation, Sanford, FL, USA), filter medium implemented in a filter cartridge, is a proprietary nanofiber filter with small diameter and high surface area (300–600 m^2/g) which can be generated by sol–gel reaction in kilogram amounts. The result is free-flowing, white powder that consists of nanofibers with a diameter of about 2 nm and a length of 10–100 nm. They are accumulated in aggregates and can be inserted into sheets of glass and cellulose and can be used in ultrafiltration.

The fibers use an electrostatic effect to attract bacteria, viruses, dirt and proteins. NanoCeram can be used for ultrapure water systems in pre-filtration mode, for commercial or industrial water treatment [102]. For the separation of organic solvents, hydrophobic nanofiber membranes could become very suitable, contributing to greater flux efficiency. Bio-nanofiber membranes based on polymers such as polylactic acid, polyurethane, and polyethylene oxide are also designed to achieve greater antifouling properties [103].

13.2.2.2 Nanocomposite Membranes

A new category of filtration materials containing mixed matrix and surface-functionalized membranes may be called nanocomposite membranes. Nanofillers which are added into a matrix material are mostly inorganic and incorporated in a matrix of polymeric or inorganic oxides and have a larger specific surface area [100, 104, 105]. The key purpose of the incorporation of metal oxide nanoparticles is to decrease fouling by increasing the membrane hydrophilicity. By introducing metal oxide nanoparticles like TiO_2, silica, alumina, zeolite, etc. to polymer membranes improves membrane hydrophilicity, water permeability or resistance to fouling [106–109]. These inorganic nanoparticles also boosted the mechanical and thermal stability of polymeric membranes, reducing the detrimental effect of heat and compaction on membrane permeability [108, 110].

By inserting ordered mesoporous carbons as nanofillers into thin film polymeric matrices, a new form of thin film nanocomposite membrane is produced. These are semi-permeable membranes containing selective coating which is used in reverse osmosis. By using ambient pressure plasma, hydrophobic mesoporous carbons turned into hydrophilic mesoporous carbons with higher solubility. In order to increase the hydrophilicity of the membrane surface, only a small percentage of hydrophilic mesoporous carbons is needed that significantly increased pure water permeability [111]. Polyamide and nano-NaX zeolite (40–150 nm) thin film nanocomposite membrane coated using trimesoyl chloride and m-phenylenediamine monomers over porous polyethersulfone by interfacial polymerization and the membrane displayed strong permeability to clean water [112].

Surface functionalization with chemical compounds able to oxidize organic pollutants and thereby preventing the creation of fouling layers can be another approach used to prevent membrane clogging. Specially formulated coatings like TiO_2, nanosilver layers, resist membrane fouling and have a disinfection effect on organic contaminants. These nanoscale surface layers need few materials because of their low thickness and retain the surface profile of the material while simultaneously immobilizing potentially hazardous nanoparticles.

13.2.2.3 Thin Film Nanocomposite Membranes

Thin film nanocomposite (TFN) membranes are a new type of composite membranes, prepared by an interfacial polymerization and to enhance the properties, nanoparticles such as nano-Ag, nano-zeolites, nano-TiO_2, and CNTs are inserted into the layer of the thin film composite membranes [29, 113]. The most commonly used dopants in thin film nanocomposite are nano-zeolites and the inclusion of nano-zeolites results in a more permeable and thicker active layer of polyamide [114]. It was found in a study that water permeability improved up to 80% compared to TFC membrane with the salt rejection retained up to 94% [115]. Doped TFN membranes at 0.2 wt% with nano-zeolites resulted in higher permeability and greater salt rejection up to 99.4% than regular RO membranes [116]. Nano-zeolites have also been used as carriers of antimicrobial agents like silver ion which imparts membrane anti-fouling properties [117]. The insertion of nano-TiO_2 into the active layer of TFC significantly improved the rejection of the membrane while preserving permeability [118]. TiO_2 can destroy organic contaminants on UV irradiation and inactivate microorganisms which help in decreasing chemical and biological fouling. Because of its antimicrobial activities, CNTs also found their use in TFN membranes. In a study covalently bound SWNTs to a TFC membrane surface is useful because it uses comparatively small volumes of nanomaterial and minimizes the active layer disruption [119]. The subsequent TFN membrane displayed mild anti-bacterial properties, theoretically minimizing or delaying biofouling of the membrane.

13.2.2.4 Nanofiltration Membranes

Nanofiltration technique is pressure-driven membrane separation in which the membrane rejects molecules and particles of size less than 0.5 to 1 nm. Nanofiltration membranes are distinguished by its unique charge-based repulsion property that allows different ions to be isolated [29, 120, 121]. Nanofiltration is turning out to be more famous nowadays, being a lower energy utilization procedure, because of the lower pressure necessities contrasted with reverse osmosis (RO) measures [122, 123]. Membranes used in nanofiltration are characterized by properties that lie between the non-porous RO membranes and porous ultrafiltration membranes and accordingly, transportation in nanofiltration happens by the process of solution diffusion [124]. The pores of the nanofiltration membranes are so small (within 1–5 nm range) that the membrane essentially excludes almost all solutes. The membrane surface charge characteristics helps the monovalent ions in hard water to migrate while holding other multivalent ions. The predominant rejection method in nanofiltration for the uncharged species is size exclusion, while both size exclusion and electrostatic interaction dismiss the ionic species [125–127].

In wastewater treatment, the use of nanofiltration is quite recent technique attracting considerable interest from different industries, such as textiles, dairy industry, in the pharmaceutical and fine chemical industries for solvent recovery, in the petrochemical industry to purify gas condensates, in desalination of sea water, in life sciences for separation of amino acids and lipids from blood and other cultures of cells etc. [128–132]. Nanofiltration is ideal for filtering out nearly all organic and inorganic pollutants, including some destructive microbes from wastewater, due to the special filtration process and availability of different forms of membranes [133, 134]. Due to their easy preparation process, high durability and low cost, most of the nanofiltration membranes are made from synthetic polymers but they have lower chemical resistance and a high fouling rate, and hence have a limited lifespan [135]. Membranes manufactured from inorganic ceramics, on the other hand, have high chemical and thermal tolerance and a longer life, but with the limitation of high production costs and a lack of flexibility. In this respect, newly designed nanomaterials can play an important role in the manufacture of nanofiltration membranes, since they can be cost-effectively synthesized and flexibly produced [136].

13.2.2.5 Aquaporin-Based Membranes

Many biological membranes are selectively permeable, which enables their use to increase membrane efficiency in polymeric membranes. Aquaporins are pore-forming protein channels which regulate the flow of water across cell membranes. *Escherichia coli* aquaporin-Z has been integrated into the amphiphilic triblock polymer, that shows water permeability with complete rejection of urea, salt, glycerol and glucose over original vesicles [137]. The first commercially available membrane with embedded aquaporin is Aquaporin Inside™ (Aquaporin A/S, Copenhagen, Denmark). This form of membrane is capable of resisting pressures of up to 10 bar and water flux >100 L/(hm^2) for example needed for desalination [105]. A study revealed that aquaporins form self-assembled polymer vesicles that form a thick hydrophobic layer of polymer. This membrane has a well-controlled selective nanostructured layer with high mechanical strength [138].

13.2.3 Metal Nanoparticles

13.2.3.1 Silver Nanoparticles

It is found that silver (Ag) nanoparticles are extremely poisonous to microorganisms, contributing to powerful microbial defense in opposition to bacteria, fungi and viruses [139–141]. Because of this Ag-nanoparticles are used in water disinfection. Ag-nanoparticles have been documented to pierce the bacterial cell wall, resulting in a shift in the composition of the cell wall and thus increasing the permeability [142].

Moreover, silver nanoparticles can interfere with sulfur and phosphorus found in DNA and demolish the entire bacterial cell. The bactericidal effects of silver nanoparticles were shown to be attributed to the production of silver ions, that associate with the thiol groups in enzymes, damage them and disrupt the normal functions of the cells [143, 144]. Due to their efficient antibacterial properties and cost-effectiveness, the incorporation of silver nanoparticles with filtration membranes has been found useful for water disinfection [145]. Unexpected depletion of microorganisms was detected, approving significant antimicrobial potency, thus providing the scope for applications for wastewater treatment [146]. Amassing silver nanoparticles with ceramic filters is a convenient way to design an efficient treatment of *Escherichia coli* from wastewater. Filters with greater porosity have been shown to eliminate a greater number of microorganisms than those with lower porosity [147].

13.2.3.2 Iron Nanoparticles

Iron nanoparticles have been commonly employed in the wastewater treatment because of its exceptional adsorption ability, greater reduction efficiency, small size, cost effectiveness and high surface area [148–151]. For the elimination of a wide variety of contaminants including nitrates, radioactive materials, toxic metals, organic dyes, antibiotics and halogenated organic molecules, various separation techniques like reduction and co-precipitation, adsorption, and oxidation were embraced by nanozerovalent iron particles [152–157]. Studies found that the rate constant for arsenic removal could be increased by 1–3 significant degrees by controlling the size of zerovalent iron from micron to nanoscale and this remarkable reactivity was attributed to the increased surface area and active sites [158].

Despite the fact that nanozerovalent iron particles have a few favorable circumstances, some disadvantages are allied with these nanoparticles, therefore, diverse coating strategies such as surface modification, doping and encapsulation have been embraced in order to enhance the performance of nanozerovalent iron particles [159].

13.2.3.3 Titanium Dioxide Nanoparticles

TiO_2 nanoparticles are the most promising photocatalysts because of its high availability, low toxicity, chemical stability, good photoactivity and cost effectiveness [33]. Photocatalysis mechanism of TiO_2 involves the production of radicals such as superoxides (O^{2-}), OH^- or other radicals that can destroy organic or inorganic pollutants in water. The photocatalytic activity of TiO_2 can be improved by doping with transition metals and non-metals such as nitrogen, sulfur, halogens etc., by coupling with narrow bandgap semiconductors, by optimizing particle size [160–167]. It was observed that TiO_2 nanotubes were more effective than TiO_2 nanoparticles to decompose organic compounds [168]. Furthermore, composites of TiO_2 nanoparticles have also demonstrated their efficacy by doping with other metal nanoparticles. In addition to TiO_2, tungsten trioxide and certain fullerene derivatives have the ability to be used in photocatalytic water remedy [169].

13.3 Conclusion

Owing to the rise in water contamination, clean water has become a competitive resource; thus wastewater treatment strategies are evolving and becoming popular. Metal ions, organic and inorganic waste from water can be regulated by conventional wastewater treatment techniques but due to the inability to effectively purify water and reuse the retentates, these strategies are energy consuming and uneconomical. Nanomaterials have unique characteristics such as high surface-to volume ratio, high adsorption, size and shape dependent catalytic properties etc., making them ideal for wastewater treatment applications. Nanomaterials can be designed to proficiently gather freely accessible solar energy and can be utilized as a photocatalyst to decontaminate water effectively. Many advanced nanotechnology-based methods for wastewater treatment are available, including adsorption based purification of water, membrane based filtration of water, photocatalytic degradation of organic contaminants, application of antimicrobial nanomaterials etc. Via the elimination of toxic metal ions, heavy metals, natural organic matter, microorganisms and organic and inorganic contaminants found in water, these highly effective advancement methods offer an efficient solution for wastewater treatment. As more advancement is made in terms of environmentally friendly and economically efficient technology development, nanomaterials will become an essential wastewater treatment framework in the future. The fate of the nanoparticles in the treatment of wastewater is very reformist, but a rapid, economical, and attainable water-treatment innovation requires joint efforts from academic and industrial resources. This chapter has shown that nanotechnology can be used as an innovative and evolving wastewater treatment system and is a promising area for current and future studies.

References

1. United Nations News Centre, Web update; March (2010). Available from: http://www.un.org/apps/news/story.asp?NewsID534150#.WlSW3FSWbIU.
2. Ahmad, A., Mohd-Setapar, S.H., Chuong, C.S., Khatoon, A., Wani, W.A., Kumar, R., Rafatullah, M., Recent advances in new generation dye removal technologies: Novel search for approaches to reprocess wastewater. *RSC Adv.*, 5, 39, 30801–30818, 2015.
3. Zhang, Y., Wu, B., Xu, H., Liu, H., Wang, M., He, Y., Pan, B., Nanomaterials-enabled water and wastewater treatment. *NanoImpact.*, 3, 22–39, 2016.
4. Ali, I. and Aboul-Enein, H.Y., *Instrumental Methods in Metal Ions Speciation: Chromatography, Capillary Electrophoresis and Electrochemistry*, Taylor & Francis Ltd, New York, 2006.
5. Ali, I., Aboul-Enein, H.Y., Gupta, V.K., *Nano Chromatography and Capillary Electrophoresis: Pharmaceutical and Environmental Analyses*, John Wiley & Sons, Hoboken, NJ, 2009.
6. Laws, E.A., *Aquatic Pollution: An Introductory Text*, third ed, John Wiley & Sons, New York, 2000.
7. Ali, I., New generation adsorbents for water treatment. *Chem. Rev.*, 112, 10, 5073–5091, 2012.
8. Carballa, M., Omil, F., Ternes, T., Lema, J.M., Fate of pharmaceutical and personal care products (PPCPs) during anaerobic digestion of sewage sludge. *Water Res.*, 41, 10, 2139–2150, 2007.
9. Ellis, J.B., Pharmaceutical and personal care products (PPCPs) in urban receiving waters. *Environ. Pollut.*, 144, 1, 184–189, 2006.
10. Mohapatra, D.P., Brar, S.K., Tyagi, R.D., Picard, P., Surampalli, R.Y., Analysis and advanced oxidation treatment of a persistent pharmaceutical compound in wastewater and wastewater sludge-carbamazepine. *Sci. Total Environ.*, 470, 58–75, 2014.

11. Briggs, A.M., Cross, M.J., Hoy, D.G., Sanchez-Riera, L., Blyth, F.M., Woolf, A.D., March, L., Musculoskeletal health conditions represent a global threat to healthy aging: A report for the 2015 World Health Organization world report on ageing and health. *Gerontologist*, 56, suppl_2, S243–S255, 2016.
12. World Health Organization, & UNICEF, *Progress on sanitation and drinking-water*, World Health Organization, 2013.
13. Vörösmarty, C.J., Green, P., Salisbury, J., Lammers, R.B., Global water resources: Vulnerability from climate change and population growth. *Science*, 289, 5477, 284–288, 2000.
14. Shannon, M.A., Bohn, P.W., Elimelech, M., Georgiadis, J.G., Marinas, B.J., Mayes, A.M., Science and technology for water purification in the coming decades, in: *Nanoscience and technology: A collection of reviews from nature Journals*, pp. 337–346, 2010.
15. Godfray, H.C.J., Beddington, J.R., Crute, I.R., Haddad, L., Lawrence, D., Muir, J.F., Toulmin, C., Food security: The challenge of feeding 9 billion people. *Science*, 327, 5967, 812–818, 2010.
16. Mulder, K., Hagens, N., Fisher, B., Burning water: A comparative analysis of the energy return on water invested. *Ambio*, 39, 1, 30–39, 2010.
17. Kemper, K.E. and Voss, C.I., Groundwater from development to management: Hydrogeology journal theme issue 12 (1). *Hydrogeol. J.*, 12, 32534, 1–132, 2004.
18. Foley, J.A., DeFries, R., Asner, G.P., Barford, C., Bonan, G., Carpenter, S.R., Snyder, P.K., Global consequences of land use. *Science*, 309, 5734, 570–574, 2005.
19. Coetser, S.E., Heath, R.G.M., Ndombe, N., Diffuse pollution associated with the mining sectors in South Africa: A first-order assessment. *Water Sci. Technol.*, 55, 3, 9–16, 2007.
20. Ritter, Solomon, K., Sibley, P., Hall, K., Keen, P., Mattu, G., Linton, B.L., Sources, pathways, and relative risks of contaminants in surface water and groundwater: A perspective prepared for the Walkerton inquiry. *J. Toxicol. Environ. Health Part A*, 65, 1, 1–142, 2002.
21. Fawell, J. and Nieuwenhuijsen, M.J., Contaminants in drinking water: Environmental pollution and health. *British Med. Bull.*, 68, 1, 199–208, 2003.
22. Rodriguez-Mozaz, S., de Alda, M.J.L., Barceló, D., Monitoring of estrogens, pesticides and bisphenol A in natural waters and drinking water treatment plants by solid-phase extraction–liquid chromatography–mass spectrometry. *J. Chromatogr. A*, 1045, 1–2, 85–92, 2004.
23. Falconer, I.R. and Humpage, A.R., Health risk assessment of cyanobacterial (blue-green algal) toxins in drinking water. *Int. J. Environ. Res. Public Health*, 2, 1, 43–50, 2005.
24. Baruah, S., Khan, M.N., Dutta, J., Perspectives and applications of nanotechnology in water treatment. *Environ. Chem. Lett.*, 14, 1, 1–14, 2016.

25. Wu, Y., Pang, H., Liu, Y., Wang, X., Yu, S., Fu, D., Wang, X., Environmental remediation of heavy metal ions by novel-nanomaterials: A review. *Environ. Pollut.*, 246, 608–620, 2019.
26. Yaqoob, A.A. and Ibrahim, M.N.M., A Review Article of Nanoparticles; Synthetic Approaches and Wastewater Treatment Methods. *Int. Res. J. Eng. Technol.*, 6, 1–7, 2019.
27. Cloete, T.E., *Nanotechnology in water treatment applications*, Caister Academic Press, 2010.
28. Libralato, G., Devoti, A.C., Ghirardini, A.V., Vignati, D.A.L., Environmental effects of nZVI for land and groundwater remediation, in: *Nanotechnologies for Environmental Remediation*, pp. 269–286, Springer, Cham, 2017.
29. Qu, X., Alvarez, P.J.J., Li, Q., Applications of nanotechnology in water and wastewater treatment. *Water Res.*, 47, 3931–3946, 2013.
30. Encyclopedia Britannica. 2014. http://global.britannica.com/EBchecked/topic/584695/taxonomy.
31. Faust, S.D. and Aly, O.M., *Chemistry of water treatment*, CRC Press, 1998.
32. Kumari, P., Alam, M., Siddiqi, W.A., Usage of nanoparticles as adsorbents for wastewater treatment: An emerging trend. *Sustainable Mater. Technol.*, 22, e00128, 2019.
33. Qu, X., Alvarez, P.J.J., Li, Q., Applications of nanotechnology in water and wastewater treatment. *Water Res.*, 47, 3931–3946, 2013.
34. Ali, I., New generation adsorbents for water treatment. *Chem. Rev.*, 112, 10, 5073–5091, 2012.
35. Dąbrowski, A., Adsorption—From theory to practice. *Adv. Colloid Interface Sci.*, 93, 1–3, 135–224, 2001.
36. Chaudhary, D.S., Vigneswaran, S., Jegatheesan, V., Ngo, H.H., Moon, H., Shim, W.G., Kim, S.H., Granular activated carbon (GAC) adsorption in tertiary wastewater treatment: experiments and models. *Water Sci. Technol.*, 47, 1, 113–120, 2003.
37. Çeçen, F. and Aktas, Ö., *Activated carbon for water and wastewater treatment: Integration of adsorption and biological treatment*, John Wiley & Sons, 2011.
38. Fairey, J.L., Speitel, G.E., Katz, L.E., Impact of natural organic matter on monochloramine reduction by granular activated carbon: The role of porosity and electrostatic surface properties. *Environ. Sci. Technol.*, 40, 13, 4268–4273, 2006.
39. Asenjo, N.G., Álvarez, P., Granda, M., Blanco, C., Santamaría, R., Menéndez, R., High performance activated carbon for benzene/toluene adsorption from industrial wastewater. *J. Hazard. Mater.*, 192, 3, 1525–1532, 2011.
40. Kadirvelu, K., Thamaraiselvi, K., Namasivayam, C., Removal of heavy metals from industrial wastewaters by adsorption onto activated carbon prepared from an agricultural solid waste. *Bioresour. Technol.*, 76, 1, 63–65, 2001.

41. Kobya, M., Demirbas, E., Senturk, E., Ince, M., Adsorption of heavy metal ions from aqueous solutions by activated carbon prepared from apricot stone. *Bioresour. Technol.*, 96, 13, 1518–1521, 2005.
42. Al-Omair, M.A. and El-Sharkawy, E.A., Removal of heavy metals via adsorption on activated carbon synthesized from solid wastes. *Environ. Technol.*, 28, 4, 443–451, 2007.
43. Tamburri, E., Orlanducci, S., Terranova, M.L., Valentini, F., Palleschi, G., Curulli, A., Rossi, M., Modulation of electrical properties in single-walled carbon nanotube/conducting polymer composites. *Carbon*, 43, 6, 1213–1221, 2005.
44. Moore, V.C., Strano, M.S., Haroz, E.H., Hauge, R.H., Smalley, R.E., Schmidt, J., Talmon, Y., Individually suspended single-walled carbon nanotubes in various surfactants. *Nano Letters*, 3, 10, 1379–1382, 2003.
45. Hu, X., Wang, T., Qu, X., Dong, S., *In situ* synthesis and characterization of multiwalled carbon nanotube/Au nanoparticle composite materials. *J. Phys. Chem. B*, 110, 2, 853–857, 2006.
46. Li, Y., Ding, J., Chen, J., Xu, C., Wei, B., Liang, J., Wu, D., Preparation of ceria nanoparticles supported on carbon nanotubes. *Mater. Res. Bull.*, 37, 2, 313–318, 2002.
47. Yang, K. and Xing, B., Adsorption of organic compounds by carbon nanomaterials in aqueous phase: Polanyi theory and its application. *Chem. Rev.*, 110, 10, 5989–6008, 2010.
48. Abbas, A., Al-Amer, A.M., Laoui, T., Al-Marri, M.J., Nasser, M.S., Khraisheh, M., Atieh, M.A., Heavy metal removal from aqueous solution by advanced carbon nanotubes: Critical review of adsorption applications. *Sep. Purif. Technol.*, 157, 141–161, 2016.
49. Pan, B., Lin, D., Mashayekhi, H., Xing, B., Adsorption and hysteresis of bisphenol A and 17α-ethinyl estradiol on carbon nanomaterials. *Environ. Sci. Technol.*, 42, 15, 5480–5485, 2008.
50. Ji, L., Chen, W., Duan, L., Zhu, D., Mechanisms for strong adsorption of tetracycline to carbon nanotubes: A comparative study using activated carbon and graphite as adsorbents. *Environ. Sci. Technol.*, 43, 7, 2322–2327, 2009.
51. Yang, K. and Xing, B., Adsorption of organic compounds by carbon nanomaterials in aqueous phase: Polanyi theory and its application. *Chem. Rev.*, 110, 10, 5989–6008, 2010.
52. Chen, W., Duan, L., Zhu, D., Adsorption of polar and nonpolar organic chemicals to carbon nanotubes. *Environ. Sci. Technol.*, 41, 24, 8295–8300, 2007.
53. Lin, D. and Xing, B., Adsorption of phenolic compounds by carbon nanotubes: Role of aromaticity and substitution of hydroxyl groups. *Environ. Sci. Technol.*, 42, 19, 7254–7259, 2008.
54. Yang, K.U.N., Wu, W., Jing, Q., Zhu, L., Aqueous adsorption of aniline, phenol, and their substitutes by multi-walled carbon nanotubes. *Environ. Sci. Technol.*, 42, 21, 7931–7936, 2008.

55. Rao, G.P., Lu, C., Su, F., Sorption of divalent metal ions from aqueous solution by carbon nanotubes: A review. *Sep. Purif. Technol.*, 58, 1, 224–231, 2007.
56. Li, Y.H., Ding, J., Luan, Z., Di, Z., Zhu, Y., Xu, C., Wei, B., Competitive adsorption of Pb2+, Cu2+ and Cd2+ ions from aqueous solutions by multiwalled carbon nanotubes. *Carbon*, 41, 14, 2787–2792, 2003.
57. Lu, C., Chiu, H., Liu, C., Removal of zinc (II) from aqueous solution by purified carbon nanotubes: Kinetics and equilibrium studies. *Ind. Eng. Chem. Res.*, 45, 8, 2850–2855, 2006.
58. De Volder, M.F., Tawfick, S.H., Baughman, R.H., Hart, A.J., Carbon nanotubes: Present and future commercial applications. *Science*, 339, 6119, 535–539, 2013.
59. Iijima, S., Helical microtubules of graphitic carbon. *Nature*, 354, 6348, 56–58, 1991.
60. Ali, I., New generation adsorbents for water treatment. *Chem. Rev.*, 112, 10, 5073–5091, 2012.
61. Gao, W., Majumder, M., Alemany, L.B., Narayanan, T.N., Ibarra, M.A., Pradhan, B.K., Ajayan, P.M., Engineered graphite oxide materials for application in water purification. *ACS Appl. Mater. Interfaces*, 3, 6, 1821–1826, 2011.
62. Lu, C., Chiu, H., Liu, C., Removal of zinc (II) from aqueous solution by purified carbon nanotubes: Kinetics and equilibrium studies. *Ind. Eng. Chem. Res.*, 45, 8, 2850–2855, 2006.
63. Hua, M., Zhang, S., Pan, B., Zhang, W., Lv, L., Zhang, Q., Heavy metal removal from water/wastewater by nanosized metal oxides: A review. *J. Hazard. Mater.*, 211, 317–331, 2012.
64. Xu, P., Zeng, G.M., Huang, D.L., Feng, C.L., Hu, S., Zhao, M.H., Liu, Z.F., Use of iron oxide nanomaterials in wastewater treatment: A review. *Sci. Total Environ.*, 424, 1–10, 2012.
65. Auffan, M., Rose, J., Proux, O., Borschneck, D., Masion, A., Chaurand, P., Bottero, J.Y., Enhanced adsorption of arsenic onto maghemites nanoparticles: As (III) as a probe of the surface structure and heterogeneity. *Langmuir*, 24, 7, 3215–3222, 2008.
66. Auffan, M., Rose, J., Bottero, J.Y., Lowry, G.V., Jolivet, J.P., Wiesner, M.R., Towards a definition of inorganic nanoparticles from an environmental, health and safety perspective. *Nat. Nanotechnol.*, 4, 10, 634–641, 2009.
67. Badruddoza, A.Z.M., Shawon, Z.B.Z., Rahman, M.T., Hao, K.W., Hidajat, K., Uddin, M.S., Ionically modified magnetic nanomaterials for arsenic and chromium removal from water. *Chem. Eng. J.*, 225, 607–615, 2013.
68. Lei, Y., Chen, F., Luo, Y., Zhang, L., Three-dimensional magnetic graphene oxide foam/Fe3O4 nanocomposite as an efficient absorbent for Cr (VI) removal. *J. Mater. Sci.*, 49, 12, 4236–4245, 2014.
69. Ngomsik, A.F., Bee, A., Talbot, D., Cote, G., Magnetic solid–liquid extraction of Eu (III), La (III), Ni (II) and Co (II) with maghemite nanoparticles. *Sep. Purif. Technol.*, 86, 1–8, 2012.

70. Tan, L., Xu, J., Xue, X., Lou, Z., Zhu, J., Baig, S.A., Xu, X., Multifunctional nanocomposite Fe3O4@ SiO2-mPD/SP for selective removal of Pb (ii) and Cr (vi) from aqueous solutions. *RSC Adv.*, 4, 86, 45920–45929, 2014.
71. Tu, Y.J., You, C.F., Chang, C.K., Wang, S.L., Chan, T.S., Arsenate adsorption from water using a novel fabricated copper ferrite. *Chem. Eng. J.*, 198, 440–448, 2012.
72. Gautam, R.K., Soni, S., Chattopadhyaya, M.C., Functionalized magnetic nanoparticles for environmental remediation, in: *Handbook of Research on Diverse Applications of Nanotechnology in Biomedicine, Chemistry, and Engineering*, pp. 518–551, IGI Global, 2015.
73. Diagboya, P.N., Olu-Owolabi, B.I., Adebowale, K.O., Synthesis of covalently bonded graphene oxide–iron magnetic nanoparticles and the kinetics of mercury removal. *RSC Adv.*, 5, 4, 2536–2542, 2015.
74. Hu, S., Shi, Q., Jing, C., Groundwater arsenic adsorption on granular TiO2: Integrating atomic structure, filtration, and health impact. *Environ. Sci. Technol.*, 49, 16, 9707–9713, 2015.
75. Srivastava, V., Sharma, Y.C., Sillanpää, M., Green synthesis of magnesium oxide nanoflower and its application for the removal of divalent metallic species from synthetic wastewater. *Ceram. Int.*, 41, 5, 6702–6709, 2015.
76. Ray, P.Z. and Shipley, H.J., Inorganic nano-adsorbents for the removal of heavy metals and arsenic: A review. *RSC Adv.*, 5, 38, 29885–29907, 2015.
77. Sankararamakrishnan, N., Jaiswal, M., Verma, N., Composite nanofloral clusters of carbon nanotubes and activated alumina: An efficient sorbent for heavy metal removal. *Chem. Eng. J.*, 235, 1–9, 2014.
78. Sharma, Y.C., Srivastava, V., Singh, V.K., Kaul, S.N., Weng, C.H., Nano-adsorbents for the removal of metallic pollutants from water and wastewater. *Environ. Technol.*, 30, 6, 583–609, 2009.
79. Hu, J., Chen, G., Lo, I.M., Selective removal of heavy metals from industrial wastewater using maghemite nanoparticle: Performance and mechanisms. *J. Environ. Eng.*, 132, 7, 709–715, 2006.
80. Yin, J. and Deng, B., Polymer-matrix nanocomposite membranes for water treatment. *J. Membr. Sci.*, 479, 256–275, 2015.
81. Lofrano, G., Carotenuto, M., Libralato, G., Domingos, R.F., Markus, A., Dini, L., Meric, S., Polymer functionalized nanocomposites for metals removal from water and wastewater: An overview. *Water Res.*, 92, 22–37, 2016.
82. Ge, F., Li, M.M., Ye, H., Zhao, B.X., Effective removal of heavy metal ions Cd2+, Zn2+, Pb2+, Cu2+ from aqueous solution by polymer-modified magnetic nanoparticles. *J. Hazard. Mater.*, 211, 366–372, 2012.
83. Kumar, V., Talreja, N., Deva, D., Sankararamakrishnan, N., Sharma, A., Verma, N., Development of bi-metal doped micro-and nano multi-functional polymeric adsorbents for the removal of fluoride and arsenic (V) from wastewater. *Desalination*, 282, 27–38, 2011.
84. Zhang, Q., Pan, B., Pan, B., Zhang, W., Jia, K., Zhang, Q., Selective sorption of lead, cadmium and zinc ions by a polymeric cation exchanger

containing nano-Zr (HPO3S) 2. *Environ. Sci. Technol.*, 42, 11, 4140–4145, 2008.
85. Pan, B., Lin, D., Mashayekhi, H., Xing, B., Adsorption and hysteresis of bisphenol A and 17α-ethinyl estradiol on carbon nanomaterials. *Environ. Sci. Technol.*, 42, 15, 5480–5485, 2008.
86. Bosman, D.A., Janssen, H.M., Meijer, E.W., About dendrimers: Structure, physical properties, and applications. *Chem. Rev.*, 99, 7, 1665–1688, 1999.
87. Tully, D.C. and Fréchet, J.M., Dendrimers at surfaces and interfaces: Chemistry and applications. *Chem. Commun.*, 14), 1229–1239, 2001.
88. Khajeh, M., Laurent, S., Dastafkan, K., Nanoadsorbents: Classification, preparation, and applications (with emphasis on aqueous media). *Chem. Rev.*, 113, 10, 7728–7768, 2013.
89. Crooks, R.M., Zhao, M., Sun, L., Chechik, V., Yeung, L.K., Dendrimer-encapsulated metal nanoparticles: Synthesis, characterization, and applications to catalysis. *Acc. Chem. Res.*, 34, 3, 181–190, 2001.
90. Diallo, M.S., Christie, S., Swaminathan, P., Johnson, J.H., Goddard, W.A., Dendrimer enhanced ultrafiltration. 1. Recovery of Cu (II) from aqueous solutions using PAMAM dendrimers with ethylene diamine core and terminal NH2 groups. *Environ. Sci. Technol.*, 39, 5, 1366–1377, 2005.
91. Breck, D.W., *Zeolite Molecular Sieves*, Wiley and Sons Publishers, New York, 1974.
92. Camblor, M.A., Corma, A., Valencia, S., Synthesis in fluoride media and characterisation of aluminosilicate zeolite beta. *J. Mater. Chem.*, 8, 9, 2137–2145, 1998.
93. Dąrowski, A., Hubicki, Z., Podkościelny, P., Robens, E., Selective removal of the heavy metal ions from waters and industrial wastewaters by ion-exchange method. *Chemosphere*, 56, 2, 91–106, 2004.
94. Bai, Y., Sun, Q., Xing, R., Wen, D., Tang, X., Removal of pyridine and quinoline by bio-zeolite composed of mixed degrading bacteria and modified zeolite. *J. Hazard. Mater.*, 181, 1–3, 916–922, 2010.
95. Kesraoui-Ouki, S., Cheeseman, C., Perry, R., Effects of conditioning and treatment of chabazite and clinoptilolite prior to lead and cadmium removal. *Environ. Sci. Technol.*, 27, 6, 1108–1116, 1993.
96. Kwakye-Awuah, B., Williams, C., Kenward, M.A., Radecka, I., Antimicrobial action and efficiency of silver-loaded zeolite X. *J. Appl. Microbiol.*, 104, 5, 1516–1524, 2008.
97. Chmielewská, E., Jesenák, K., Gáplovská, K., Arsenate and chromate removal with cationic surfactant-loaded and cation-exchanged clinoptilolite-rich tuff vs montmorillonite. *Collect. Czech. Chem. Commun.*, 68, 4, 823–836, 2003.
98. Qu, X., Brame, J., Li, Q., Alvarez, P.J., Nanotechnology for a safe and sustainable water supply: Enabling integrated water treatment and reuse. *Acc. Chem. Res.*, 46, 3, 834–843, 2013.
99. Li, D. and Xia, Y., Electrospinning of nanofibers: Reinventing the wheel? *Adv. Mater.*, 16, 14, 1151–1170, 2004.

100. Wegmann, M., Michen, B., Graule, T., Nanostructured surface modification of microporous ceramics for efficient virus filtration. *J. Eur. Ceram. Soc.*, 28, 8, 1603–1612, 2008.
101. Ramakrishna, S., Fujihara, K., Teo, W.E., Yong, T., Ma, Z., Ramaseshan, R., Electrospun nanofibers: Solving global issues. *Mater. Today*, 9, 3, 40–50, 2006.
102. Karim, M.R., Rhodes, E.R., Brinkman, N., Wymer, L., Fout, G.S., New electropositive filter for concentrating enteroviruses and noroviruses from large volumes of water. *Appl. Environ. Microbiol.*, 75, 8, 2393–2399, 2009.
103. Rana, D. and Matsuura, T., Surface modifications for antifouling membranes. *Chem. Rev.*, 110, 4, 2448–2471, 2010.
104. Feng, C., Khulbe, K.C., Matsuura, T., Tabe, S., Ismail, A.F., Preparation and characterization of electro-spun nanofiber membranes and their possible applications in water treatment. *Sep. Purif. Technol.*, 102, 118–135, 2013.
105. Gehrke, I., Geiser, A., Somborn-Schulz, A., Innovations in nanotechnology for water treatment. *Nanotechnol. Sci. Appl.*, 8, 1, 2015.
106. Maximous, N., Nakhla, G., Wong, K., Wan, W., Optimization of Al2O3/PES membranes for wastewater filtration. *Sep. Purif. Technol.*, 73, 2, 294–301, 2010.
107. Bottino, A., Capannelli, G., D'asti, V., Piaggio, P., Preparation and properties of novel organic–inorganic porous membranes. *Sep. Purif. Technol.*, 22, 269–275, 2001.
108. Pendergast, M.T.M., Nygaard, J.M., Ghosh, A.K., Hoek, E.M., Using nanocomposite materials technology to understand and control reverse osmosis membrane compaction. *Desalination*, 261, 3, 255–263, 2010.
109. Bae, T.H. and Tak, T.M., Effect of TiO2 nanoparticles on fouling mitigation of ultrafiltration membranes for activated sludge filtration. *J. Membr. Sci.*, 249, 1–2, 1–8, 2005.
110. Ebert, K., Fritsch, D., Koll, J., Tjahjawiguna, C., Influence of inorganic fillers on the compaction behaviour of porous polymer based membranes. *J. Membr. Sci.*, 233, 1–2, 71–78, 2004.
111. Kim, E.S. and Deng, B., Fabrication of polyamide thin-film nano-composite (PA-TFN) membrane with hydrophilized ordered mesoporous carbon (H-OMC) for water purifications. *J. Membr. Sci.*, 375, 1–2, 46–54, 2011.
112. Fathizadeh, M., Aroujalian, A., Raisi, A., Effect of added NaX nano-zeolite into polyamide as a top thin layer of membrane on water flux and salt rejection in a reverse osmosis process. *J. Membr. Sci.*, 375, 1–2, 88–95, 2011.
113. Lau, W.J., Gray, S., Matsuura, T., Emadzadeh, D., Chen, J.P., Ismail, A.F., A review on polyamide thin film nanocomposite (TFN) membranes: History, applications, challenges and approaches. *Water Res.*, 80, 306–324, 2015.
114. Lind, M.L., Ghosh, A.K., Jawor, A., Huang, X., Hou, W., Yang, Y., Hoek, E.M., Influence of zeolite crystal size on zeolite-polyamide thin film nanocomposite membranes. *Langmuir*, 25, 17, 10139–10145, 2009.

115. Jeong, B.H., Hoek, E.M., Yan, Y., Subramani, A., Huang, X., Hurwitz, G., Jawor, A., Interfacial polymerization of thin film nanocomposites: A new concept for reverse osmosis membranes. *J. Membr. Sci.*, 294, 1–2, 1–7, 2007.
116. Lind, M.L., Eumine Suk, D., Nguyen, T.V., Hoek, E.M., Tailoring the structure of thin film nanocomposite membranes to achieve seawater RO membrane performance. *Environ. Sci. Technol.*, 44, 21, 8230–8235, 2010.
117. Lind, M.L., Jeong, B.H., Subramani, A., Huang, X., Hoek, E.M., Effect of mobile cation on zeolite-polyamide thin film nanocomposite membranes. *J. Mater. Res.*, 24, 5, 1624–1631, 2009.
118. Lee, H.S., Im, S.J., Kim, J.H., Kim, H.J., Kim, J.P., Min, B.R., Polyamide thin-film nanofiltration membranes containing TiO2 nanoparticles. *Desalination*, 219, 1–3, 48–56, 2008.
119. Tiraferri, A., Vecitis, C.D., Elimelech, M., Covalent binding of single-walled carbon nanotubes to polyamide membranes for antimicrobial surface properties. *ACS Appl. Mater. Interfaces*, 3, 8, 2869–2877, 2011.
120. Jagadevan, S., Jayamurthy, M., Dobson, P., Thompson, I.P., A novel hybrid nano zerovalent iron initiated oxidation—Biological degradation approach for remediation of recalcitrant waste metalworking fluids. *Water Res.*, 46, 7, 2395–2404, 2012.
121. Sharma, V. and Sharma, A., Nanotechnology: An emerging future trend in wastewater treatment with its innovative products and processes. *Nanotechnology*, 1, 2, 1–8, 2012.
122. Cadotte, J., Forester, R., Kim, M., Petersen, R., Stocker, T., Nanofiltration membranes broaden the use of membrane separation technology. *Desalination*, 70, 1–3, 77–88, 1988.
123. Bellona, C., Drewes, J.E., Oelker, G., Luna, J., Filteau, G., Amy, G., Comparing nanofiltration and reverse osmosis for drinking water augmentation. *J. Am. Water Works Assoc.*, 100, 9, 102–116, 2008.
124. Bowen, W.R. and Mukhtar, H., Characterisation and prediction of separation performance of nanofiltration membranes. *J. Membr. Sci.*, 112, 2, 263–274, 1996.
125. Choi, J.H., Dockko, S., Fukushi, K., Yamamoto, K., A novel application of a submerged nanofiltration membrane bioreactor (NF MBR) for wastewater treatment. *Desalination*, 146, 1–3, 413–420, 2002.
126. Schafer, A.I., *Nanofiltration*, Elsevier Advanced Technology, 2005.
127. Verliefde, A.R., Cornelissen, E.R., Heijman, S.G.J., Verberk, J.Q.J.C., Amy, G.L., Van der Bruggen, B., Van Dijk, J.C., The role of electrostatic interactions on the rejection of organic solutes in aqueous solutions with nanofiltration. *J. Membr. Sci.*, 322, 1, 52–66, 2008.
128. Mohammad, A.W., Teow, Y.H., Ang, W.L., Chung, Y.T., Oatley-Radcliffe, D.L., Hilal, N., Nanofiltration membranes review: Recent advances and future prospects. *Desalination*, 356, 226–254, 2015.
129. Bowen, W.R., Welfoot, J.S., Williams, P.M., Linearized transport model for nanofiltration: Development and assessment. *AIChE J.*, 48, 4, 760–773, 2002.

130. Bowen, W.R. and Welfoot, J.S., Modelling the performance of membrane nanofiltration—Critical assessment and model development. *Chem. Eng. Sci.*, 57, 7, 1121–1137, 2002.
131. Yangali-Quintanilla, V., Sadmani, A., McConville, M., Kennedy, M., Amy, G., Rejection of pharmaceutically active compounds and endocrine disrupting compounds by clean and fouled nanofiltration membranes. *Water Res.*, 43, 9, 2349–2362, 2009.
132. Yangali-Quintanilla, V., Maeng, S.K., Fujioka, T., Kennedy, M., Amy, G., Proposing nanofiltration as acceptable barrier for organic contaminants in water reuse. *J. Membr. Sci.*, 362, 1–2, 334–345, 2010.
133. Thanuttamavong, M., Oh, J.I., Yamamoto, K., Urase, T., Comparison between rejection characteristics of natural organic matter and inorganic salts in ultra low pressure nanofiltration for drinking water production. *Water Sci. Technol.: Water Supply*, 1, 5-6, 77–90, 2001.
134. Thanuttamavong, M., Yamamoto, K., Oh, J.I., Choo, K.H., Choi, S.J., Rejection characteristics of organic and inorganic pollutants by ultra low-pressure nanofiltration of surface water for drinking water treatment. *Desalination*, 145, 1–3, 257–264, 2002.
135. Van der Bruggen, B., Mänttäri, M., Nyström, M., Drawbacks of applying nanofiltration and how to avoid them: A review. *Sep. Purif. Technol.*, 63, 2, 251–263, 2008.
136. Endo, M., Muramatsu, H., Hayashi, T., Kim, Y.A., Terrones, M., Dresselhaus, M.S., 'Buckypaper' from coaxial nanotubes. *Nature*, 433, 7025, 476–476, 2005.
137. Kumar, M., Grzelakowski, M., Zilles, J., Clark, M., Meier, W., Highly permeable polymeric membranes based on the incorporation of the functional water channel protein Aquaporin Z. *Proc. Nat. Acad. Sci.*, 104, 52, 20719–20724, 2007.
138. Xie, W., He, F., Wang, B., Chung, T.S., Jeyaseelan, K., Armugam, A., Tong, Y.W., An aquaporin-based vesicle-embedded polymeric membrane for low energy water filtration. *J. Mater. Chem. A*, 1, 26, 7592–7600, 2013.
139. Krishnaraj, C., Ramachandran, R., Mohan, K., Kalaichelvan, P.T., Optimization for rapid synthesis of silver nanoparticles and its effect on phytopathogenic fungi. *Spectrochim. Acta Part A*, 93, 95–99, 2012.
140. Kalhapure, R.S., Sonawane, S.J., Sikwal, D.R., Jadhav, M., Rambharose, S., Mocktar, C., Govender, T., Solid lipid nanoparticles of clotrimazole silver complex: An efficient nano antibacterial against *Staphylococcus aureus* and MRSA. *Colloids Surf. B*, 136, 651–658, 2015.
141. Borrego, B., Lorenzo, G., Mota-Morales, J.D., Almanza-Reyes, H., Mateos, F., López-Gil, E., Bogdanchikova, N., Potential application of silver nanoparticles to control the infectivity of Rift Valley fever virus *in vitro* and *in vivo*. *Nanomed. Nanotechnol. Biol. Med.*, 12, 5, 1185–1192, 2016.
142. Sondi, I. and Salopek-Sondi, B., Silver nanopartiklar som antimikrobiellt medel: En fallstudie på *E. coli* som modell för gramnegativa bakterier. *J. Colloid Interface Sci.*, 275, 1, 177–182, 2004.

143. Dhanalekshmi, K.I. and Meena, K.S., DNA intercalation studies and antimicrobial activity of Ag@ ZrO2 core–shell nanoparticles *in vitro*. *Mater. Sci. Eng. C*, 59, 1063–1068, 2016.
144. Prabhu, S. and Poulose, E.K., Silver nanoparticles: Mechanism of antimicrobial action, synthesis, medical applications, and toxicity effects. *Int. Nano Lett.*, 2, 1, 32, 2012.
145. Biesinger, M.C., Payne, B.P., Grosvenor, A.P., Lau, L.W., Gerson, A.R., Smart, R.S.C., Resolving surface chemical states in XPS analysis of first row transition metals, oxides and hydroxides: Cr, Mn, Fe, Co and Ni. *Appl. Surf. Sci.*, 257, 7, 2717–2730, 2011.
146. Ferreira, A.M., Roque, É.B., Fonseca, F.V.D., Borges, C.P., High flux microfiltration membranes with silver nanoparticles for water disinfection. *Desalin. Water Treat.*, 56, 13, 3590–3598, 2015.
147. Kallman, E.N., Oyanedel-Craver, V.A., Smith, J.A., Ceramic filters impregnated with silver nanoparticles for point-of-use water treatment in rural Guatemala. *J. Environ. Eng.*, 137, 6, 407–415, 2011.
148. Fu, F., Dionysiou, D.D., Liu, H., The use of zero-valent iron for groundwater remediation and wastewater treatment: A review. *J. Hazard. Mater.*, 267, 194–205, 2014.
149. Matheson, L.J. and Tratnyek, P.G., Reductive dehalogenation of chlorinated methanes by iron metal. *Environ. Sci. Technol.*, 28, 12, 2045–2053, 1994.
150. Wang, X., Zhu, M., Liu, H., Ma, J., Li, F., Modification of Pd–Fe nanoparticles for catalytic dechlorination of 2, 4-dichlorophenol. *Sci. Total Environ.*, 449, 157–167, 2013.
151. Wang, Y., Liang, S., Chen, B., Guo, F., Yu, S., Tang, Y., Synergistic removal of Pb (II), Cd (II) and humic acid by Fe3O4@ mesoporous silica-graphene oxide composites. *PloS one*, 8, 6, e65634, 2013.
152. Ryu, A., Jeong, S.W., Jang, A., Choi, H., Reduction of highly concentrated nitrate using nanoscale zero-valent iron: Effects of aggregation and catalyst on reactivity. *Appl. Catal. B*, 105, 1–2, 128–135, 2011.
153. Fang, Z., Chen, J., Qiu, X., Qiu, X., Cheng, W., Zhu, L., Effective removal of antibiotic metronidazole from water by nanoscale zero-valent iron particles. *Desalination*, 268, 1–3, 6—67, 2011.
154. Ling, L., Pan, B., Zhang, W.X., Removal of selenium from water with nanoscale zero-valent iron: Mechanisms of intraparticle reduction of Se (IV). *Water Res.*, 71, 274–281, 2015.
155. Ling, L., Pan, B., Zhang, W.X., Removal of selenium from water with nanoscale zero-valent iron: Mechanisms of intraparticle reduction of Se (IV). *Water Res.*, 71, 274–281, 2015.
156. Hoag, G.E., Collins, J.B., Holcomb, J.L., Hoag, J.R., Nadagouda, M.N., Varma, R.S., Degradation of bromothymol blue by 'greener' nano-scale zero-valent iron synthesized using tea polyphenols. *J. Mater. Chem.*, 19, 45, 8671–8677, 2009.
157. Liang, D.W., Yang, Y.H., Xu, W.W., Peng, S.K., Lu, S.F., Xiang, Y., Nonionic surfactant greatly enhances the reductive debromination of polybrominated

diphenyl ethers by nanoscale zero-valent iron: Mechanism and kinetics. *J. Hazard. Mater.*, 278, 592–596, 2014.

158. Kanel, S.R., Greneche, J.M., Choi, H., Arsenic (V) removal from groundwater using nano scale zero-valent iron as a colloidal reactive barrier material. *Environ. Sci. Technol.*, 40, 6, 2045–2050, 2006.
159. Stefaniuk, M., Oleszczuk, P., Ok, Y.S., Review on nano zerovalent iron (nZVI): From synthesis to environmental applications. *Chem. Eng. J.*, 287, 618–632, 2016.
160. Graciani, J., Nambu, A., Evans, J., Rodriguez, J.A., Sanz, J.F., Au ↔ N synergy and N-doping of metal oxide-based photocatalysts. *J. Am. Chem. Soc.*, 130, 36, 12056–12063, 2008.
161. Ji, G., Gu, Z., Lu, M., Zhou, J., Zhang, S., Chen, Y., First principles calculations of N:H co-doping effect on energy gap narrowing of ZnO. *Physica B*, 405, 24, 4948–4950, 2010.
162. Rehman, S., Ullah, R., Butt, A., Gohar, N.D., Strategies of making TiO2 and ZnO visible light active. *J. Hazard. Mater.*, 170, 2–3, 560–569, 2009.
163. Fujihara, S., Ogawa, Y., Kasai, A., Tunable visible photoluminescence from ZnO thin films through Mg-doping and annealing. *Chem. Mater.*, 16, 15, 2965–2968, 2004.
164. Tao, Z., Yu, X., Fei, X., Liu, J., Zhao, Y., Wu, H., Yang, L., Synthesis and optical properties of halogen-doped ZnO phosphor. *Mater. Lett.*, 62, 17–18, 3018–3020, 2008.
165. Wu, L., Jimmy, C.Y., Fu, X., Characterization and photocatalytic mechanism of nanosized CdS coupled TiO2 nanocrystals under visible light irradiation. *J. Mol. Catal. A: Chem.*, 244, 1–2, 25–32, 2006.
166. Ho, W. and Jimmy, C.Y., Sonochemical synthesis and visible light photocatalytic behavior of CdSe and CdSe/TiO2 nanoparticles. *J. Mol. Catal. A: Chem.*, 247, 1–2, 268–274, 2006.
167. Jianhua, L.I.U., Rong, Y.A.N.G., Songmei, L.I., Preparation and characterization of the TiO2-V2O5 photocatalyst with visible-light activity. *Rare Met.*, 25, 6, 636–642, 2006.
168. Macak, J.M., Zlamal, M., Krysa, J., Schmuki, P., Self-organized TiO2 nanotube layers as highly efficient photocatalysts. *Small*, 3, 2, 300–304, 2007.
169. Meng, Z.D., Zhu, L., Choi, J.G., Park, C.Y., Oh, W.C., Preparation, characterization and photocatalytic behavior of WO 3-fullerene/TiO 2 catalysts under visible light. *Nanoscale Res. Lett.*, 6, 1, 1–11, 2011.

Index

Also of Interest

Check out these published and forthcoming related titles from Scrivener Publishing

Sustainable Practices in the Textile Industry
Edited by Luqman Jameel Rather, Mohd Shabbir and Aminoddin Haji
Published 2021. ISBN 978-1-119-81888-5

Innovative and Emerging Technologies for Textile Dyeing and Finishing
Edited by Luqman Jameel Rather, Aminoddin Haji and Mohd Shabbir
Published 2021. ISBN 978-1-119-71014-1

Recycling from Waste in Fashion and Textiles
A Sustainable & Circular Economic Approach
Edited by Pintu Pandit, Shakeel Ahmed, Kunal Singha and Sanjay Shrivastava
Published 2020. ISBN 978-1-119-62049-5

Frontiers of Textile Materials
Polymers, Nanomaterials, Enzymes, and Advanced Modification Techniques
Edited by Mohd Shabbir, Shakeel Ahmed and Javed N. Sheikh
Published 2020. ISBN 978-1-119-62037-2

Advanced Functional Textiles and Polymers
Edited by Shahid-ul-Islam and B.S. Butola
Published 2019. ISBN 978-1-119-60579-9

Textiles and Clothing
Environmental Concerns and Solutions
Edited by Mohd Shabbir
Published 2019. ISBN 978-1-119-52631-5

Smart Textiles
Wearable Nanotechnology
Edited by Nazire D. Yilmaz
Published 2019. ISBN 978-1-119-46022-0

Nanomaterials in the Wet Processing of Textiles
Edited by Shahid-ul-Islam and B.S. Butola
Published 2018. ISBN 978-1-119-45980-4

Advanced Textile Engineering Materials
Edited by Shahid-ul-Islam and B.S. Butola
Published 2018. ISBN 978-1-119-48785-2

Handbook of Renewable Materials for Coloration and Finishing
Edited by Mohd Yusuf
Published 2018. ISBN 978-1-119-40775-1

Textile Finishing: Recent Developments and Future Trends
Edited by K.L. Mittal and Thomas Bahners
Published 2017. ISBN 978-1-119-42676-9

Chemicals and Methods for Conservation and Restoration
Paintings, Textiles, Fossils, Wood, Stones, Metals, and Glass
By Johannes Karl Fink
Published 2017. ISBN 978-1-119-41824-5

Plant-Based Natural Products
Derivatives and Applications
Edited by Shahid-ul-Islam
Published 2017. ISBN 978-1-119-42383

Printed and bound by CPI Group (UK) Ltd, Croydon, CR0 4YY

17/09/2026

14964247-0005